NUMERICAL MODELLING OF CONSTRUCTION PROCESSES IN GEOTECHNICAL
ENGINEERING FOR URBAN ENVIRONMENT

BALKEMA – Proceedings and Monographs
in Engineering, Water and Earth Sciences

PROCEEDINGS OF THE INTERNATIONAL CONFERENCE ON NUMERICAL SIMULATION OF CONSTRUCTION PROCESSES IN GEOTECHNICAL ENGINEERING FOR URBAN ENVIRONMENT, 23/24 MARCH 2006, BOCHUM, GERMANY

Numerical Modelling of Construction Processes in Geotechnical Engineering for Urban Environment

Edited by

Th. Triantafyllidis
Ruhr University Bochum, Institute of Foundation Engineering and Soil Mechanics, Bochum, Germany

Taylor & Francis
Taylor & Francis Group

LONDON/LEIDEN/NEW YORK/PHILADELPHIA/SINGAPORE

Published by: A.A. Balkema Publishers, a member of Taylor & Francis Group plc
P.O. Box 447, 2300 AK Leiden, The Netherlands
e-mail: Pub.NL@tandf.co.uk
www.balkema.nl, www.tandf.co.uk, www.crcpress.com

ISBN 10: 0 415 39748 0   ISBN 13: 9 78 0 415 39748 3

Numerical Modelling of Construction Processes in Geotechnical Engineering for Urban Environment – Triantafyllidis (ed)
© 2006 Taylor & Francis Group, London, ISBN 0 415 39748 0

Table of Contents

Numerical Modelling of Construction Processes in Geotechnical Engineering for
Urban Environment – Triantafyllidis (ed)
© 2006 Taylor & Francis Group, London, ISBN 0 415 39748 0

Preface

This volume contains papers that have been selected by the scientific committee for presentation at the Inter-national Conference on Numerical Simulation of Construction Processes in Geotechnical Engineering in Urban Environment, held in Bochum on the 23–24 March 2006. The main scope of this conference was to bring together scientists from all over the world working on the specific subject simulation of construction methods in geotechnical engineering in order to summarize old and new approaches in this field, to provide a platform for discussion between the active scientists, to exchange experiences and to identify new fields for the future activities and research in this field.

The construction methods in geotechnical engineering cover a wide field from retaining structures, deep foundation elements and soil improvement, whereby different installation methods and techniques have been applied in the past with several variations thereof. Especially in an urban environment the application of an installation method in geotechnical engineering affects the serviceability of the existing structures and utilities. Quite often an installation technique or construction method is required in order to protect cultural heritage or old structures and the demand for simulation tools is high in this field in order to predict the impact of the considered techniques on the area of interest. In the meantime, very sophisticated constitutive relations have been developed, which realistically describe the soil behaviour and the further development in the information technology and hardware components facilitate the development of computer codes, which can be used in order to predict the impact of the one or the other construction method in the existing infrastructure.

In this volume, approaches are presented for modelling deep excavations starting from methods for the determination of soil parameters, case studies of excavation support systems and modelling including predictions for real projects. Very close to this topic are the simulations of the diaphragm wall construction methods and comparisons between monitoring data on construction sites with the respective predictions.

Further interesting topics refer to deep vibro-compaction simulations, where very promising results and simulation techniques are presented for the first time. In an adjacent topic referring to ground freezing constitutive relations, numerical simulations of freezing and applications of the analysis to several projects are presented. Numerical methods and strategies for the analysis of tunnel construction processes specially in urban environment are discussed.

A separate section is referring to pile installation techniques, like jacking and driving as well as to the installation of bored and screwed piles. Very close to this topic are the simulation methods for the installation of rammed aggregate piers and the backanalysis of a failure of a micro-pile wall during construction.

All the abstracts and the papers selected for this conference have been reviewed by the members of the international scientific committee and the editor is deeply indebted to all members of the committee for their efforts. Furthermore we like to express our gratitude and special thanks to all our keynote speakers, Prof. Finno (Northwestern University, USA), Prof. Phoon and Prof. Lee (National University of Singapore) for their valuable contributions and their spontaneous acceptance of our invitation.

For the organization of the conference, the efforts of the local organizing committee are gratefully acknowl-edged herewith and my special thanks are dedicated to my co-workers Dr. Diethard König, Dr. Andrzej Niemunis and Ms. Anke Wellmann.

Last but not least we also like to gratefully acknowledge the financial support of the German Research Council (Deutsche Forschungsgemeinschaft, DFG) for this conference. The German Ministry of Education and Research is highly acknowledged for the financial support of the research in this field.

The international scientific committee, the local organizing committee and the editor hope that this conference volume will be frequently used and appreciated as a basis and reference for future scientific work in this field.

Theodoros Triantafyllidis
March 2006

Numerical Modelling of Construction Processes in Geotechnical Engineering for Urban Environment – Triantafyllidis (ed)
© 2006 Taylor & Francis Group, London, ISBN 0 415 39748 0

Organization

Numerical Modelling of Construction Processes in Geotechnical Engineering for Urban Environment

Papers presented at the International Conference "Numerical Simulation of Construction Processes in Geotechnical Engineering for Urban Environment" held in Bochum, Germany on 23/24 March 2006

Organized by
Ruhr University Bochum
Faculty of Civil Engineering
Institute of Foundation Engineering and Soil Mechanics
Block IA – 4/126
D-44780 Bochum, Germany

E-mail:office-gub@rub.de
Internet: http://www.gub.ruhr-uni-bochum.de
Phone: +49-(0)234-32-26135
Fax: +49-(0)234-32-14150

Supported by
German Research Foundation (DFG)
Federal Ministry of Education and Research (BMBF), Germany

Scientific Committee
Anagnostou, G., Zurich, Switzerland
Burland, J., London, UK
Finno, R.J., Evanston, USA
Grabe, J., Hamburg, Germany
Gudehus, G., Karlsruhe, Germany
Houlsby, G.T., Oxford, UK
Kastner, R., Lyon, France
Mandolini, A., Naples, Italy
Mróz, Z., Warsaw, Poland
Phoon, K.K., Singapore
Savidis, S., Berlin, Germany
Schanz, T., Weimar, Germany
Vermeer, P., Stuttgart, Germany
Vogt, N., Munich, Germany
Wu, W., Vienna, Austria

Local Organizing Committee
Triantafyllidis, Th. – Chairman
König, D.
Niemunis, A.

1. Deep excavations

Numerical Modelling of Construction Processes in Geotechnical Engineering for
Urban Environment – Triantafyllidis (ed)
© 2006 Taylor & Francis Group, London, ISBN 0 415 39748 0

Selected topics in numerical simulation of supported excavations

Richard J. Finno & Xuxin Tu

Department of Civil and Environmental Engineering, Northwestern University, Evanston, IL USA

ABSTRACT: This paper describes factors that affect the selection of the type of finite element formulation, the initial stress conditions with emphasis on urban environments, the importance a faithful representation of the construction process, and factors affecting the selection of the constitutive model. To illustrate the importance of small strain non-linearity, responses measured in laboratory specimens cut from hand-carved block samples of soft to medium clay obtained from an excavation in Chicago are summarized. To model these responses, a tangential stiffness model is described and used to represent the constitutive relationship of the clays in a finite element simulation of an instrumented tied-back excavation. The results of these simulations and those made assuming responses modeled by a Mohr-coulomb plasticity-based model are compared to observed results to illustrate the differences in the capabilities of the two types of constitutive models.

1 INTRODUCTION

A successful numerical prediction of the performance of a supported excavation contains many key elements, including proper finite element formulation, accurate integration of the material models, adequate representation of the constitutive responses of the geomaterials and structural elements, accurate definition of initial conditions and reasonable numerical simulations of the construction activities at the excavation site. A complete presentation is well beyond the scope of a single paper, necessitating a selection of topics to be discussed within the confines of this paper. This paper will focus on simulating geotechnical construction activities and selected aspects of constitutive modeling of soils. These topics are chosen because they are linked in the sense that the same computed result can be attained by a number of combinations of these factors. This paper can be thought as an effort to dispel the old saying "any accurate prediction in geotechnical engineering is a result of compensating errors" – an adage that is particularly applicable to numerical predictions of the performance of supported excavations.

This paper describes factors that affect the selection of the type of finite element formulation, the initial stress conditions with emphasis on urban environments, the importance a faithful representation of the construction process, and factors affecting the selection of the constitutive model. The discussions include examples culled from the authors' experiences with recent excavations made in Chicago.

2 TYPE OF FINITE ELEMENT FORMULATION

A key to a successful finite element simulation is representing faithfully in the numerical simulation what happens in the field during construction. To do this, one must match the expected drainage conditions, which impacts the formulation required. For example, for undrained conditions, one can employ either a coupled or mixed finite element formulation where both displacements and pore water pressures are solved for explicitly (e.g. Carter et al. 1979) or a penalty formulation (e.g., Hughes 1980; Borja 1990) wherein the bulk modulus of water – or in fact any large number can be used that depends on the accuracy of computation – is added to the diagonal terms in the element stiffness matrix during global matrix assembly. This additional term constrains the volumetric strain to nearly zero, i.e., undrained. In both these approaches, the constitutive response of the soil is defined in terms of effective stress parameters. During the stress recovery phase of the penalty formulation calculation, the bulk modulus of water is not included in the stiffness terms so that the effective stress responses can be determined.

A simpler, alternate approach is to define the constitutive response in terms of total stress parameters, with care being taken to make the diagonal terms of the element stiffness matrix large, typically by using a Poisson's ratio close to 0.5. In this case, a Young's modulus corresponds to an undrained value and failure is expressed in terms of an undrained shear strength, S_u ($\varphi = 0$ and $c = S_u$).

Mana (1978) and field data have shown that for excavations through saturated clays with typical excavation periods of several months, the clays remain essentially undrained with little dissipation of excess pore pressures. However, there may be cases (i.e., O'Rourke and O'Donnell 1997) where substantial delays during construction occur and excess pore pressures partially dissipate, and in these cases one must use a mixed formulation to account for the pore water effects.

For cases where the water table is below the bottom of the excavation and when one does not expect any excess pore water pressures to develop during excavation, a displacement-based formulation is appropriate. Alternatively, in cases with high water levels, pore water pressures may need to be computed based on a fluid flow analysis wherein the free boundary of the seepage can be found, depending on the boundary conditions, presence/absence of a dewatering system, wall type and hydraulic conductivity of the soil (e.g., Brezis et al. 1978; Borja and Kishnani 1991). Generally a steady state analysis is adequate in cases with high hydraulic conductivity.

3 INITIAL CONDITIONS

A reasonable prediction of the ground response to construction of a deep excavation starts with a good estimate of the initial stress conditions, in terms of both effective stresses and pore water pressures. The effective stress conditions for excavations in well-developed urban areas rarely correspond to at-rest conditions because of the myriad past uses of the land. Existence of deep foundations and/or basements from abandoned buildings and nearby tunnels changes the effective stresses from at-rest conditions prior to the start of excavation. For example, Calvello and Finno (2003) showed that an accurate computation of movements associated with an excavation could only be achieved when all the pre-excavation activities affecting the site were modeled explicitly. They used the case of the excavation for the Chicago-State subway renovation project (Finno et al. 2002). In this project, construction of both a tunnel and a school impacted the ground stresses prior to the subway renovation project. Ignoring these effects made a difference of a factor of 3 in the computed lateral movements.

One must also take care when defining the initial ground water conditions. Even in cases where the ground water level is not affected by near surface construction activities, non-hydrostatic conditions can exist for a variety of reasons. For example, Finno et al. (1989) showed pneumatic piezometer data that indicated a downward gradient within a 20 m thick sequence of saturated clays. This downward flow arose from a gradual lowering of the water level in the upper rock aquifer in the area as a result of municipal use and the effects of the presence of a deep tunnel system constructed in the 1970s and 1980s for the purposes of storm water control and treatment. A non-hydrostatic water condition affects the magnitude of the effective stresses at the start of an excavation project.

An engineer has two choices to define such conditions – to measure the *in situ* conditions directly or to simulate all the past construction activities at a site starting from appropriate at-rest conditions. Because both approaches present challenges in their own right, it is advantageous to do both to provide some redundancy in the input. In any case, careful evaluation of the initial conditions is required when numerically simulating supported excavation projects.

4 REPRESENTING CONSTRUCTION

While supported excavations commonly are simulated by modeling cycles of excavation and support installation, it is necessary to simulate all aspects of the construction process that affect the stress conditions around the cut in order to obtain an accurate prediction of behavior. This may involve simulating installation of the supporting wall and any deep foundation elements, as well as the removal of cross-lot supports or detensioning of tieback ground anchors. Furthermore, issues of time effects caused by hydrodynamic effects or material responses may be important.

4.1 *Wall installation*

Many times the effects of the installation of a wall is ignored and the wall is *wished into place* with no change in the stress conditions in the ground or any attendant ground movements. However, there is abundant information that shows ground movements may arise during installation of the wall.

O'Rourke and Clough (1990) present data that summarize observed settlements that developed during installation of 5 diaphragm walls. They noted settlements as large as 0.12% of the depth of the excavation. These effects can be evaluated by 3-dimensional modeling of the construction process (e.g., Ng and Yan 1999; Gourvenec and Powrie 1999). For these analyses, selecting a value of specific gravity for the supporting fluid in the trench is not as straightforward as it first may seem. The specific gravity of the supporting fluid usually varies during excavation of a panel as a result of excavated solids becoming suspended – increasing the specific gravity above the value of the water and bentonite mixture – and subsequently decreasing when the slurry is cleaned prior to the concrete being tremied into place. Consequently, it is difficult to select one value that represents an average condition. Furthermore the effects of the fluid concrete on the stresses in the surrounding soil depend how fast the concrete hardens relative to its placement rate. Some guidance in selecting the fresh concrete pressure is provided by Lings et al. (1994).

It is less straightforward when considering diaphragm wall installation effects in a more conventional plane strain analysis because the arching caused by the geometry of the excavation of individual panels cannot be taken directly into account. To approximate the effects of the arching that is not present in the plane strain analysis when making such an analysis, an equivalent fluid pressure must be applied to the walls of the trench to maintain stability that is generally higher than the level of the fluid during construction. It is more difficult to simulate in a plane strain analysis the effects of installation of a secant pile wall when the individual piles comprising the wall are drilled without any slurry as was the case at the Chicago-State subway excavation (Finno et al. 2002). In either case, some degree of empiricism is required to consider these effects in plane strain analyses. One can back-calculate an equivalent fluid pressure corresponding to the observed ground response if good records of lateral movements close to the wall are recorded during construction. More data of this type is needed before any recommendations can be made regarding magnitudes of appropriate equivalent pressures.

The effects of installing a sheet pile wall are different than those of a diaphragm wall, yet the effects on observed responses also can be significant. In this case, both settlements due to transient vibrations developed during driving or vibrating the piles into place and ground movements due to the displacement of the ground by the piles may develop. The former is of practical importance if installing the piles through loose to medium dense sands, and can be estimated by procedures proposed by Clough et al. (1989). However, these effects are not included in finite element simulations. The latter effects in clays were illustrated by Finno et al. (1988). In this case, the soil was displaced away from the sheeting as it was installed and was accompanied by an increase in pore water pressure and a ground surface heave. As these excess pore water pressures dissipated, the ground settled. The maximum lateral movement and surface heave was equal to one-half the equivalent width of the sheet pile wall, defined as the cross-sectional area of the sheet pile section per unit length of wall. For a PZ-40 section, this is 25 mm.

Simulating sheet-pile installation can be accomplished in plane strain by representing a section of sheeting as indicated in Figure 1. Insertion of the wall can be modeled by displacing two rows of nodes away from one another by the amount equal to the equivalent width of the sheet-pile section (1a). Thereafter, the sheet-pile elements are activated by removing one row of nodes and the vertical and horizontal restraints for the other displaced row and initializing the beam

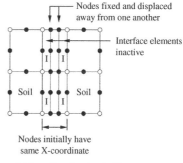

(a) Inserting the sheet-piles

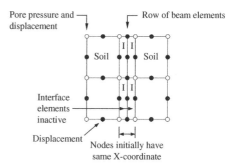

(b) Before start of excavation simulation

Figure 1.   Simulation of sheet-pile installation.

element representing the sheet-pile (1b). The interface elements are activated by assigning shear stiffness representative of the soil-wall interface and assuming no slip has occurred.

In addition to these effects, installing the walls can have a large influence on subsequent movements, especially if the walls are installed relatively close to each other, as may be the case in cut-and-cover excavation for a tunnel. Sabatini (1991) conducted a parametric study wherein the effects of sheet-pile wall installation in clays were compared with simulations where the walls were *wished into place* as a function of the depth, H, to width, B, of the excavation. A mixed finite element formulation was employed to allow pore water pressure generation and dissipation in the clays. The results of the study are shown in Figure 2 where the computed maximum lateral movements, $\delta_{H(max)}$, are plotted versus H/B. The value of H was kept constant at 12.2 m, the depth to an underlying firm layer was a constant value of 6.1 m whereas the values of B were varied. Three levels of cross-lot braces provided support for the walls.

Two types of simulations are reported in Figure 2, a centerline symmetric geometry and a full mesh where the centerline symmetry is not assumed. In

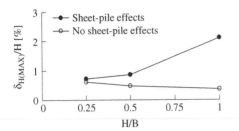

Figure 2. Effects of sheet-pile installation on computed lateral movements.

the former case, it is implicitly assumed that both sheet-pile walls are installed at the same time, whereas sequential installation effects can be simulated in the latter arrangement. Installing the sheets at the same time maximizes the effects because the sheet-pile installation-induced pore water pressures have no time to dissipate. For a time frame of up to 3 months between installation, differences less than 5% in $\delta_{H(max)}$ were noted and thus are not shown explicitly on the figure.

The results show that for the *wished in place* case when the sheet-pile installation effects are ignored, the lateral movements are larger for wider excavations, a similar trend reported by Mana and Clough (1981). The installation procedure has two main effects: the soil adjacent to the excavation is preloaded and the shear strength on the passive side is (partially) mobilized prior to the beginning of the cycles of excavation. Wall installation tends to preload the soil on the active side of the excavation as a result of the reduction in shear stress at approximately constant mean normal effective stress. This mechanism provides the soil outside the walls with more available shearing resistance when the cycles of excavation start. However, the soil between the walls has less available passive resistance as a result of the preloading and this promotes the larger movements during excavation as compared to the case of ignoring the sheet-pile effects. It is apparent for wide excavations, the decision to include installation effects in a simulation is not critical. However, these effects become pronounced for narrow excavations and should be explicitly considered.

4.2 *Excavation and support cycles*

Ghaboussi and Pecknold (1985) indicated that the correct solution to the incremental excavation problem involves satisfying at any step n:

$$R_{(n)} = F_{ext(n)} - F_{int(n)} = \int_{\Omega^e} \gamma_t N^T f \, d\Omega^e - \int_{\Omega^e} B^T \sigma \, d\Omega^e \quad (1)$$

where R is the residual force vector, F_{ext} is the external force vector, F_{int} is the internal force vector, γ^t is the total unit weight of the soil, N is the shape function

matrix, f is the unit body force vector, Ω^e is the element domain, B is the strain-displacement matrix, σ is the vector of total stresses and the superscript T implies a matrix transposition. For equilibrium R approaches zero. In (1), the total gravity loads are balanced by total internal stresses distributed over the excavated surface. Equation (1) implies that whenever water is present and the constitutive response is represented by an effective stress model, both total and effective stress vectors must be saved. Other approaches for simulating the excavation process (e.g., Christian and Wong 1973; Clough and Mana 1976) are approximate and can lead to errors in the solution simply as a result of applying incorrect loads.

Care must be taken when specifying water levels in conjunction with excavation loading. In some cases, the physics of the solution removes any potential ambiguity when handling the water levels. For example, for excavations through high permeability soils, the excavation must be dewatered prior to removing the soil in the field and the water levels in the numerical simulation must account for the dewatered state. In contrast, when excavating through saturated clays wherein the constitutive responses are represented by an effective stress model, one must be sure that the σ in (1) is indeed total stresses and that the water levels are correctly manipulated throughout the simulation. If a mixed formulation is used, the phreatic surface inside the excavation must be specified as the excavated surface, but ideally there should be small enough elements so sharp gradients can exist near the excavated surface and little dissipation of excess pore water pressure occurs during the normal durations of excavations. Commercially available codes handle the question of pore water pressures and excavation forces in various ways, and thus excavation procedures can vary from code to code.

Representing lateral support elements in a finite element simulation in plane strain conditions is accomplished by dividing the actual support stiffness by the support spacing. For cross lot bracing, this is a direct procedure. In this case, the effect of the waler is assumed to uniformly spread the load to the wall. When a wall is supported by tiebacks, several options are available to an analyst. One can model the ground anchor and the tierod explicitly, thereby resulting in a complicated mesh. The ground anchor can consist of a solid element surrounded by interface elements and the tierod can be represented by a bar element tied to the wall and the end of the anchor. The benefits of going to these extremes are not necessarily clear, given the simplifications inherent in the model, e.g., the effects of drilling and grouting pressures are not included in the analysis. Clough and Tsui (1974) suggested that the tiebacks could be represented by bar elements attached to the wall oriented along the line of the tieback with its stiffness equal to the tierod stiffness

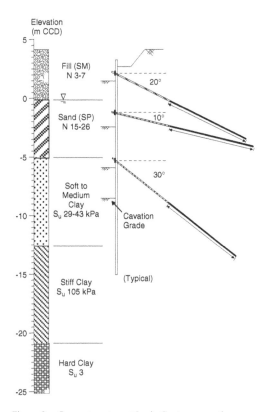

Figure 3. Support system at Lurie Center excavation.

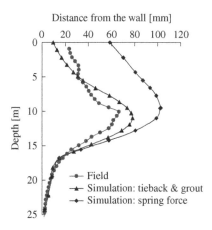

Figure 4. Effects of tieback simulation.

The effects of how one models the anchors are quite clear. The more complicated approach produces much higher lateral displacements in the form of a rotation of the displaced profile, as a result of movements of the ground anchors. This is not observed in the computed results for the simplified approach, which more faithfully represent the observed movements. This admittedly is an extreme case because of the orientation of the anchors. The finite element discretization near the ground anchors was such that only one row of soil elements were located between the anchors, exaggerating the effects of the load transfer from the anchors to the soil, and resulting in excessive lateral movements near the top of the wall.

divided by its spacing. In this simplified approach, it is assumed that the ground anchor is unyielding. If performance test data are available, one can use the stiffness determined from the unloading portion of the test from the maximum load to the lock-off load.

Sometimes it is necessary to model tieback anchors in one way because of the geometry of the installation. Figure 3 shows a section of the support system for the Lurie Center excavation (Finno and Roboski 2005). The top two rows of ground anchors were supported in the same sand stratum. The horizontal spacing between the anchors was 1.5 m, but the anchors at levels 1 and 2 were staggered so that theoretically there was at least 0.75 m between each anchor. When making a plane strain analysis of this excavation, if one explicitly models the entire anchor, there will be very little space between the two rows of anchors.

To illustrate the effect this can have on computed displacements, Figure 4 shows horizontal displacements computed by considered the anchor system explicitly and the simplified approach. The simulation was made with ABAQUS and the same soil parameters were used in each simulation. Also shown is inclinometer data obtained at the end of the excavation.

5 CONSTITUTIVE MODELS

When one undertakes a numerical simulation of a deep supported excavation, one of the key decisions made early in the process is the selection of the constitutive model. This selection should be compatible with the objectives of the analysis. For example, for preliminary analysis, a simple model with few parameters is appropriate, but one's expectations of the results of the simulation must be tempered with the realization that the model is inherently limited in its ability to reflect all aspects of the complicated response of natural soils. For simulations that form the basis of the design of the system, particularly a case where control of ground movements is a key design consideration, the constitutive model must be able to reproduce the key elements of response to the expected loadings. The following sections describe some considerations that are applicable specifically to excavations made in relatively soft cohesive soils, but also are applicable in general to other subsurface conditions.

5.1 Model selection

The response of soil at small strain levels is an important input for many geotechnical problems where the design depends on limiting deformations to an acceptably small level. Real soils are neither linear elastic nor elastic-perfectly plastic, but exhibit complex behavior characterized by zones of high constant stiffness at very small strains, followed by decreasing stiffness with increasing strain. This behavior under static loading initially was realized through back-analysis of foundation and excavation movements in the United Kingdom from the late 1960s to the early 1980s (Burland 1989), primarily because of significantly smaller measured deformations than predicted by analytical, empirical, and finite element methods. The recognition of zones of high initial stiffness under typical field conditions was followed by efforts to measure this behavior in the laboratory environment (Burland and Symes 1982; Jardine et al. 1984). While this initial work was based on responses of stiff clays, laboratory data shows small strain non-linearity for relatively soft clays (e.g., Clayton and Heymann 2001; Santagata et al. 2005; Callisto and Calebresi 1998), and indeed is present in all soils.

Burland (1989) suggested that working strain levels in soil around well-designed tunnels and foundations were on the order of 0.1%. If one uses data collected with conventional triaxial equipments to discern the soil responses, one can reliably measure strains 0.1% or higher. This limitation led to the development of on-sample instrumentation and subsequent evaluation of strain responses less than 0.1% (e.g. Jardine et al. 1984). Thus in many practical situations, it is not possible to accurately incorporate site-specific small strain non-linearity into a constitutive model based on conventionally derived laboratory data.

When using simpler, elasto-plastic models, the key decision to select the elastic parameters that are representative of the secant values that correspond to the strain levels in the soil mass. Examples of the strain levels behind a wall for an excavation with lateral wall movements of 29 and 57 mm are shown in Figure 5.

These strain levels were computed based on the results of displacement-controlled simulations where the lateral wall movements and surface settlements were incrementally applied to the boundaries of a finite element mesh. The patterns of movements were typical of excavations through clays and were based on those that were observed at the excavation for the Ford Design Center made through Chicago clays (Finno and Blackburn 2005). Because the simulations were displacement-controlled, the computed strains do not depend on the assumed constitutive behavior.

As can be seen in Figure 5, the maximum shear strains correspond to about 0.3% for 29 mm maximum wall lateral movement, and represent good control of ground movements in these soft soils. Shear strains

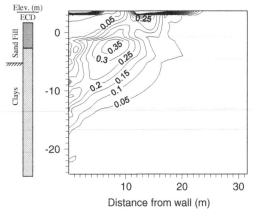

(a) 29 mm maximum lateral wall movement

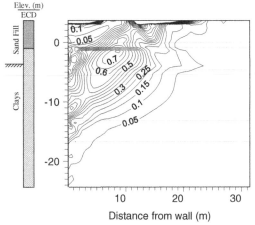

(b) 57 mm maximum lateral wall movement

Figure 5. Shear strain levels behind excavation.

as high as 0.7% occur when 57 mm of maximum wall movement develop. These strain levels can be accurately measured in conventional triaxial testing, and thus if one can obtain sufficiently undisturbed specimens, then secant moduli corresponding to these strain levels can be determined. Because the maximum horizontal wall displacement can be thought of as a summation of the horizontal strains behind a wall, the maximum wall movements can be accurately calculated with a selection of elastic parameters that corresponds to the expected strain levels. In this case, the fact that small strain non-linearity is not explicitly considered will not have a large impact on the computed horizontal displacements because they are dominated by the larger strains in the soil mass. However, if one needs to have an accurate representation of the distribution of settlements with distance from the wall, then this approach of selecting strain-level appropriate elastic parameters will not work. The small strain non-linearity must be explicitly considered to find the *extent* of the settlement because the strains in the area of interest vary from the maximum value to zero. As a consequence, many cases reported in literature indicate computed wall movements agree reasonably well with observed values, but the results from the same computations do not accurately reflect the distribution of settlements. This latter result can be obtained only if the small stain non-linearity of the soil is adequately represented in the constitutive model.

There are a number of models reported in literature wherein the variation of small strain nonlinearity can be represented, for example, a three-surface kinematic model develop for stiff London clay (Stallebrass and Taylor 1997), MIT-E3 (Whittle and Kavvadas 1994) and hypoplasticity models (e.g., Viggiani and Tamagnini 1999; Kolymbas 1991). The authors have tried somewhat unsuccessfully to apply these approaches to excavations in Chicago, although this fact does not necessarily indicate that the models cannot be useful in these cases, but rather indicate the shortcomings of the particular users.

To illustrate some difficulties and to highlight the main advantage of incorporating small strain non-linearity into a constitutive model, the remainder of this paper will be used to report on the results of an extensive, on-going experimental program aimed at defining the behavior of compressible Chicago glacial clays from strain ranges of 0.001% and higher, briefly describe a tangent stiffness model that incorporates many aspects of the observed laboratory behavior and present a comparison of the results of a finite element simulation with both conventional elastoplastic and the tangent stiffness models with field observations made at the Lurie Research Center excavation.

5.2 *Example of small strain nonlinearity – compressible clay*

Sample disturbance is the greatest impedance to accurate measurement of very and small strain shear moduli for soft clays. Disturbance is caused by many factors, including but not limited to borehole instability, reduction of effective stress, shear strains induced by tube penetration, and sample extrusion and trimming (Hight 2001). Disturbance causes a variety of different effects on soft soils, but typically, results in a flatter modulus degradation curve due to progressive destructuring and shrinkage of the yield surface (Clayton et al. 1992). However, it is also possible for the moduli of a somewhat disturbed sample to become greater after reconsolidation due to significant decreases in void ratio. Using high quality block samples and careful handling of triaxial specimens can minimize the effects of disturbance.

Table 1. Summary of average block sample properties.

Parameter	Block LB2	Block LB3
Water content (%)	29.2 (0.5)	28.5 (0.6)
Liquid limit (%)	38	37
Plasticity index (%)	19	19
Void ratio	0.82 (0.01)	0.79 (0.02)
Unit weight (kN/m^3)	18.9 (0.1)	19.0 (0.1)
Limit pressure (kPa)	186	190
Overconsolidation ratio	1.5	1.5

Note: standard deviations are shown in ( ).

To this end, compressible Chicago glacial clay samples were obtained from high quality block samples obtained at several excavations in the Chicago area. Results presented herein are taken from those blocks cut from the excavation for the Robert H. Lurie Cancer Research Center in downtown Chicago (Finno and Roboski, 2005). Hand-cut block samples, approximately 0.3 m in each dimension were removed from a depth of about 10.4 m below street level. The results presented herein were obtained from experiments on two blocks, designated LB2 and LB3. Table 1 presents a summary of the average properties of these blocks, and illustrates the degree of uniformity of the clay within the blocks.

5.2.1 *Experimental program*

Triaxial specimens were hand-trimmed from the block samples with a nominal diameter of 71 mm and a height to diameter ration between 2.1 and 2.3. Each specimen was reconsolidated under K_0 conditions to the in-situ vertical effective stress σ'_{v0} of 134 kPa, and then subjected to a 36 hour K_0 creep cycle. Bender element (BE) tests were conducted during the reconsolidation and stress probing portions of the test for each specimen. The results are presented in detail by Holman (2005).

Following the K_0 creep phase, the specimens were subjected to directional stress probes under drained axisymmetric conditions. The internal deformation measurements made by subminiature LVDTs mounted directly on the specimen were used to calculate raw axial and radial strain values using the measured axial gage length and sample diameter, respectively. The axial load was measured using an internal load cell and corresponding axial stresses were calculated using the measured axial load and the current sample area from the measured radial deformation. Cell and pore pressures were measured using external differential pressure transducers. Internal stress and strain measurements were made at 5 to 20 second intervals by an automated data acquisition and control system.

The readings of the axial LVDTs were averaged to produce a single axial deformation response, assumed

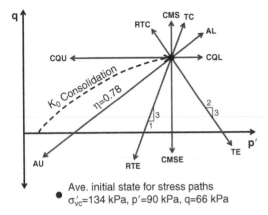

Ave. initial state for stress paths
σ'_{vc}=134 kPa, p'=90 kPa, q=66 kPa

Figure 6.   Schematic diagram of drained directional stress probes.

to be representative of the centerline deformations within the zone of local measurement. Smoothed values of data collected by each transducer and load cell were used to calculate the local axial strain ε_a, local radial strain ε_r, shear stress, $q = \sigma'_v - \sigma'_r$, and mean normal effective stress, $p' = (\sigma'_v + 2\sigma'_r)/3$. The local triaxial shear strain, ε_s, secant shear modulus, G_{sec}, and secant bulk modulus, K_{sec}, were computed from these data as:

$$\varepsilon_s = \frac{2}{3}(\varepsilon_a - \varepsilon_r) \qquad (2)$$

$$G_{sec} = \frac{\Delta q}{3\varepsilon_s} \qquad (3)$$

$$K_{sec} = \frac{\Delta p'}{\varepsilon_a + 2\varepsilon_r} \qquad (4)$$

The use of these quantities for secant stiffness, while strictly applicable to isotropic linear elasticity, are employed to present the data in a conventional fashion and is not intended to imply either isotropic or elastic responses.

Figure 6 illustrates the stress probe directions in q-p' space and Table 2 summarizes the notation used to describe each path. All stress probes were carried out at a stress rate of 1.2 kPa/hour to minimize accumulation of excess pore water pressure within a specimen. Duplicate tests were conducted for the majority of the stress probes.

5.2.2   Results of stress probe tests

The stress-strain data from the directional stress probes was processed to allow for examination of the directional stiffness as a function of strain. Data are presented in terms of secant shear and bulk moduli as defined in equations 3 and 4.

Table 2.   Notation for drained stress probes.

Label	Description
AL	Anisotropic loading ($\eta = 0.78$)
TC	Triaxial compression
CMS	Constant mean normal stress
RTC	Reduced triaxial compression
CQL	Constant shear loading
AU	Anisotropic unloading ($\eta = -0.78$)
RTE	Reduced triaxial
CMSE	Reduced constant mean
TE	Triaxial extension
CQU	Constant shear unloading

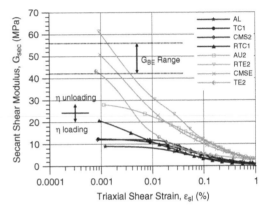

Figure 7.   Secant shear modulus degradation curves for directional stress probes.

5.2.2.1   Secant shear modulus

The secant shear modulus was plotted versus triaxial shear strain in Figure 7 for selected natural specimens whose stress probes involved changes in the shear stress q. The stress probes wherein q and the stress ratio, $\eta = q/p'$, is increased ("η loading") exhibit a zone of near-constant G_{sec} for strains up to 0.002 to 0.005%. Various researchers have referred to the strain level beyond which the modulus degrades rapidly as an elastic threshold strain. By referring to this strain level as elastic, it is implied that the behavior is completely reversible. However, as no cyclic load-unload cycles were performed during the stress probes, there is no direct evidence that the behavior of compressible Chicago glacial clays is truly recoverable, even at strain levels less than 0.005%.

In contrast, cases where q and η initially decrease ("η unloading"), do not contain obvious zones of constant G_{sec} at shear strains greater than 0.001%, and thus no elastic threshold is observed in these data for strain levels.

Differences in shear modulus degradation for deviatoric loading and unloading stress probes at small

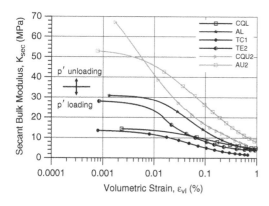

Figure 8. Bulk modulus degradation for directional stress probes.

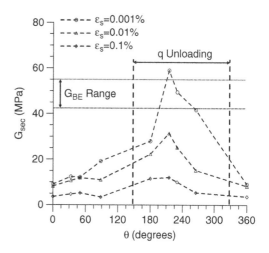

Figure 9. Secant shear moduli as a function of recent stress history.

shear strains are clearly illustrated in Figure 7. The value of G_{sec} at 0.001% strain, hereafter referred to as $G_{0.001}$, varies from 9 MPa to 19 MPa for the loading probes and ranges from 28 to 59 MPa for the unloading probes. The secant shear modulus degradation is more pronounced for the unloading probes than for the loading probes. The rates of shear modulus degradation in the small strain range can be represented as $G_{0.001}/G_{0.1}$. The mean values of $G_{0.001}/G_{0.1}$ are 3.3 for the loading probes and 5.0 for the unloading probes. Thus the unloading stress paths resulted in a greater decrease in shear modulus within the small strain range.

5.2.2.2 Secant bulk modulus

The secant bulk modulus K_{sec} was plotted versus the volumetric strain ε_{vl} in Figure 8 for stress probes on natural specimens involving changes in p'. Most stress probes resulted in the characteristic reverse S-shaped degradation curve. Like the shear modulus degradation curves for increasing η in Figure 7, the bulk modulus degradation curves for increasing p' in Figure 8 exhibit a relatively constant K_{sec} or threshold volumetric strain for strains as high as 0.005%.

The modulus degradation curves shown in Figure 9 demonstrate the differences in behavior as a function of p' loading or unloading, where loading implies an increase in p' during the stress probe. The value of, K_{sec} at 0.001% strain, or $K_{0.001}$, varies from 13.4 MPa to 31 MPa for the loading probes. $K_{0.001}$ varies from 28 to 74 MPa for the unloading probes.

5.2.2.3 Effects of recent stress history on stiffness

Experimental results published by a number of investigators, (e.g., Smith et al. 1992; Atkinson et al. 1990) showed the effects that the recent or previous stress history has on the stress-strain behavior of clays. These researchers have presented data that show any stress path rotations from the previous path, denoted the recent stress history, result in changes in soil

moduli. Atkinson et al. (1990) defined stress path rotation in terms of the angular difference θ (positive counterclockwise herein) between the previous stress path direction and a new stress path direction. The recent stress history of the compressible Chicago clays employed in this experimental evaluation is assumed to be defined by the K_0 reconsolidation path. However, the reconsolidation path is nonlinear in p'-q space, complicating an exact definition of recent stress history in terms of a constant η stress path direction. For simplicity, the η representing the recent stress history was therefore assumed to be 0.74 and probe AL is assumed to represent an extension of the recent stress history path. Even though the K_0 creep cycle resulted in a nearly 90° change in stress path direction at the end of consolidation, the length of the creep stress path was generally less than 10 kPa and was not considered large enough to have any significant influence on the directionally-dependent stiffness.

The clays from which the blocks were cut are ice margin deposits which after deposition about 8000 years ago were subjected to a large drop in the water table with a subsequent recovery thereof. In the late 1800s, about 4 m of fill consisting of debris from the Chicago fire of 1871 was placed to raise the grade to its current level. Thus the most recent event that affected the *in situ* clay's stress history prior to the excavation for the Lurie Building can be represented as a one-dimensional loading of fill, similar to the path of the reconsolidated clay in the laboratory.

The recent stress history effects on initial shear moduli and degradation behavior are examined in Figure 9. The angle θ was calculated for each of the stress probes presented in Figures 7 and 8 and plotted against the values of G_{sec} and K_{sec} for 0.001%, 0.01%, and 0.1% strain. Data points for path AL are shown at 0° and 360°. Zones of θ corresponding to reduced shear stress (unloading) also are delineated. The variation in G_{sec} is nonlinear with respect to stress path rotation angle θ for all small strain levels. The dependence of shear modulus on θ is large at 0.001% and 0.01% ε_{sl}, but the apparent directional dependence decreases as the strains reach and exceed 0.1%. The majority of the variation in G_{sec} occurs in the unloading region of the plot. There are clear differences in the behavior demonstrated by each G_{sec}–θ curve for $0° < \theta < 120°$ and $160° < \theta < 300°$. Between 0 and 120°, the trend of the G_{sec}–θ data is nearly constant. Between 160° and 300°, the variation of G_{sec} with respect to θ is nonlinear. The peak G_{sec} for each curve occurs within this range with a maximum occurring at θ of about 210°. The secant shear modulus is technically indeterminate at $\theta = 143.5°$ and $\theta = 323.5°$ because there is no change in q for those probes. The curves should be discontinuous at these points, but the data trend is very strong to either side. The existence of shear strain development at these values of θ can be attributed to coupling between shear and volumetric behavior.

Unlike the $G_{0.001}$ from the directional stress probes, the dynamic modulus G_{BE} is primarily dependent on the mean normal effective stress p'. G_{BE} is commonly assumed to represent the maximum shear modulus G_{max} at very small strains between 0.0001 and 0.001% (Dyvik and Madshus, 1985), similar to the assumed strain levels for field geophysical testing such as seismic cone penetration (SCPT). The mean V_{BE} was 161 m/s, resulting in an average G_{BE} value of 51 MPa. Figure 9 indicates that G_{BE} approximates the $G_{0.001}$ value occurring at about 210°, near the CMSE and RTE paths where q decreases at constant or nearly constant mean normal effective stress.

The secant bulk modulus, K_{sec}, also varies nonlinearly with strain level and θ, as shown in Figure 10. The general shapes of each K_{sec}–θ curve are similar to the secant shear moduli in Figure 9 in that most variation occurs in the unloading range. However, the peak value of K_{sec} occurs at about 130° for the CQU path suggesting that the stiffest bulk modulus response is observed in the p' unloading direction. The secant bulk modulus is technically indeterminate at $\theta = 53.5°$ and $\theta = 233.5°$ because there is no change in p' for those probes. The curves should be discontinuous at these points. The development of volumetric strains at these values of θ can also be attributed to coupled shear and volumetric behavior.

In summary, the experimental data clearly indicate that the stress-strain response of these compressible Chicago clays is directionally dependent. The stiffness of the clay generally evolves continuously and nonlinearly as a function of loading direction, recent stress history and strain level. This variability in response shows these clays are incrementally nonlinear.

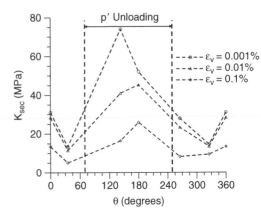

Figure 10. Secant bulk moduli as a function of recent stress history.

5.3 Tangent stiffness constitutive relation

Based on triaxial stress probe tests on Chicago clay, the material response can be separately analyzed in terms of deviatoric and volumetric components using the following relation:

$$\begin{cases} \delta\varepsilon_v = \delta p'/K + \delta q/J_v \\ \delta\varepsilon_s = \delta p'/J_s + \delta q/(3G) \end{cases} \tag{5}$$

Eq. 6 provides definitions for individual stiffness indices and triaxial tests that can be used to determine the values.

$$\begin{array}{ll} K = \partial p'/\partial\varepsilon_v; & CQL, CQU \\ J_v = \partial q/\partial\varepsilon_v; & CMC, CME \\ G = \partial q/3\partial\varepsilon_s; & any\ stress\ path \\ J_s = \partial p'/\partial\varepsilon_s; & implicitly\ derived \end{array} \tag{6}$$

As shown in Figures 9 and 10, there generally are two distinct types of response, loading and unloading, for each stiffness index. Irrespective of path, all stiffness indices tend to decrease rapidly when the stress changes from the initial stress state. The rapid degradation rate slows after a certain stress change from the initial stress state. This characteristic stress change outlines a boundary in stress space that can be used to define the region of the small strain behaviors.

5.3.1 Shear modulus G and coupling modulus J_s

Experimental data have shown that when $\delta p' < 0$ and $\delta\eta = 0$ (stress probe path – AE) virtually no shear strain is developed. This observation suggests that for most stress paths leading to failure, ε_s is more likely due to change in the quantity η instead of changes in the

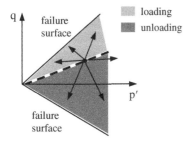

Figure 11. Loading–unloading criterion for G.

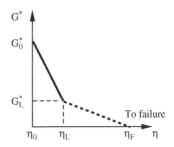

Figure 12. Stiffness variations for G^*.

deviatoric stress q. Thus a new stiffness index G^* is introduced to describe the relation between ε_s and η.

$$\delta\varepsilon_s = \frac{\delta\eta}{G^*} = \frac{\delta q/p'}{G^*} - \frac{(q/p'^2)\delta p'}{G^*} = \frac{\delta q}{p'G^*} - \frac{\delta p'}{(p'/\eta)G^*} \quad (7)$$

The values of G and J_s can be derived from G^* in the following way:

$$G = p'G^*/3$$
$$J_s = -(p'/\eta)G^* \quad (8)$$

Figure 11 shows the loading–unloading criterion for the shear modulus G. It can be expressed as:

$$loading:\ \ \delta\eta > 0 \ \ or \ \ \frac{\delta q}{p} - \frac{q\delta p}{p^2} > 0$$
$$unloading:\ \ \delta\eta < 0 \ \ or \ \ \frac{\delta q}{p} - \frac{q\delta p}{p^2} < 0 \quad (9)$$

According to eq. (8), the coupling modulus J_s has the same criterion as G. Note that this loading–unloading criterion is an idealization of the laboratory observation. Test results show that the stiffness index G^* varies continuously as the direction of the stress path changes. However, the variation is relatively minor within either zone, as indicated in Figure 8.

For stress paths leading to the failure surface, the quantity η is used to describe the boundary of the small strain region. As shown in Figure 12, η_0 refers to the initial condition and $(\eta_L - \eta_0)$ defines the range of the small strain region.

5.3.2 Bulk modulus, K, and coupling modulus, J_v

As indicated in eq. (6), only two stress probe tests can be used to calculate the bulk modulus K. The loading-unloading criterion shown in Figure 13 is a hypothesis expressed as:

$$loading:\ \ \delta p' > 0$$
$$unloading:\ \ \delta p' < 0 \quad (10)$$

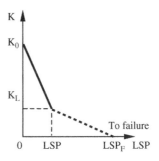

Figure 13. Loading–unloading criterion for K.

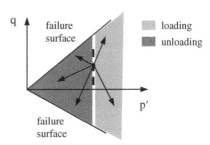

Figure 14. Variation for K with LSP.

This hypothesis works reasonably well in simulating soil response in other stress probe paths. Compared to the shear modulus G, K in most cases is a secondary concern. Thus, the directional dependence of K can be simplified with this loading–unloading criterion in the numerical model. Figure 14 shows the variation of K in terms of the length of stress path, LSP, defined as:

$$LSP = \sum \sqrt{(\Delta p_i)^2 + (\Delta q_i)^2} \quad (11)$$

where $(\Delta p_i, \Delta q_i)$ is the ith stress increment. The summation is made from the initial stress state to the current stress state. LSP and η are both stress-derived

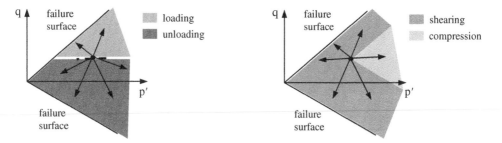

Figure 15. Loading–unloading criterion for J_v.

Figure 17. Shearing and compression zones.

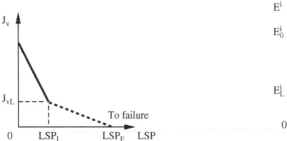

Figure 16. Variation of J_v with LSP.

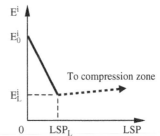

Figure 18. Stiffness variation in compression zone.

5.3.4 Incremental formulation

It is assumed that in situ Chicago glacial clay is a cross-anisotropic material with the following compliance matrix:

quantities. Using them to describe the stiffness degradation curve yields a similar pattern, as shown in both Figures 12 and 14. For stress paths leading to the failure surface, η is preferred, while for paths leading to the so-called compression zone using LSP is more reasonable. The definition of the compression zone is made in the next section.

Like K, the value of J_v can be obtained in only two stress probe tests. As shown in Figure 15, the loading-unloading criterion is defined as:

$$
\begin{aligned}
loading : \quad & \delta q > 0 \\
unloading : \quad & \delta q < 0
\end{aligned} \tag{12}
$$

Figure 16 shows the variation of J_v with respect to LSP. This hypothesis is reasonable as shown by successful simulation of other stress probe tests (Tu 2006).

5.3.3 Compression zone

The relation of stiffness variation for each stiffness index shown before is for stress paths leading to the failure surface. These stress paths fall in the shearing zone indicated in Figure 17. If a stress path does not intersect with the failure surface, it falls in the compression zone. In this zone, the stiffness variation follows the general trend shown in Figure 18. In this figure, E_i represents any of the stiffness indices.

$$
\begin{Bmatrix} \delta\varepsilon_x \\ \delta\varepsilon_y \\ \delta\varepsilon_z \\ \delta\gamma_{xy} \\ \delta\gamma_{yz} \\ \delta\gamma_{zx} \end{Bmatrix} = \begin{bmatrix} 1/E_h & -v_{hh}/E_h & -v_{vh}/E_v & 0 & 0 & 0 \\ -v_{hh}/E_h & 1/E_h & -v_{vh}/E_v & 0 & 0 & 0 \\ -v_{hv}/E_h & -v_{hv}/E_h & 1/E_v & 0 & 0 & 0 \\ 0 & 0 & 0 & 1/G_{hh} & 0 & 0 \\ 0 & 0 & 0 & 0 & 1/G_{vh} & 0 \\ 0 & 0 & 0 & 0 & 0 & 1/G_{vh} \end{bmatrix} \begin{Bmatrix} \delta\sigma_x' \\ \delta\sigma_y' \\ \delta\sigma_z' \\ \delta\tau_{xy} \\ \delta\tau_{yz} \\ \delta\tau_{zx} \end{Bmatrix} \tag{13}
$$

where subscripts x and y indicate the two horizontal axes and z indicates the vertical axis. Under axisymmetric load conditions, G_{hh} and G_{vh} cannot be investigated. They will be discussed later. For a hyper-elastic material, the matrix in (13) should be symmetric. For an inelastic material like Chicago glacial clay, the compliance matrix generally is asymmetric, and (13) can be written as:

$$
\begin{Bmatrix} \delta\varepsilon_x \\ \delta\varepsilon_y \\ \delta\varepsilon_z \end{Bmatrix} = \begin{bmatrix} A & B & C \\ B & A & C \\ D & D & E \end{bmatrix} \begin{Bmatrix} \delta\sigma_x' \\ \delta\sigma_y' \\ \delta\sigma_z' \end{Bmatrix} \tag{14}
$$

There are 5 independent unknowns in eq. (14), A, B, C, D and E. For axisymmetric conditions, the incremental shear and volumetric strains can be written as:

$$
\begin{cases} \delta\varepsilon_s = 2(2D + E - A - B - C)\delta p/3 + 2(-2D + 2E + A + B - 2C)\delta q/9 \\ \delta\varepsilon_v = (2D + E + 2A + 2B + 2C)\delta p + 2(-D + E - A - B + 2C)\delta q/3 \end{cases} \tag{15}
$$

Comparing eq. (13) with eq. (6), one can derive the following set of linear equations:

$$\begin{cases} 2(A+B)+2C+2D+E=1/K \\ 2[-(A+B)+2C-D+E]/3=1/Jv \\ 2[-(A+B)-C+2D+E]/3=1/Js \\ 2[(A+B)-2C-2D+2E]/9=1/(3G) \end{cases} \quad (16)$$

Solutions to these equations can be expressed as:

$$\begin{cases} A+B=(-6/J_s+4/K+3/G-6/J_v)/18 \\ C=(6/J_v-3/J_s+2/K-3/G)/18 \\ D=(2/K-3/G+6/J_s-3/J_v)/18 \\ E=(3/G+1/K+3/J_s+3/J_v)/9 \end{cases} \quad (17)$$

The solutions given in eq. (17) are incomplete mappings of stiffness indices from $p'-q$ space to general stress space. The values of A and B cannot be calculated simply based on results from triaxial tests. Either an extra test or measurement must be made or a value for ν_{hh} should be assumed. Furthermore, G_{hh} and G_{vh} are still unknown. One simplification is to assume that all these shear stiffness indices are the same. With such assumptions, mapping from eq. (6) to eq. (13) is complete. Experimental work is continuing to define these quantities for the compressible Chicago clays.

Failure is defined in terms of the Matsuoka-Nakai surface. This surface coincides with the corners of the Mohr-Coulomb failure surface for both compression and extension stress paths, and can be expressed as:

$$f = I_3 + \frac{\cos^2 \phi_c}{9-\sin^2 \phi_c} I_1 I_2 = 0 \quad (18)$$

where

$$I_1 = \sigma_{kk} = \sigma_{11}+\sigma_{22}+\sigma_{33}$$
$$I_2 = (\sigma_{ij}\sigma_{ij}-\sigma_{kk}^2)/2 = \sigma_{12}^2+\sigma_{23}^2+\sigma_{31}^2-\sigma_{11}\sigma_{22}-\sigma_{22}\sigma_{33}-\sigma_{33}\sigma_{11}$$
$$I_3 = \det(\sigma_{ij}) = \sigma_{11}\sigma_{22}\sigma_{33}+2\sigma_{12}\sigma_{23}\sigma_{31}-\sigma_{11}\sigma_{23}^2-\sigma_{22}\sigma_{31}^2-\sigma_{33}\sigma_{12}^2$$

and ϕ_c is the friction angle in triaxial compression.

When the cohesion c is considered, the failure surface is formulated for a modified stress:

$$\bar{\sigma}_{ij} = \sigma_{ij} - c\cot\phi\delta_{ij} \quad (19)$$

For implementation within a finite element code, the compliance matrix is inverted. Details can be found in Tu (2006).

5.3.5 Input material parameters

A typical set of parameters include G_0^*, G_L^*, K_0, K_L, J_{v0}, J_{vL}, η_L or LSP_L, friction angle, ϕ, compression index, λ, and OCR.

Table 3. Parameters for tangent stiffness model.

Parameter	Loading	Unloading
G_0^*	90(CMS)	330(CME)
G_L^*	330(CMS)	1000(CME)
LSP_L (kPa)	8.1(CMS)	18(CME)
K_0^*	11000(CQL)	60000(CQU)
K_L^*	5000(CQL)	13000(CQU)
LSP_L (kPa)	14(CQL)	14(CQU)
ϕ_c	28° (TC)	
λ^*	0.04 (oedometer)	
OCR	1.5 (oedometer)	

The parameters G_0^* and G_L^* are functions of the stress path direction. One approach is to internally specify these functions such that G_0^* and G_L^* can be obtained in any single stress probe test, and their values in other directions can be derived from the internal function. In the version of the model implemented in the computations presented in this paper, a numerical interpolation scheme was developed so that values at different angles θ could be obtained. A more general function will be determined as more data are available. Values of K_0, K_L, J_{v0}, and J_{vL} have either a loading or unloading value. To further reduce the large number of parameters, one can assume that the ratio of stiffness parameters is constant, i.e, the ratio of $K_0^{loading}$ to $K_0^{unloading}$ is constant. Similarly, the ratio of K_0 to G_0 could be assumed constant, as was assumed herein based on available experimental results.

The values of the parameters based on the drained stress probes for specimens cut from the block samples obtained at the Lurie Center excavation are shown in Table 3.

5.4 Finite element simulation of Lurie Center excavation

5.4.1 Lurie Center description

The Robert H. Lurie Medical Research Building included a 12.8 m deep cut for two basement levels. This building is supported on caissons with a basement slab-on-grade. A plan view of the approximately 80 m × 68 m area is shown in Figure 18. Numerous timber piles and pile caps existed in the northeastern portion of the excavation. Old concrete caissons were present throughout the site. Steam, water, sewer, gas and electric lines run parallel with the excavation beneath the three streets around the cut. A pedestrian and utilities tunnel borders the northwest corner of the excavation. A ramp was moved between several locations on the south wall to provide access to the excavation.

Figure 3 previously presented when discussing effects of simulating tieback anchors showed the support system in relation to the stratigraphy. The

15

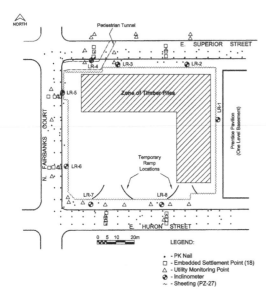

Figure 19. Plan view of Lurie excavation site.

Table 4. M-C parameters for Lurie Center simulation.

Stratum	E (MPa)	υ	$\phi°$	c (kPa)	$\psi°$
Fill	51	0.2	30	0	2
Sand	29	0.39	35	0	5
Stiff clay	171	0.49	0	105	0
Hard clay	677	0.49	0	383	0
Soft clay (settlement matched)	6.5	0.49	0	40	0
Soft clay (wall movement matched)	16	0.49	0	40	0

excavation is supported by a hot-rolled PZ-27 sheet pile wall on all sides. Two levels of tieback ground anchors are installed on the east wall due to the presence of the basement of the Prentice Pavilon. Three levels of tieback anchors provided lateral support on the other three walls. Both the first and second level ground anchors are founded in the beach sand.

Details of the construction process are found in Finno and Roboski 2005. Prior to installation of the sheeting, the excavator "pot-holed" the site to remove large obstructions such as pile caps and building rubble. PZ-27 sheets of length 18.3 meters were installed by a vibratory hammer. Ground water within the site was removed by dewatering wells. Care was taken to ensure the interlocks were kept intact to minimize leakage of ground water into the site. Surface water and leakage through the sheeting into the excavation was controlled by sump pits and pumps.

Excavation of the site and tieback installation took place simultaneously within the site. However, four distinct excavation stages were defined: excavation to elevations +1.5 m CCD, −2.5 m CCD, −5.8 m CCD, and −8.5 m CCD, corresponding to levels immediately below tieback elevations and the final excavated grade. Excavation was limited to a distance of 0.6 to 1.2 meters below the tieback installation elevation, depending on the angle of the tieback installation.

To monitor the ground response to excavation activities, 150 surface survey points, 18 embedded settlement points and 30 utility points were installed on three surrounding streets prior to wall installation (Figure 19). Measurements of both lateral and vertical ground surface movements were obtained. In addition to the optical survey data, seven inclinometers were installed at distances from 1 to 2.4 m from the sheet-pile wall.

5.4.2 *Finite element description*

The finite element simulation of the excavation for the Lurie Center was made using the commercial program ABAQUS. The purpose of this presentation is to show the effects small strain nonlinearity on computed results. Simulations were made in the same way for all cases, only the assumed constitutive responses were changed.

The glacial clays were simulated by either a Mohr-Coulomb (M-C) model or by the tangent stiffness model described in section 5.3. The M-C model was that implemented in ABAQUS (Mohr-Coulomb Plasticity) while the user-defined material option permitted use of the tangent stiffness model.

The M-C parameters for each soil stratum (Figure 3) are shown in Table 4. The values for the fill, sand, stiff clay and hard clay are based on results of site investigations made in conjunction with foundation design studies for the Lurie Center structure, and on past experience with excavations in the Chicago area. Experience has shown that the response of excavations in Chicago primarily depends on the responses in the soft clays. Note that the M-C parameters shown for the soft clay are those that were obtained by matching the computed responses with either the settlement or inclinometer data. The parameters used for the tangent stiffness model were those based on the drained stress probes on block samples, as summarized in Table 3.

The plane strain finite element simulation consisted of twelve steps and represented the construction history near inclinometer LR-8 (Figure 18). The simulation consisted of initializing stresses using by a "gravity calculation," activating the wall element, and cycles of excavation, tieback ratio less than 0.2. Because of the proximity of the first two levels of ground anchors (Figure 3), the tiebacks were simulated as springs with axial stiffness determined from performance tests.

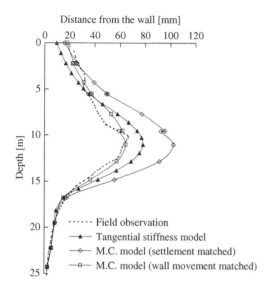

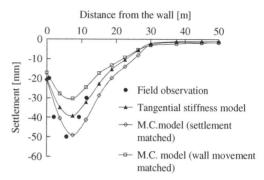

Figure 21. Computed and observed settlements at end of excavation.

Figure 20. Computed and observed lateral deformations at end of excavation.

The results of the calculations are shown in Figures 20 and 21. Figure 20 shows the lateral displacements computed based on the tangent stiffness model, two M-C based models – one with the settlements matched and one with the lateral movements matched, and the field observations based at inclinometer LR-8. The settlements computed with the same 3 models and the field observations are shown in Figure 20. Note that the results are based on the conditions at the end of the excavation.

It is clear that the M-C models can be used to repli-cate the observed results, but at a price that the elastic must be altered to provide such a match. This is a result of using constant elastic parameters that cannot take into account the small strain non-linearity of the compressible Chicago clays. In contrast, the tangent stiffness model based on the laboratory test results does a reasonable job of replicating both the lateral movements and settlements observed at this location.

The comparisons shown in Figures 20 and 21 illus-trate that even simple models can be used to provide reasonable comparisons with at least one set of obser-vations, if one adjusts the key parameters. Note that the variation in the moduli is Table 4 is 250%, making this approach as rather unreliable predictive tool. These results show use of simple models in numerical simu-lations is most applicable to parametric studies during the design stages of a project. If one wants to make predictions of behavior, particularly of the distribution of settlements, then one must employ a constitutive model that incorporates small strain nonlinearity.

Also note that the agreement of computed and observed results in Figures 19 and 20 was only possible

because the tiebacks were simulated as springs applied at the wall. The initial conditions and expected con-struction procedures must be incorporated correctly into any finite element simulation for reasonable results to be obtained.

6 CONCLUDING REMARKS

This paper discusses several aspects of making finite element simulations of supported excavations and emphasizes the need to accurately represent all pre-construction and construction activities that signifi-cantly affect the stresses in the soil affected by the cut. Examples were drawn from excavation from recent excavations through saturated, compressible glacial clays in Chicago. In addition to reasonably represent-ing pertinent construction activities, proper consider-ation of small strain non-linearity must be included in the constitutive model representing the soil clays if one needs to evaluate both lateral wall movements and settlements behind a wall.

ACKNOWLEDGMENTS

Financial support for this work was provided by National Science Foundation grant CMS-0219123 and the Infrastructure Technology Institute (ITI) of Northwestern University. The support of Dr. Richard Fragaszy, program director at NSF, and Mr. David Schulz, ITI's director, is greatly appreciated.

REFERENCES

Atkinson, J.H., Richardson, D. and Stallebrass, S.E. (1990). Effect of recent stress history on the stiffness of overcon-solidated soil. *Geotechnique*, Vol. 40(4), 531–540.
Borja, R.I. (1990). Analysis of incremental excavation based on critical state theory. *Journal of Geotechnical Engineer-ing*, ASCE, Vol. 116(6), 964–985.

Borja, R.I. and Kishnani, S.S. (1991). On the solution of elliptical free-boundary problems via Newton's method. *Computer Methods and Applications in Mechanical Engineering*, Vol. 88(3). 341–361.

Brezis, H., Kinderlehrer, D. and Stampacchia G. (1978). Sur une nouvelle formulation due probleme de l'ecoulement a travers une digue. *C.R. Acad. Sci.* Paris, France, 287(ser A) 711–714.

Burland, J.B. (1989). 'Small is beautiful' – the stiffness of soils at small strains: Ninth Laurits Bjerrum Memorial Lecture. *Canadian Geotechnical Journal*, Vol. 26, 499–516.

Burland, J.B. and Symes, M.J. (1982). A simple axial displacement gauge for use in the triaxial apparatus. *Geotechnique*, Vol. 32(1), 62–65.

Carter, J.P., Booker, J.R. and Small, J.C. (1979). The analysis of finite elasto-plastic consolidation. *International Journal for Numerical and Analytical Methods in Geomechanics*, Vol. 3, 107–129.

Calisto, L. and Calebresi, G. (1998). Mechanical behavior of a natural soft clay. *Geotechnique*, Vol. 48(4), 495–513.

Calvello, M. and Finno, R.J. (2002). Calibration of soil models by inverse analysis. *Proc. International Symposium on Numerical Models in Geomechanics*, NUMOG VIII, Balkema, p. 107–116.

Calvello, M. and Finno, R.J. (2003). Modeling excavations in urban areas: effects of past activities. *Italian Geotechnical Journal*, 37(4), 9–23.

Calvello, M. and Finno, R.J. (2004). Selecting parameters to optimize in model calibration by inverse analysis. *Computers and Geotechnics*, Elsevier, Vol. 31, 5, 2004, 411–425.

Christian, J.T. and Wong, I.H. (1973). Errors in simulation of excavation in elastic media by finite elements. *Soils and Foundations*, Vol. 13(1), 1–10.

Chung, C.K. and Finno, R.J. (1992). Influence of Depositional Processes on the Geotechnical Parameters of Chicago Glacial Clays. *Engineering Geology*, Vol. 32, p. 225–242.

Clayton, C.R.I., and Heymann, G. (2001). Stiffness of geomaterials at very small strains. *Geotechnique* Vol. 51(3), 245–255.

Clayton, C.R.I., Hight, D.W. and Hopper, R.J. (1992). Progressive destructuring of Bothkennar clay: implications for sampling and reconsolidation procedures. *Geotechnique* Vol. 42(2), 219–239.

Clough, G.W. and Mana, A.I. (1976). Lessons learned in finite element analysis of temporary excavations. *Proceedings, 2nd International Conference on Numerical Methods in Geomechanics*, ASCE, Vol. I, 496–510.

Cough, G.W. and Tsui, Y. (1974). Performance of tied-back walls in clay. *Journal of the Geotechnical Engineering Division*, ASCE, Vol. 100 (12), 1259–1274.

Clough, G.W., Smith, E.M. and Sweeney, B.P. (1989). Movement control of excavation support systems by iterative design. *Current Principles and Practices, Foundation Engineering Congress*, Vol. 2, ASCE, 869–884.

Finno, R.J., Atmatzidis, D.K. and Nerby, S.M. (1988). Ground response to sheet-pile installation in clay, *Proceedings, Second International Conference on Case Histories in Geotechnical Engineering*, St. Louis, MO.

Finno, R.J., Atmatzidis, D.K. and Perkins, S.B. (1989). Observed Performance of a Deep Excavation in Clay, *Journal of Geotechnical Engineering*, ASCE, Vol. 115 (8), 1045–1064.

Finno, R.J. and Blackburn, J.T. (2005). Automated monitoring of supported excavations, *Proceedings, 13th Great Lakes Geotechnical and Geoenvironmental Conference, Geotechnical Applications for Transportation Infrastructure*, GPP 3, ASCE, Milwaukee, WI., 1–12.

Finno, R.J., Bryson, L.S. and Calvello, M. (2002). Performance of a stiff support system in soft clay. *Journal of Geotechnical and Geoenvironmental Engineering*, ASCE, Vol. 128, No. 8, p. 660–671.

Finno, R.J. and Roboski, J.F., (2005). "Three-dimensional Responses of a Tied-back Excavation through Clay," *Journal of Geotechnical and Geoenvironmental Engineering*, ASCE, Vol. 131, No. 3, March, 273–282.

Ghaboussi, J. and Pecknold, D.A. (1985). Incremental finite element analysis of geometrically altered structures. *International Journal of Numerical Methods in Engineering* Vol. 20(11), 2061–2064.

Gourvenec, S.M. and Powrie, W. (1999). Three-dimensional finite element analysis of diaphragm wall installation. *Geotechnique*, Vol. 49(6), 801–823.

Hight, D.W. (2001). Sampling effects on soft clay: An update on Ladd and Lambe (1963). *Proceedings of the Symposium on Soil Behavior and Soft Ground Construction*, ASCE Geotechnical Special Publication No. 119, Cambridge, 86–121.

Holman, T.P. (2005). Small strain behavior of compressible Chicago glacial clay. PhD thesis, Northwestern University, Evanston, IL.

Hughes, T.J.R. (1980). Generalization of selective integration procedures to anisotropic and nonlinear media. *International Journal of Numerical Methods in Engineering*, Vol. 15(9), 1413–1418.

Jardine, R.J., Symes, M.J. and Burland, J.B. (1984). The measurement of small strain stiffness in the triaxial apparatus. *Geotechnique*, Vol. 34(3), 323–340.

Lings, M.L., Ng, C.W.W. and Nash, D.F.T. (1994). The lateral pressure of wet concrete in diaphragm wall panels cast under bentonite. *Proceedings of the Institution of Civil Engineers, Geotechnical Engineering*, 107, 163–172.

Kolymbus, D. (1991). An outline of hypoplasticity. *Archive of Applied Mechanics*, Vol. 61, 143–151.

Mana, A.I. (1979). Finite element analysis of deep excavation behavior in soft clay. PhD dissertation, Stanford University, Stanford, CA.

Mana, A.I. and Cough, G.W. (1981). Prediction of movements for braced cut in clay. *Journal of Geotechnical Engineering*, ASCE, New York, Vol. 107, No. 8, pp 759–777.

Ng, C.W.W. and Yan, R.W.M. (1999). Three-dimensional modeling of a diaphragm wall construction sequence. *Geotechnique*, Vol. 49(6), 825–834.

O'Rourke, T.D. and Clough, G.W. (1990). Construction induced movements of insitu walls. *Proceedings, Design and Performance of Earth Retaining Structures*, Lambe, P.C. and Hansen L.A. (eds). ASCE, 439–470.

O'Rourke, T.D. and O'Donnell, C.J. (1997). Deep rotational stability of tiedback excavations in clay, *Journal of Geotechnical Engineering*, ASCE, Vol. 123(6), 506–515.

Rampello, S., Viggiani, G.M.B. and Amorosi, A. (1997). Small-strain stiffness of reconstituted clay compressed along constant triaxial effective stress paths. *Geotechnique*, Vol. 47(3), 475–489.

Santagata, M., Germaine, J.T. and Ladd, C.C. (2005). Factors Affecting the Initial Stiffness of Cohesive Soils. *Journal of Geotechnical and Geoenvironmental Engineering*, ASCE, Vol. 131(4), 430–441.

Sabatini, P.J. (1991). Sheet-pile installation effects on computed ground response for braced excavations in soft to medium clays. MS thesis, Northwestern University, Evanston, IL.

Smith, P.R., Jardine, R.J. and Hight, D.W. (1992). The yielding of Bothkennar clay, *Geotechnique*, Vol. 42(2), 257–274.

Stallebrass, S.E. and Taylor, R.N. (1997). The development and evaluation of a constitutive model for the prediction of ground movements in overconsolidated clay. *Geotechnique*, Vol. 47(2), 235–253.

Tu, X.X. (2006). Tangent stiffness model for clays including small strain non-linearity. PhD thesis, Northwestern University, Evanston, IL.

Viggiani, G. and Tamagnini, C. (1999). Hypoplasticity for modeling soil non-linearity in excavation problems. *Prefailure Deformation Characteristics of Geomaterials*, M. Jamiolkowski, M, Lancellotta, R. and Lo Presti, D. (eds.), Balkema, Rotterdam, 581–588.

Whittle, A.J. and Kavvadas, M.J. (1994). Formulation of MIT-E3 constitutive model for overconsolidated clays. *Journal of Geotechnical and Geoenvironmental Engineering*, ASCE, Vol. 120(1), 173–198.

*Numerical Modelling of Construction Processes in Geotechnical Engineering for
Urban Environment – Triantafyllidis (ed)
© 2006 Taylor & Francis Group, London, ISBN 0 415 39748 0*

Determination of soil parameters for modeling of deep excavations utilizing an inverse approach

Martin M. Zimmerer
VAROCON – Software & Engineering for Geotechnical Application

Tom Schanz
Laboratory of Soil Mechanics, Bauhaus-Universität Weimar

ABSTRACT: Recently we observe a significant increase in importance of complex numerical procedures in order to fulfill requirements of the latest generation of standards in geotechnical engineering.

In this study, we focus on the determination of soil stiffness parameter using both oedometer and triaxial data. Different optimization schemes as genetic and gradient approaches are used. Arguments are given, which optimization procedures are to be used to solve successfully the above mentioned problem by FEM. The next issues in our contribution are the proof of the solution uniqueness and the demonstration of the importance of the initial guess. Finally, we discuss the minimum amount of measurements required for a successful inverse parameter identification and/or result interpretation.

1 INTRODUCTION

The new generation of the "German standard code" follows mainly the European agreements. In the standard code there is a requirement to verify the ultimate bearing capacity. This verification is supplemented by proving the serviceability, i.e. the compatibility of the deformations. Consequently, there is a strong need in understanding the in situ nonlinear soil behavior, both in terms of compressibility and strength. The procedures for determination of constitutive parameters and the assessment strategy to the geotechnical characteristics based on the former generation of standard codes are no longer sufficient. Existing Finite Element codes provide many different material models to be selected by the user. These models range from simple to highly complex formulations. The correct use of the existing FE codes requires high competence in the code application and a good knowledge and experience with complex material laws. Instead of using the proper material model, mostly linear material laws are used, which represent the material behavior only in an insufficient approximation and gives wrong or insufficient predictions. However, often it is indispensable to use advanced material laws, especially for complex geotechnical constructions with large deformation potential.

The quality of the analysis as well as of the given proof are strongly depending on the used soil parameters. Based on a numerical experiment (modeling of a deep excavation with a double anchored retaining wall and analysis of wall deflections), the following task is analyzed

- Basic discussion on the developed model and the simulation technique of a double anchored retaining wall
- Which constitutive parameters are of importance concerning relevant wall deflections (sensitivity analysis)?
- Is it possible to use mathematical procedure, viz. optimization procedure, to derive relevant constitutive parameters?

To assess which measurements are required for successful inverse procedure, we did our analysis in two steps. In a first step, for the definition of the objective function for the inverse problem only the first stage of excavation and the horizontal displacements in several points of the retaining wall and the vertical deformation at the upper edge behind the retaining structure are used. In the second step, all stages of construction are considered in the objective function.

2 CONCEPTUAL MODEL

To do an acceptable calibration of a FE-model it is reasonable to follow the following pattern:

1. geometrical modeling of the problem (deep excavation)

2. choice of the required soil model (Soil model for soft soils, for soils with hardening plasticity, elastic soil models, ...)
3. selection of reasonable measurements (vertical and horizontal deformations, pore pressures, ...)
4. sensitivity analysis for the specific problem to detect the influence of soil parameters as well as the necessity of measurements
5. from the sensitivity analysis derive a selection of reasonable parameters to optimize and a proposal of the essential measurement equipment
6. selection of an adequate optimization technique for the specific problem
7. perform the optimization to calibrate the model during real construction

3 ANALYSIS

3.1 Idea of the inverse analysis

In inverse analysis, a given model has to be calibrated by iteratively changing the input parameters until the calculated or simulated results match together with the observed or measured data (observations). Figure 1 shows a scheme of an inverse analysis procedure. The first step of the procedure involves the estimation of the input parameters or setting boundaries for the parameters by conventional means. In the next step, a numerical simulation of the geotechnical problem or any engineering problem in general has to be done based on the estimated input parameters in the previous step. After the calculation is finished, the calculated results are compared with the observed or measured data. The difference between this data is called the objective function.

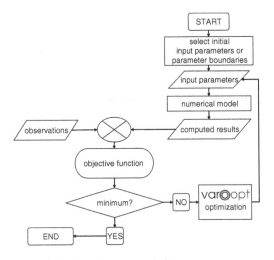

Figure 1. Scheme of inverse analysis.

When the simulation agrees with the measurements viz the result of the objective function is zero, the realistic soil parameters have been found, else an optimization routine (VARO²PT) is called to modify the input parameters for a next numerical simulation. This step is repeated automatically until the result of the objective function based on the observations and simulation is minimized.

This kind of inverse analysis allows the simultaneous calibration of multiple input parameters. However, identifying the most important parameters to be included in the inverse analysis or to be optimized can be problematic. The number and type of input parameters that can be optimized depends on many factors such as the used soil model, how the parameters are combined with the stiffness matrix, the site, the number and the type of the observations, the characteristics of the simulated system and the required computational time.

Hence, another step has to be added to the beginning of the scheme of the inverse analysis procedure to define the parameters to optimize. This step involves the selection of the parameters relevant to the problem and performing a sensitivity analysis to choose the parameters that will be optimized in order to find their values based on the observations.

The main advantage of the inverses analysis is its ability to automatically calculate parameter values based on the comparison between the observed and the computed results. It has also some other advantages, such as saving the time over traditional trial-and-error methods, providing statistics that quantify the quality of the calibration and reliability. The main disadvantages of this method are complexity, question of existence and uniqueness of the solution and maybe instability. Complexity of real nonlinear systems can sometimes lead to problems of insensitivity when the observations do not contain enough information to support the analysis. Non uniqueness may happen when different combinations of parameters match the observations very well. Instability may happen when slight changes in the model may radically change the inverse model results.

3.2 Optimization techniques

Because gradient based optimization techniques always are strongly dependent on the starting values and very sensitive regarding local minima, in this case study we focus on the use of an evolutionary algorithm called *Particle Swarm Optimizer* (parameters are constrained between fixed boundaries).

The Particle Swarm Optimization (PSO) technique has been developed by Eberhard and Kennedy (Kennedy and Eberhard 1995) for continuous nonlinear functions.

It is a simple evolutionary algorithm which was discovered through simulation of a simplified social model. PSO simulates a social behavior such as bird flocking or fish schooling to a promising position or region for food or other objectives in an area or space. Like evolutionary algorithm, PSO conducts a search using a population, which is called swarm, of individuals, which are called particles. Each particle represents a candidate position or solution to the problem. During searching for optima, each PSO particle adjusts its trajectory towards its own previous best position, and towards the best previous position attained by any member of its neighborhood (i.e., the whole swarm). Thus, global sharing of experience or information takes place and particles profit from the discoveries of themselves (i.e., local best) and previous experience of all other companions (i.e., global best) during search process. PSO is initialized with a population of M random particles and then searches for best position (solution or optimum) by updating generations until getting a relatively steady position or exceeding the limit of iteration number (i.e., T). In every iteration or generation, the local bests and global bests are determined through evaluating the performances, i.e., fitness values or objectives, of the current population of particles. Each particle is treated as a point in an N-dimensional space. Two factors characterize a particle status on the search space: its position and velocity. The N-dimensional position for the ith particle in the tth generation (i.e., iteration) can be denoted as $X_i(t) = \{x_1(t), x_2(t), \ldots, x_N(t)\}$. Similarly, the velocity (i.e., distance change), also an N-dimensional vector, for the ith particle in the tth generation can be described as $V_i(t) = v_1(t), v_2(t), \ldots, v_N(t)$.

The following equations represent the updating mechanism of a population of particles' status from the ones of the last generation during search process:

$$
\begin{aligned}
V_i(t) &= w(t)V_i(t-1) \\
&+ c_1 r_1 \left(X_i^L - X_i(t-1) \right) \\
&+ c_2 r_r \left(X^G - X_i(t-1) \right)
\end{aligned}
\tag{1}
$$

$$
X_i(t) = V_i(t) + X_i(t-1)
\tag{2}
$$

where $i = 1, 2, \ldots, M$, $t = 1, 2, \ldots, T$, $X_i^L = \{x_1^L, x_2^L, \ldots, x_N^L\}$ represents the local best (position or solution) of the ith particle associated with the best fitness encountered after $t-1$ iterations, $X^G = \{x_1^G, x_2^G, \ldots, x_N^G\}$ represents the global best among all the population of particles achieved so far, c_1 and c_2 are positive constants (namely learning factors) and r_1 and r_2 are random numbers between 0 and 1 and $w(t)$ is the inertia weight used to control the impact of the previous velocities on the current velocity, influencing

the trade-off between the global and local exploration abilities during search. Equation (1) is used to calculate the particles' new velocity according to its previous velocity and the distances of its current position from its own best experience or position and the groups' best experience or position. Then the particle "flies" toward a new position according to the equation (2).

It can be noticed that the PSO has the following properties:

- it generates a random initial population,
- the algorithm searches for optima by updating generations or iterations, by evaluating a fitness or objective for possible solutions
- each PSO particle shares its own search experience (local best)
- and its companions' search experience (global best).

4 EXAMPLES

4.1 The synthetic model

Based on the example of a double anchored diaphragm wall for an excavation the inverse analysis has to be examined. For the modeling, the finite element code PLAXIS (Vermeer and Brinkgreve 1995) has been used.

Below the surface of the site three soil layers have been modeled. The first soil layer consists of Fill material (3 m), the second of Sand (12 m) and the third of Loam material. All layers have been modeled with the Hardening Soil Model (Schanz 1998).

The diaphragm wall has been modeled using elastic plate elements. To simulate an accurate behavior of the soil-plate interaction, interface elements have to be applied to the plate element. They simulate a reduced strength between soil and diaphragm wall.

Both of the anchors are prestressed (1st-Anchor 120 kN, 2nd-Anchor 200 kN). The grout bodies are modeled using geotextile elements as they simulate realistic load transfer between soil and grout body.

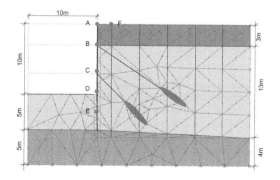

Figure 2.   The FE-Model.

23

Table 1. Soil Model Parameters.

Parameter	Fill	Sand	Loam
E_{oed}^{ref} [MPa]	20.5	38.0	20.0
E_{50}^{ref} [MPa]	20.5	38.0	20.0
Power m [-]	0.5	0.5	0.5
E_{ur}^{ref} [MPa]	61.5	115.5	60
Cohesion [kPa]	1	1	8
ϕ [^{o}]	30.0	34.0	29.0
ψ [^{o}]	0.0	4.0	0.0
γ_{unsat} [$\frac{kN}{m^3}$]	16	17	17
γ_{sat} [$\frac{kN}{m^3}$]	20	20	19
ν [-]	0.2	0.2	0.2
R_{inter} [-]	0.65	0.7	1
Material type	Drained	Drained	Drained

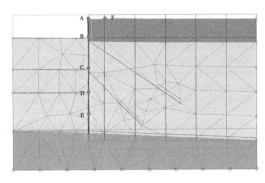

Figure 3. The deformed mesh of the FE-model (1st stage of construction) and the selected measurement points.

The model has been developed using plane strain condition, 15 nodes triangular elements. The main soil parameters used in this experiment are shown in table 1.

Three meters below surface we have saturated soil condition due to groundwater level.

Because the model is part of the objective function of the inverse analysis and each calculation can be very time consuming, a relatively coarse mesh has been applied to the FE-model.

During the excavation stages, the phreatic level in the excavation is always below the actual excavation surface. The dry excavation needs a groundwater flow calculation in each excavation phase to generate the new water pressure distribution. Because in reality measurements are mainly time-series, the following time schedule has been adapted:

1st part of excavation and:
installation of diaphragm wall: 2 days

installation and prestressing of
1st anchor and grout body (geogrid):   1 day

2nd part of excavation:
and updating the groundwater condition   2 days

installation and prestressing of
2nd anchor and grout body (geogrid)   1 day

3rd part of excavation:
and updating the groundwater condition   2 days

Due to drained material behavior all calculation phases are modeled plastic. A plastic calculation type is selected to carry out the elastic-plastic deformation. In this case, it is not necessary to consider the decay of excess pore pressures with time (Vermeer and Brinkgreve 1995). The results show that we have a maximum horizontal wall deflection of ~7 cm (point C) and a maximum vertical deformation in point F of ~6.5 cm (see figures 4 and 5)

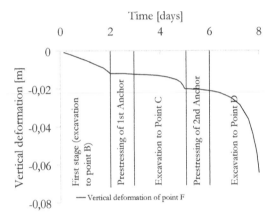

Figure 4. Stages of construction, vertical deformations of Point F.

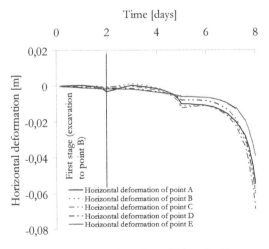

Figure 5. Horizontal deformations of Points A to E.

24

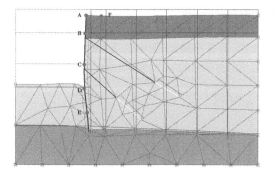

Figure 6. The deformed mesh of the FE-model (all stages of construction).

4.2 The model sensitivity

To evaluate the relative importance of each parameter, a sensitivity analysis has to be done for the different input parameters for each layer. The result of this analysis can be used to determine the number of relevant and uncorrelated parameters for each layer. The characteristics of the used soil model and the type of the observations influence the sensitivity of the observations to the changes in the input parameters (Calvello and Finno 2002; Calvello and Finno 2004; Finno and Calvello 2005).

The model sensitivity has been analyzed using a scaled sensitivity analysis. The scaled sensitivities indicate the amount of information provided by the observations i for the estimation of parameter j or the opposite way.

$$\frac{\partial y_i}{\partial x_j} \qquad (3)$$

For the sensitivity analysis the vertical and horizontal deformation in points A to F (see figure 2) have been analysed.

The sensitivity of the model has been investigated regarding cohesion c, friction ϕ, stiffness E_{ur}^{ref} and E_{oed}^{ref} for all layers. For all values a range of $\pm 10\%$ has been applied and the difference of deflections in each points where calculated.

The sensitivity shows that the friction angle ϕ of the sand layer has major influence on the wall and soil deflections, followed by the E_{oed}^{ref} of the sand layer.

It has been seen that the horizontal wall deflection in points A to E and the vertical deformation in point F have major influence on the sensitivity. Therefore in the later inverse analysis only the horizontal deformation in points A to E (on the diaphragm wall) and the vertical deformation in point F (on the surface behind the diaphragm wall) are considered.

To show the influence of the E_{oed}^{ref} and ϕ of the sand layer to the objective function (equation (6)),

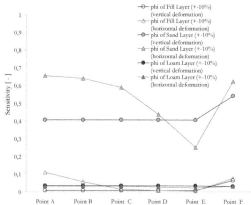

Figure 7. The scaled sensitivity ϕ.

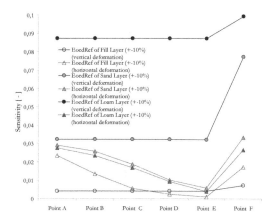

Figure 8. The scaled sensitivity E_{oed}^{ref}.

calculations have been done ($E_{oed,orig}^{ref} \pm 10\%$, $\phi_{orig} \pm 10\%$) within an equidistant grid. These calculations have been done for the 1st stage of construction (see figure 9) and for all stages of construction (see figure 10).

Each knot of the grid represents a combination of different parameters. A further comparison between the objective function values of these individual sections (points) can be done to determine the minimum value.

The grid method shows that the influence of ϕ of the sand layer increases with ongoing excavation, while the influence of E_{oed}^{ref} of the sand layer decreases (see figure 10).

4.3 The optimization

For the inverse analysis the program package VARO²PT (Zimmerer and Lobers 2005) has been

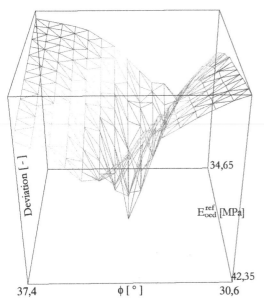

Figure 9. Relation between ϕ of Sand Layer and E_{oed}^{ref} of Sand Layer – 1st excavation part.

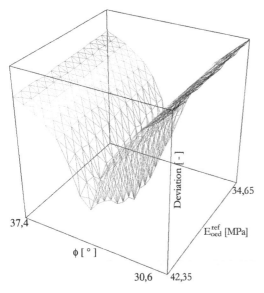

Figure 10. Relation between ϕ of Sand Layer and E_{oed}^{ref} of Sand Layer – all stages of construction.

used. With this program environment several optimization techniques (gradient based and evolutionary algorithms) can be used to do inverse studies on complex problems.

In our case, the inverse analysis aims to minimize the difference between experimental data (laboratory or field tests) and numerically computed results. The

Table 2. Input parameters for the optimization technique.

	Std. Value	Bound. min	Bound. max	Final generation
First stage				
Sand				
E_{oed}^{ref} [MPa]	38.5	34.65	42.35	
ϕ [^{o}]	34	30.6	37.4	100
All stages				
Sand				
E_{oed}^{ref} [MPa]	38.5	34.65	42.35	
ϕ [^{o}]	34	30.6	37.4	130

difference is called objective function. The calibration can be done using optimization techniques which are used to find the minimum value of the objective function. In general, objective function is defined using the following equation:

$$F\left(\mathbf{x}\right) = \sum_{i=1}^{n} \left| f_{calc}^{i}\left(\mathbf{x}\right) - f_{meas}^{i} \right| \tag{4}$$

where

$$\mathbf{x} = (x_1, x_2, \ldots, x_m) \tag{5}$$

where m is the number of parameters to be optimized, n is the number of measurements $f_{calc}^{i}(\mathbf{x}) - f_{meas}^{i}$ is the difference between the observations and the simulation based on the m optimized parameters $x_1, x_2, \ldots, x_m$. The objective function can be, in the scene of the least square method, expressed as

$$F\left(\mathbf{x}\right) =$$

$$\left(\frac{1}{n} \sum_{i=1}^{n} \frac{\left(f_{calc}^{i}\left(\mathbf{x}\right) - f_{meas}^{i}\right)^2}{(1-\alpha) f_{meas}^{i\,2} + \alpha f_{meas}^{i\,2} \frac{1}{n} \sum_{j=1}^{n} f_{meas}^{j\,2}} \right)^2 \tag{6}$$

where n is the number of measurements, f_{meas} are the measurements and f_{calc} is the associated numerical result. The coefficient α has to be chosen between 0 and 1 ($\alpha = 1$: $F(\mathbf{x})$ refers to a rate of relative error, $\alpha = 0$: $F(\mathbf{x})$ refers to a rate of absolute error) (Malecot et al. 2004).

The Criterion for stopping the calculation was set to the deviation of the objective function exceeding the value of $1 \cdot 10^{-6}$. Whenever the deviation exceeded this value, the program Optimization Interface was automatically terminated.

4.4 The results

In a first analysis it has been tried whether it is possible to determine the "right" soil parameters of the

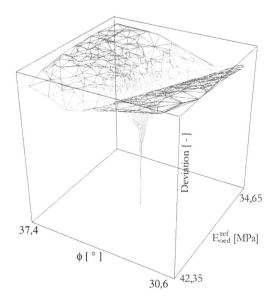

Figure 11. Updated relation between ϕ of Sand Layer and E_{oed}^{ref} of Sand Layer using Delaunay triangulation- 1st excavation part.

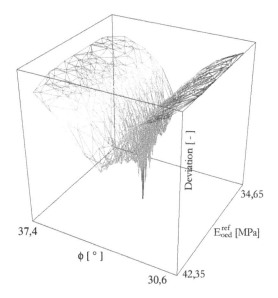

Figure 12. Updated relation between ϕ of Sand Layer and E_{oed}^{ref} of Sand Layer using Delaunay triangulation – all stages of construction.

soil layer, which has the main influence on the wall deformations due to the installation of the diaphragm wall and the first part of the excavation. Therefore, calculations for the first calculation phase have been done in several grid knots shown in figure 9 to analyze whether there exists a minimum. In a second step, we

Table 3. Optimization parameters.

Properties of Particle Swarm Optimizer	
Number of neighbors	20
w_{max}	0.9
w_{min}	0.5
c_1	1.4
c_2	1.4
min deviation	$1 \cdot 10^{-6}$

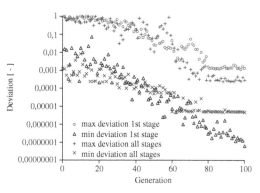

Figure 13. Minimum and maximum deviation of each generation of particles for 1st and 2nd analysis.

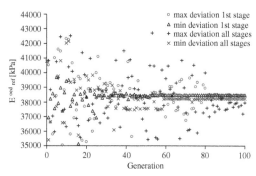

Figure 14. Best and worst E_{oed}^{ref}-Value of each generation of particles for 1st and 2nd analysis.

tried to let the Particle Swarm Optimizer find the minimum within the boundaries shown in table 2. After 100 generations an acceptable parameter combination was found ($\phi = 33,99$ [^{o}] and $E_{oed}^{ref} = 38496$ [kPa]).

In a second analysis, we investigated whether a longer observation of soil deformation would lead to a different result. Just as in the first analysis, we first run a raster calculation to get an idea of the global minimum of the specific parameter combination ϕ and E_{oed}^{ref} of the sand layer (see figure 10). In the second step, we let the particle swarm optimization technique find

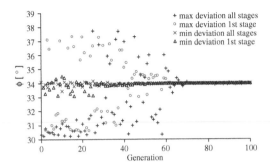

Figure 15. Best and worst ϕ-Value of each generation of particles for *I*st and 2nd analysis.

the minimum within the constraints shown in table 2 and the optimization parameters shown in table 3.

After 130 generations with 20 neighbors an acceptable parameter combination has been found ($\phi = 33, 98$ [^{o}] and $E_{oed}^{ref} = 38490$ [kPa]).

The decreasing influence of E_{oed}^{ref} with ongoing construction (viz excavation, installation of anchors, prestressing of anchors) leads to a need o more generations to reach acceptable values for E_{oed}^{ref}.

We could conclude the following:

1. It is possible to determine relevant soil parameters witch have great influence on the deformation of a diaphragm wall by just taking into account the first deformation of the construction.
2. It would be possible to give a prediction of the possible total final deformation of the diaphragm wall by using the investigated soil parameters. Based on these predictions, the right prestressing force for the anchors could be determined to keep the final deformation of the construction as small as possible.
3. It is sufficient to monitor only the horizontal deformation of the diaphragm wall and the vertical deformation on the surface behind the diaphragm wall in order to identify a complete set of realistic material parameters.
4. The updated relation between E_{oed}^{ref} and ϕ (figures 11 and 12) for the sand layer in both cases shows, that an initial raster method has to be applied carefully. Sometimes the real global minimum does not even lie close to the minimum obtained by a raster method.

5 CONCLUSIONS

Based on a synthetic experiment it has been tried whether it is possible to get model parameters from

"observations" by back calculation or to calibrate FE-models by means of field measurements.

Based on a scaled sensitivity analysis it has been concluded, that the friction angle ϕ shows the most significant influence on the vertical and horizontal deformation of the subsoil and therefore the diaphragm wall. In this study we focused on the determination of these selected parameters, hence these parameters have been detected by inverse analysis.

It can been shown, that with suitable optimization strategies it is possible to investigate soil model parameters from observations. Moreover, it has been shown that the deformation due to the first construction phase is sufficient to detect the model parameter with acceptable accuracy.

Based on these results, a reasonable number of anchors and prestressing forces could be acquired and predictions to the final deformation could be done.

It has also been shown that, with ongoing construction (excavation), the influence of the Oedometer reference stiffness E_{oed}^{ref} decreases while the influence of the friction angle ϕ increases, suggesting a transformation from a deformation problem towards a stability problem.

REFERENCES

Calvello, M. and R. J. Finno (2002). Calibration of soil models by inverse analysis. In G. N. Pande and S. Pietruszczak (Eds), *Numerical Models in Geomechanics NUMOG VIII*. A. A. Balkema.

Calvello, M. and R. J. Finno (2004). Selecting parameters to optimize in model calibration by inverse analysis. *Computers and Geotechnics 31*, 411–425.

Finno, R. J. and M. Calvello (2005). Supported excavations: Observational method and inverse modeling. *Journal of Geotechnical and Geoenvironmental Engineering 137*(7), 826–836.

Kennedy, J. and R. C. Eberhard (1995). Particle swarm optimization. In *International Conference on Neuronal Networks*, Perth – Australia.

Malecot, Y., E. Flavigny, and M. Boulon (2004, 25. June). Inverse Analysis of Soil Parameters for Finite Element Simulation of Geotechnical Structures: Pressuremeter Test and Excavation Problem. In R. Brinkgreve (Ed.), *Geotechnical Innovations: Studies in Honour of Professor Pieter A. Vermeer on Occasion of his 60th Birthday and Proceedings of the Symposium held at Stuttgart*, Stuttgart.

Schanz, T. (1998). *Zur Modellierung des mechanischen Verhaltens von Reibungsmaterialien*. Universität Stuttgart: Mitteilung 45 des Instituts für Geotechnik.

Vermeer, P. A. and R. Brinkgreve (1995). *PLAXIS: Finite element code for soil and rock analyses (Version 6.3)*. Rotterdam: Balkema.

Zimmerer, M. M. and S. Lobers (2005). *VAROPT- Software for inverse analysis*. Weimar: VAROCON – Software and engineering for geotechnical application.

Numerical Modelling of Construction Processes in Geotechnical Engineering for Urban Environment – Triantafyllidis (ed)
© 2006 Taylor & Francis Group, London, ISBN 0 415 39748 0

Comparison and feasibility of three dimensional finite element modelling of deep excavations using non-linear soil models

H.C. Yeow, D.P. Nicholson & B. Simpson
Arup Geotechnics, London, England

ABSTRACT: Construction of the proposed Crossrail Moorgate station adjacent to the Moorhouse development involves a 40 m deep excavation in Central London. Full two- and three-dimensional finite element analyses were undertaken as part of numerical modelling to predict ground movements associated with the proposed construction of the station. This paper presents a comparison of ground movement predictions, both 2D and 3D, by two small strain, non-linear soil models of *BRICK* using *Oasys* programs and the Jardine type model using the Imperial College Finite Element Program. It also investigates the feasibility of undertaking such complex analyses in a full three-dimensional model using the *Oasys* LS-Dyna program.

1 INTRODUCTION

The development of Moorhouse, which is within the safeguarded zone of the proposed Crossrail alignment, required the designer to address the complex interaction problems of future ground movements during Crossrail construction on the foundations of the new development (Torp-Peterson *et al.* 2003), see Figure 1. With the footprint of the new development lying within the influence zone of the potential future ground movements, the base grouted bored piles supporting the new structure have been designed to tolerate the additional ground forces induced by the movements.

The ground movement prediction undertaken by the designer using a two-dimensional (2D) axi-symmetric finite element (FE) model was found to be conservative after comparing to three-dimensional (3D) analyses undertaken by Crossrail, who also have the role to ensure that the foundation scheme proposed for the new Moorhouse development is adequate for the safe construction of the proposed new railway. The analyses undertaken by Crossrail were based on small strain, non-linear Jardine type soil model (referred to as the ICFEP model in this paper, Jardine *et al.* 1986). However, such 3D analyses were not available to the designer at the time when the design was made.

As part of the research development of the Numerical Skills Team within Arup Geotechnics, full 3D analyses were undertaken using the Arup in-house *Oasys* LS-Dyna program to compare ground movement predicted for the proposed excavation using the non-linear *BRICK* soil model (Simpson 1992). This paper summarises the computed wall deflections and ground movements from these two non-linear soil models. It also discusses feasibilities of using

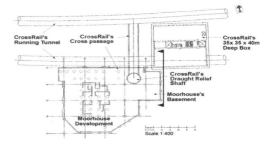

Figure 1. Moorhouse development and the proposed Crossrail tunnel and access box.

such 3D numerical modelling tools to study complex soil-structure interaction problems.

2 GEOTECHNICAL USE OF 3D NUMERICAL MODELLING

Despite the advance in computer technology in recent years, geotechnical use of full 3D FE analyses have been restricted to highly sophisticated geotechnical problems. Such limitation has not prevented many attempts of 3D modelling by academic researchers and designers (Easton *et al.* 1999, Burd *et al.* 2000 and Torp-Peterson *et al.* 2003). In order to make the computation possible, simplifications of the problems or the FE meshes were introduced. Recent attempts by MIT (Hsieh and Whittle, 2004) using the FETI (Finite Element Tearing and Interconnecting) method allowed a complicated soil-structure interaction problem to be broken down into several more manageable components which were then analysed using several parallel processors. This has substantially improved the processing time. However, these analyses are still taking many hours if not days to compute.

The approach described in this paper, on the other hand, utilizes the benefits of an explicit computing approach in *Oasys* LS-Dyna program. The program is originated from Livermore Software Technology Corporation. It performs small time stepping computations which have been found to offer significant advantages in terms of computing cost with substantially larger FE models (Yeow 2004). Its geotechnical features are incorporated by *Oasys* as an on-going effort within Arup to provide its staff advanced technical software to tackle complex geotechnical engineering problems. To date the largest 3D model created for ground movement assessment for the work surrounding Channel Tunnel Rail Link's London Terminal at St Pancras has a staggering 600,000 single point 8-noded hexahedral elements.

3 FE ANALYSES

Full 2D, axi-symmetry and plane strain, and 3D analyses were undertaken by the designer and the Crossrail team to predict potential ground movements of the proposed 40 m deep excavation. The 2D models, analysed using *Oasys* SAFE program were based on the original *BRICK* soil parameters (Simpson 1992), which was found to be conservative. This was subsequently revisited using revised *BRICK* parameters (Pillai 1996) in order to compare with the prediction of Crossrail using ICFEP model. This work allows the comparison of 2D and 3D predictions from two advanced small strain, non-linear soil models regularly used for over-consolidated London Clay.

3.1 Stratigraphy, geometry and excavation sequence

The site is underlain by 3.7 m of Made Ground over 3.5 m of Terrace Gravel which in turn is underlain by 28.3 m of London Clay, 18.2 m of Lambeth Beds, 13 m of Thanet Sand with Chalk at a depth of about 64 m below ground level, see Figure 2. Groundwater is at +7.1 mOD with sub-hydrostatic water pressures in the underlying clay strata underdrained by the more permeable Thanet Sand and Chalk. Beneath the clays is a deep aquifer in the Thanet Sand and Chalk with a water level at −35 mOD, as shown in Figure 2.

The proposed Crossrail station is a 35 × 35 m, 40 m deep box to be constructed within a 1.2 m thick diaphragm wall retaining structure. A bottom-up excavation sequence is currently anticipated with seven levels of temporary props at the levels shown in Figure 2. This allows flexibilities in construction method of the proposed scheme, although it was recognised

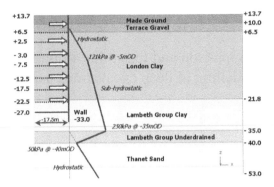

Figure 2. Geology and groundwater conditions of the Moorhouse site.

that a top-down construction method would be more appropriate. A total of eight stages of construction had been modelled in the FE analyses, with excavation following propping 0.5 m above preceding excavated level in seven stages of the excavation.

3.2 Constitutive models

Crossrail assessments were performed using the following constitutive models:

- elastic-plastic Mohr-Coulomb model for the Made Ground
- small strain, non-linear ICFEP model with Mohr-Coulomb yield criteria for all other strata.

The small strain, non-linear *BRICK* (Simpson 1992, Pillai 1996) model was used by the authors for the London Clay and Lambeth Beds soils while an elastic-plastic Mohr Coulomb model was used for the Made Ground, Terrace Gravel and the Thanet Sand. Properties of these soil models are given in Appendix A.

3.3 2D and 3D meshes

Figure 3 shows the 2D mesh used for the axi-symmetry and plane strain analyses. The mesh consisted of 571 8-noded quadilateral isoparametric elements while and props were modelled as horizontal springs.

Figure 4a shows the 3D mesh which is similar to the $\frac{1}{8}$ model used by Crossrail. The soil was modelled with 8,700 single integration point hexahedral 8-noded solid elements while the wall and the props were modelled as fully integrated shell elements. In area close to the wall eight-points solid elements were used to model the soils. Two sets of shells were used to incorporate the non-linear behaviour of the retaining structure with full stiffness when the wall spans vertically but very small stiffness in the horizontal direction. Parameters for the retaining wall can be found in Appendix A of this paper. A $\frac{1}{4}$ model was also created to study the efficiency of a larger 3D mesh and also to study the

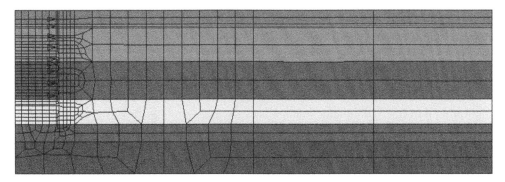

Figure 3. FE mesh for axisymmetry and plane strain analyses.

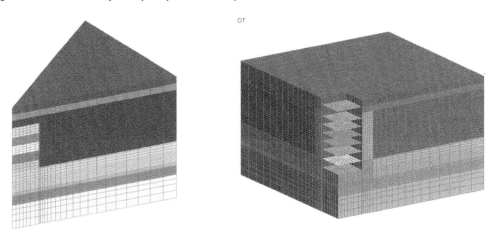

Figure 4. 3D $\frac{1}{8}$ and $\frac{1}{4}$ FE meshes.

effects of the capping beam on the performance of the retaining wall.

4   COMPARISON OF THE RESULTS

4.1   *2D analyses*

Before engaging in full 3D modelling, a 2D model was created using *Oasys* LS-Dyna program to repeat and compare with the preliminary plane strain analysis using the *Oasys* SAFE program with original *BRICK*(Simpson 1992) input parameters. Figure 5 shows the comparison of the computed wall deflection at the first stage and the final stage of the excavation. The computed cantilever (1st stage) wall deflection using the LS-Dyna program is about 35 mm compared with the SAFE prediction of about 29 mm and the difference in the maximum computed deflection at the last stage of excavation is only about 3%. Therefore the overall difference in computed wall deflection is considered to be insignificant.

Both axi-symmetry and plane strain analyses were undertaken using revised *BRICK* parameters (Pillai

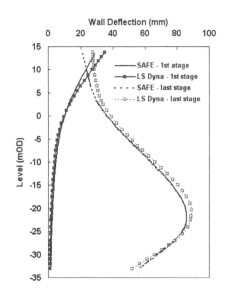

Figure 5. Comparison of computed wall deflections for SAFE and LS-Dyna plane strain model.

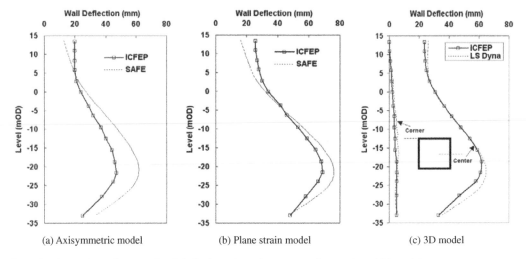

(a) Axisymmetric model   (b) Plane strain model   (c) 3D model

Figure 6. Comparison of computed wall deflection using axi-symmetry, plane strain and 3D models.

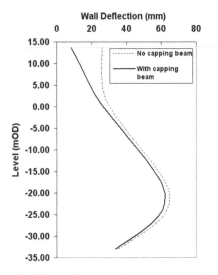

Figure 7. Effects of capping beam on computed wall deflection.

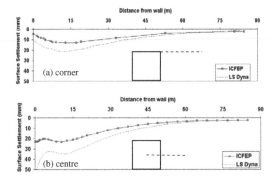

Figure 8. Comparison of computed ground movements for 3D models.

the revised *BRICK* model is about 5% higher than the model using ICFEP soil model.

From the wall deflection comparison in Figure 6, it is interesting to observe a difference of up to 10% in computed deflection in the plane strain and 3D models while a much higher 30% of difference in the axi-symmetry model. This is possibly due to the very small hoop stiffness of the wall assumed in the Cross-rail axi-symmetry model while zero stiffness has been assumed in the authors' model.

Figure 7 shows that when the $\frac{1}{4}$ model was used to incorporate the benefits of a 1.8 m deep by 1.5 m think capping beam, the cantilever deflection of the wall was more than halved but the overall wall deflection has not changed significantly.

4.3 Surface settlement

The computed surface settlements at corner and centre of the wall from the 3D model are shown in Figure 8.

1996) and compared with the results from the ICFEP model. The wall movements predicted are compared in Figures 6(a) and 6(b). Wall deflection computed with the revised *BRICK* model is about 30% higher than that with the ICFEP model for the axi-symmetry model while the corresponding difference in the plane strain model is only about 10%.

4.2 3D analyses

Figure 6(c) shows the comparison of the computed wall deflections of the FE analyses undertaken with the two soil models. The computed wall deflection using

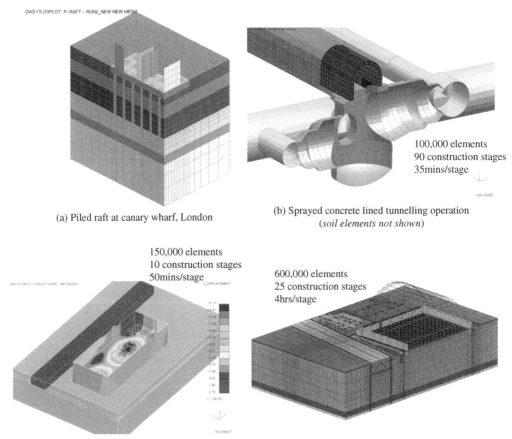

14,300 elements
2.5hrs

100,000 elements
90 construction stages
35mins/stage

(a) Piled raft at canary wharf, London

(b) Sprayed concrete lined tunnelling operation
(*soil elements not shown*)

150,000 elements
10 construction stages
50mins/stage

600,000 elements
25 construction stages
4hrs/stage

(c) Basement excavation and raft
foundation, London

(d) Combined excavation and shallow and deep foundation
at St Pancras station, London

Figure 9.   Other examples of full 3D models (project names are listed where appropriate).

LS-Dyna 3D model consistently predict slightly over 50% larger settlement. The majority of the computed displacements was in the form of undrained movements in the London Clay and Lambeth Beds, and detailed investigation into the computed volume of London Clay's surface settlement, wall deflection and base heave in the LS-Dyna model indicated consistent undrained behaviour. However, no detailed information is available for such a study for the ICFEP model.

5   FEASIBILITY OF 3D GEOTECHNICAL MODELLING

The use of efficient software has given the Numerical Skills Team in Arup Geotechnics the opportunity to undertake full 3D non-linear geotechnical modelling

as part of routine design. Some examples of the full 3D models used in recent projects are shown in Figure 9 and they consist of large scale modelling of full geotechnical problems. The efficiency provided by the LS-Dyna software makes these types of analyses practical in design offices and practitioners are no longer worried about unrealistic design programmes imposed by such complex computation. The full 3D modelling also provides the designers with full insight of the complex soil-structure interaction previously not possible using more conventional methods of design analyses. One such example is the beneficial effects of the ground bearing raft of a piled raft foundation where the raft shares some of the building load and enhances the effective stress of the ground which in turn improves the long term pile capacity of the foundation piles (Nicholson *et al.* 2002).

6 CONCLUSIONS

The comparisons presented in this paper show that wall deflections and ground movements computed using two advanced non-linear soil models (i.e. ICFEP and *BRICK*) are comparable under plane strain and full 3D conditions, with a difference of up to 10% in computed wall deflection. However, the difference is significant when comparing the axisymmetry model, with the computed wall deflection for the ICFEP model about 30% less than the associated figure using the *BRICK* model.

It is now feasible to undertake full 3D modelling of non-linear soil-structure interaction problems involving many thousands of elements in a design office. Continued development and advances in computer technology and computing methods will inevitably improve the processing time which will make this design technique more attractive to general practitioners. However, the software is at present costly and not generally available. More importantly, its use requires significant expertise both in finite element technology and in modelling of material behaviour.

ACKNOWLEDGEMENT

This paper has been published at the 11th International Conference of Computer Methods and Advances in Geomechanics (IACMAG) in Turin 2005. The authors acknowledge the kind permission given by the publisher and editors of the IACMAG conference to publish this paper in this International Conference on "Numerical Simulation of Construction Processes in Geotechnical Engineering for Urban Environment" (NSC06).

APPENDIX A

Made ground, $E = 10,000\,kPa$, $\nu = 0.2$, $\phi' = 25°$, $K_o = 0.57$

Terrace gravel, $E = 50,000\,kPa$, $\nu = 0.2$, $\phi' = 35°$, $K_o = 0.43$

BRICK parameters used for London Clay and Lambeth Beds are listed below:

$\lambda = 0.1$, $\kappa = 0.02$, $\iota = 0.0019$, $\nu = 0.2$, $\beta_\phi = 4$, $\beta_g = 4$

Strain	G/G_{max}
3.04e-005	0.92
6.08e-005	0.75
0.000101	0.53

Strain	G/G_{max}
0.000121	0.29
0.00082	0.13
0.00171	0.075
0.00352	0.044
0.00969	0.017
0.0222	0.0035
0.0646	0

λ, κ, ι, β_ϕ and β_g = constants in *BRICK* soil model
ν = Poisson's ratio
G_t = tangent shear modulus
$G_{max} = G_t$ at very small strain.

Linear elastic properties of the retaining wall used in the analyses are listed below:
Wall spanning vertically, Young's modulus, $E = 28\,GPa$
Wall spanning horizontally, Young's modulus, $E = 2.8\,GPa$
In plane horizontal direction, Young's modulus, $E = 0$.

REFERENCES

Burd H.J. *et al.* (2000), *"Modelling of tunnelling-induced settlement of masonry buildings"*, Proceeding of Civil Engineers, Geotechnical Engineering 143, 17–29.

Easton M.R. *et al.* (1999), *"Design guidance on soil berms as temporary support for embedded retaining walls"* TRL Report 398, TRL Ltd. Crowthorne.

Hsieh Y-M and Whilltle A.J (2004) *"Finite element modelling of stacked-drift cavern construction"* Numerical Models in Geomechanics – NUMOG IX, Balkema, pp 279–284.

Jardine R.J, Potts D.M, Fourie A.B and Burland J.B (1986), *"Studies of the influence of non-linear stress-strain characteristics in soil-structure interaction"*, Geotechnique 36, No. 3, 377–396.

Nicholson D.P., Morrison P.R.J. and Pillai, A.K. (2002), Piled Raft Design for High Rise Buildings in East London, UK.

Pillai A.K. (1996), Review of the BRICK of soil behaviour. MSc dissertation, Imperial College, London.

Simpson B. (1992), 32nd Rankine Lecture: Retaining structures – displacement and design. Geotechnique, 42, 4, 539–576.

Torp-Peterson *et al.* (2003), *"The prediction of ground movements associated with the construction of deep station boxes"*, Tunnelling World Congress.

Yeow H. (2004), *"Three-dimensional modelling of geotechnical problems – its feasibility as a routine analysis tool"* Ground Engineering Sept 2004.

Numerical Modelling of Construction Processes in Geotechnical Engineering for
Urban Environment – Triantafyllidis (ed)
© 2006 Taylor & Francis Group, London, ISBN 0 415 39748 0

Lessons learned from case studies of excavation support systems through Chicago glacial clays

Richard J. Finno & Cecilia Rechea

Department of Civil and Environmental Engineering, Northwestern University, Evanston, IL USA

ABSTRACT: This paper summarizes the lessons learned from three detailed case studies of supported excavations made through soft to medium glacial clays. In particular, the use of inverse analysis techniques with finite element simulations are shown to be an effect way to calibrate soil parameters, if sufficient detail is included in the numerical simulation of the excavation. The large effect of past construction activities at an urban site on computed displacements is illustrated by results from all three excavations. Previous tunneling operations, building construction, site grading and wall installation affects the stresses that exist in the ground just prior to start of excavation. Not properly accounting for these factors results in either numerical results that do not match field observations or if a reasonable match does result, then the soil parameters are not representative of the natural soils. Furthermore, the combined use of lateral movements and forces in internal bracing are shown to provide a higher sensitivity to the observations that either type of measurement alone.

1 INTRODUCTION

When making numerical simulations of deep excavations in urban areas, one must both adequately represent soil behavior with an appropriate constitutive model and faithfully represent the construction sequence and procedures. In complex problems like deep excavations, one must always relate computed results to carefully-measured field responses because the "real" part of the solution is the field observations. The core of the well-known observational approach to underground construction is to be able to relate ground and structural responses to construction activities. This can only be accomplished if data obtained is evaluated in light of the construction activities that impacted the observed response, a task that requires much attention to detail. This paper summarizes the lessons learned concerning parameter identification based on field data from three detailed case studies of supported excavations in Chicago. The excavations were made through soft to medium glacial clays (Chung and Finno 1992) with support systems consisting of a secant pile wall supported by internal bracing and ground anchors (Chicago-State excavation), a sheet-pile wall supported by tiebacks (Lurie Center excavation) and a sheet-pile wall supported by internal bracing (Ford Center excavation). In all cases, detailed observations were obtained throughout construction so that observed responses are correlated with construction activities.

Herein, inverse analysis techniques are used to relate field observations to results of numerical computations. This paper first summarizes three excavation projects where detailed field responses were observed. The inverse analysis approach employed is summarized, and its use with finite element simulations is shown to be an effect way to calibrate soil parameters, if sufficient detail is included in the numerical simulation of the excavation. This is illustrated with results of the Chicago-State excavation supported by a secant pile wall and cross-lot braces and ground anchors. Soil parameters calibrated at an early stage of excavation were used with subsequent computations to provide good agreement with observed movements measured by inclinometers on both sides of the excavation. The usefulness of this approach is illustrated by the results of numerical simulations of the two sheet-pile supported excavations made using the optimized parameters from the Chicago-State excavation with no additional optimization. In both cases, reasonable agreement was noted between the computed results and field observations of lateral movements.

Some of the practical difficulties of defining responses for a given stage of excavation are shown by examining the detailed record of construction activities at the Chicago-State project.

The large effect of past construction activities at an urban site on computed results is illustrated at a site where previous activities had altered the initial conditions from conventionally-assumed k_0 values. Previous tunneling operations, building construction and wall installation impact the stresses that exist in the ground just prior to start of excavation. Not properly accounting for these factors results in either numerical results that do not match field observations or if a reasonable match does result, then the soil parameters are not representative of the natural soils. This latter effect is elucidated by results of inverse analysis techniques.

2 OVERVIEW OF PROJECTS

Detailed performance data have been collected at three projects in the Chicago area, and these are reviewed herein.

2.1 Chicago-State excavation project

The Chicago Ave. and State St. subway renovation project involved the excavation of 12.2 m of soft to medium clay within 2 m of a school supported on shallow foundations (Finno et al., 2002). Figure 1 shows a section of the stratigraphy and the excavation support system. The support system consisted of a secant pile wall with three levels of support, which included pipe struts (1st level) and tieback anchors (2nd and 3rd levels). Ground and school movements were recorded during construction to monitor the effects of the excavation on the integrity of the school.

The subsurface conditions at all three sites generally consist of a fill deposit overlying a sequence of glacial clays deposited during the Wisconsin stage of the Pleistocene period. At the Chicago-State site, the fill is mostly medium dense sand, but also contains construction debris. Beneath the fill lie four strata associated with the repetitive process of advance and retreat of the Wisconsin glacier. The upper three are ice margin deposits deposited underwater, and are distinguished by water content and undrained shear strength (Chung and Finno, 1992). With the exception of a clay crust in the upper layer, these deposits are lightly overconsolidated as a result of lowered groundwater levels after deposition and/or aging. Stratigraphy is shown in terms of Chicago City Datum (CCD) elevation.

Lateral movements of the soil behind the secant pile wall were recorded using five inclinometers located around the site. Vertical movements were obtained from optical survey points located along the outside walls of the school, on the roof, and on eight interior columns. Measurements of the different instruments were taken before the installation of the wall, and at frequent intervals during construction. The observations for the inverse analysis described herein are derived from inclinometer data obtained on opposite sides of the excavation at five stages of construction.

Figure 2 summarizes deformation responses to excavation and support. Both lateral movements and settlements are shown. The movements that occurred as the secant pile wall was installed extend through all compressible layers. This is important when using these observations to calibrate parameters using inverse techniques in that these movements occur at an early stage of the excavation, and hence contain information that can be used to optimize parameters in all layers that can be useful to predict movements at subsequent stages of excavation.

Very little movements beyond those that occurred during wall installation were observed until the excavation was lowered below elev. −1.4 m CCD; a

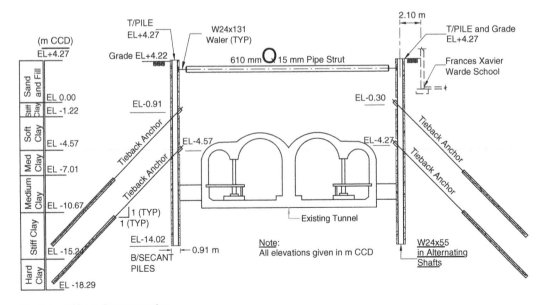

Figure 1. Chicago-State excavation.

maximum of 4 mm additional lateral movement occurred as a result of excavating to this elevation. This behavior suggests that the upper clays initially are relatively stiff, and provide field indications of the small strain nonlinearity of these soils. After wall installation, the secant pile wall incrementally moved toward the excavation in response to excavation-induced stress relief. When the excavation reached final grade, the maximum lateral movement was 28 mm. The school settled as the secant pile wall moved laterally. The maximum settlement at the school at the end of excavation was also 28 mm. A more complete record of performance can be found in Finno et al. (2005).

2.2 Lurie Center excavation project

The excavation for the Lurie Research Center is supported by a PZ-27 sheet pile wall on all sides. As indicated in Figure 3, three levels of tieback anchors provided lateral support on the other three walls. Both the first and second level ground anchors are founded in the beach sand. The anchors were proof tested to 1.2 times their design loads and performance tests were carried to 1.33 times the design load. All anchors were locked off to a pre-stress of 100% of the design load. To meet these requirements, the tiebacks typically were re-grouted. On the fourth side of the cut, only two levels of anchors were used due to the presence of a caisson-supported excavation with 2 basement levels.

Figure 3 also summarizes the stratigraphy at the site. Beneath the surficial medium dense to dense rubble fill lies a loose to medium dense beach sand. This beach sand is not present in the other two case studies presented herein. The granular soils overlie a sequence of glacial clays of increasing shear strength with depth. Undrained shear strengths in the soft to stiff clays in Figure 3 are based on results of vane shear tests. The excavation averages approximately 12.8 m deep and bottoms out in the medium stiff clay. The ground water level perched in the granular soils is related to the water level of nearby Lake Michigan. This excavation and the observed performance are described by Finno and Roboski (2005).

Figure 4 summarizes deformation responses to the excavation and support process for inclinometers located behind the walls of the excavation. Similar patterns of movements are observed in each inclinometer, except smaller movements are noted along the east wall due to the presence of a building

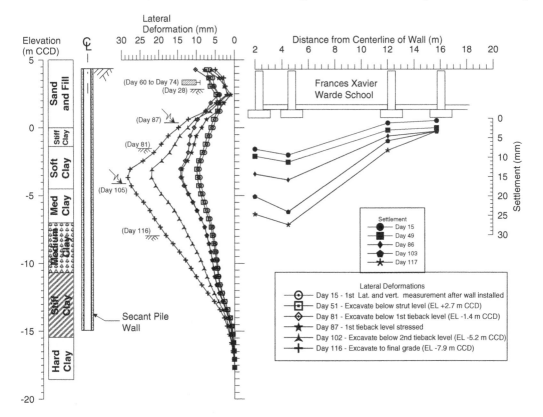

Figure 2.   Lateral movements and settlements at Chicago-State excavation.

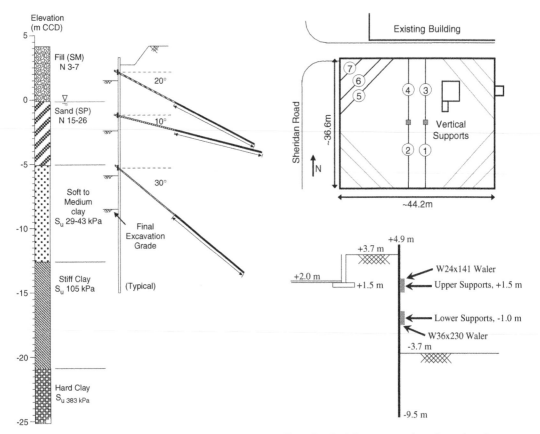

Figure 3. Lurie Center excavation.

Figure 5. Ford Center excavation: plan and section.

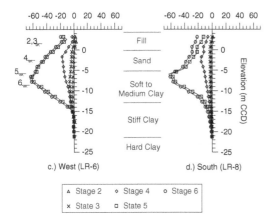

Figure 4. Lateral movements at Lurie Center excavation.

with 2 basement levels supported by deep foundations. Little movements were observed in the clay until the excavation progressed to stage 5 when it cut through the clay crust. Thereafter relatively large movements were observed, suggesting the importance of the small strain nonlinearity of these soft clays.

2.3 Ford Center excavation project

The excavation for the Ford Engineering Design Center (FEDC) on the Evanston campus of Northwestern University undercut the adjacent Technological Institute (Tech), as indicated in Figure 5. The excavation was 10 m deep and included a sheet-pile wall supported by two levels of internal braces. Complete descriptions of the monitoring efforts, observed performance and finite element simulations of this excavation are found in Blackburn (2005).

The excavation support system consisted of a XZ85 sheet pile wall supported by two levels of internal bracing. The ground surface was lower at the north side of the excavation between the existing building and the excavation. The bracing at each level consisted of 3 diagonal braces in each corner and 2 cross-lot braces, consisting of 0.46–0.61 m diameter pipes with 12.7 mm wall thickness. The shortest corner braces consisted of wide flange sections. Soil conditions were similar to those at the Chicago-State excavation.

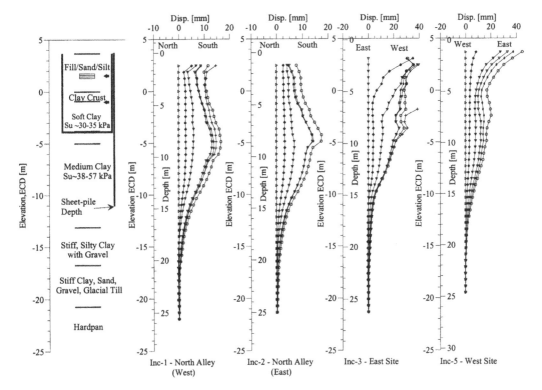

Figure 6. Lateral movements at Ford Center.

Large walers were employed for the FEDC excavation; however they were not uniformly attached to the sheet pile walls. Steel plates were welded to connect the walers to the sheeting at locations where the wale did not abut against the sheeting as a result of normal driving tolerances. Internal bracing was installed by positioning the strut on the waler and partially welding it, preloading the strut with two hydraulic jacks, and fully welding the strut to the waler after removing the preload. The middle and large diagonal supports were preloaded to fifty percent of their design load; however the preload was not sustained during the welding and the method of preload in the diagonals imposed bending moment rather than axial force. The cross-lot braces were supported by a steel frame located at their midpoint.

Figure 6 summarizes the lateral movements at the site. The lateral movements near the ground surface reflect the higher pre-excavation grade on the east side of the excavation. While slopes were cut so the top elevation of the sheetpile wall was the same around the excavation, the additional loading resulted in a more pronounced cantilever movements at the sides with the higher ground surface. While there were significant differences in the movements in the upper parts of the observed responses, the differences in the softer clays were much smaller. These observations emphasize the

importance of recognizing three dimensional effects when interpreting excavation responses under certain conditions.

3 INVERSE MODELING: UPDATING DESIGN PREDICTIONS USING FIELD OBSERVATIONS

In model calibration, various parts of the model are changed so that the measured values are matched by equivalent computed values until the resulting calibrated model accurately represents the main aspects of the actual system. In practice, numerical models typically are calibrated using trial-and-error methods. Inverse analysis works in the same way as a non-automated calibration approach: parameter values and other aspects of the model are adjusted until the model's computed results match the observed behavior of the system. Despite their apparent utility, however, inverse models are used for this purpose much less than one would expect because, perhaps, of the difficulties of implementing an inverse analysis and the complexity of the simulated systems.

Use of an inverse model provides results and statistics that offer numerous advantages in model analysis and, in many instances, expedites the process of

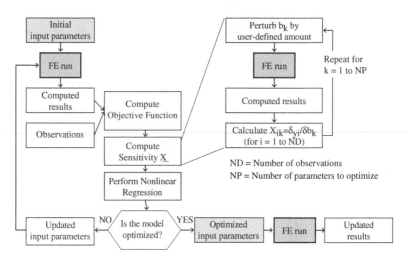

Figure 7. Flow chart for inverse analysis.

adjusting parameter values. The fundamental benefit of inverse modeling is its ability to calculate automatically parameter values that produce the best fit between observed and computed results. The main difficulties inherent to inverse modeling algorithms are complexity, non-uniqueness, and instability. Complexity of real, non-linear systems sometimes leads to problems of insensitivity when the observations do not contain enough information to support estimation of the parameters. Non-uniqueness may result when different combinations of parameter values match the observations equally well. Instability can occur when slight changes in model variables radically change inverse model results. Although these potential difficulties make inverse models imperfect tools, recent work in related civil engineering fields (Poeter and Hill, 1997; Ou and Tang, 1994; Keidser et al., 1991) demonstrate that inverse modeling provides capabilities that help modelers significantly, even when the simulated systems are very complex.

In the work described herein, model calibration by inverse analysis is conducted using UCODE (Poeter and Hill, 1998), a computer code designed to allow inverse modeling posed as a parameter estimation problem. Macros can be written in a windows environment to couple UCODE with any finite element software.

Figure 7 shows a detailed flowchart of the parameter optimization algorithm used in UCODE. With the results of a finite element prediction in hand, the computed results are compared with field observations in terms of weighted least-squares objective function, $S(b)$:

$$S\left(\underline{b}\right)=\left[\underline{y}-\underline{y}'\left(\underline{b}\right)\right]^T \underline{\omega}\left[\underline{y}-\underline{y}'\left(\underline{b}\right)\right]=\underline{e}^T \underline{\omega e} \quad (1)$$

where b is a vector containing values of the parameters to be estimated; y is the vector of the observations being matched by the regression; $y'(b)$ is the vector of the computed values which correspond to observations; ω is the weight matrix wherein the weight of every observation is taken as the inverse of its error variance; and e is the vector of residuals. This function represents a quantitative measure of the accuracy of the predictions.

A sensitivity matrix, X, is then computed using a forward difference approximation based on the changes in the computed solution due to slight perturbations of the estimated parameter values. This step requires multiple runs of the finite element code. Regression analysis of this non-linear problem is used to find the values of the parameters that result in a best fit between the computed and observed values. In UCODE, this fitting is accomplished with the modified Gauss-Newton method, the results of which allow the parameters to be updated using:

$$\left(\underline{C}^T \underline{X}^T_r \underline{\omega} \underline{X}_r \underline{C}+\underline{I}m_r\right)\underline{C}^{-1}\underline{d}_r = \underline{C}^T \underline{X}^T_r \underline{\omega}\left(\underline{y}-\underline{y}'\left(\underline{b}_r\right)\right) \quad (2)$$

$$\underline{b}_{r+1} = \rho_r\underline{d}_r + \underline{b}_r \quad (3)$$

where d_r is the vector used to update the parameter estimates b; r is the parameter estimation iteration number; X_r is the sensitivity matrix ($X_{ij} = \partial y_i/\partial b_j$) evaluated at parameter estimate b_r; C is a diagonal scaling matrix with elements c_{ij} equal to $1/\sqrt{(X^T \omega X)_{ij}}$; I is the identity matrix; m_r is the Marquardt parameter (Marquardt 1963) used to improve regression performance; and ρ_r is a damping parameter, computed as the change in consecutive estimates of a parameter normalized by its initial value, but it is restricted to values less than 0.5.

40

3.1 Chicago-State excavation

The finite element software PLAXIS 7.11 (Brinkgreve and Vermeer, 1998) was used to compute the response of the soil around the excavation. The problem was simulated assuming plane-strain conditions. The soil stratigraphy was assumed to be uniform across the site. Eight soil layers were considered: a fill layer overlaying a clay crust, a compressible clay deposit in which 5 distinct clay layers were modeled and a relatively incompressible hard silty clay stratum, locally known as hardpan. These 5 clay layers are noted as 1 through 5 in subsequent discussion. The side boundaries of the mesh (total size 183 m × 28.7 m) were established beyond the zone of influence of the settlements induced by the excavation (e.g. Roboski, 2004). The finite element mesh boundary conditions were set using horizontal restraints for the left and right boundaries and total restraints for the bottom boundary.

Table 1 shows the calculation phases and the construction stages used in the finite element simulations. Note that the tunnel tubes and the school adjacent to the excavation were explicitly modeled in the first 12 phases of the simulation to take into account the effect of their construction on the soil surrounding the excavation. Stages 1, 2, 3, 4 and 5 refer to the construction stages for which the computed results were compared to inclinometer data taken from two inclinometers on opposite sides of the excavation. Construction steps not noted as "consolidation" on Table 1 were modeled as undrained. Consolidation stages were included after the tunnel, school and wall installation calculation phases to permit excess pore water pressures to equilibrate. Secant pile wall installation in the field is a three-dimensional process. To simulate this construction in the plane strain analysis, elements representing the wall were excavated and a hydrostatic pressure equivalent to a water level located at the ground surface was applied to the face of the resulting trench (calculation phase 13 in Table 1). After computing the movements associated with this process, the excavated elements were replaced by elements with the properties of the secant pile wall (calculation phase 14). Details about the definition of the finite element problem, the calculation phases and the model parameters used in the simulation described herein can be found in Calvello (2002).

Figure 8 is presented to illustrate some of the challenges of using field observations to calibrate numerical models of any kind, even when detailed records exist. This figure summarizes the construction progress at the Chicago-State excavation in terms of excavation surface and support installation on one of the walls of the excavation for selected days after construction started. Also shown are the locations of two inclinometers placed several meters behind the wall. If one is making a computation assuming plane

Table 1. FE simulation of construction.

Phase	Construction step
0	Initial conditions
1–4	Tunnel construction (1940)
5	Consolidation
6–10	School construction (1960)
11–12	Consolidation
13	Drill secant pile wall (1999)
14	Place concrete in wall **Stage 1**
15	Consolidation (20 days)
16	Excavate and install struts **Stage 2**
17	Excavate below first tieback level
18	Prestress first level of tiebacks **Stage 3**
19	Excavate below second tieback level
20	Prestress second level of tiebacks **Stage 4**
21	Excavate to final grade **Stage 5**

strain conditions, then it is clear that one must judiciously select a data set so that planar conditions would be applicable to a set of inclinometer data. If one is using an integrated approach wherein data is collected and compared with numerical predictions in almost real time (Finno and Blackburn 2005; Hashash and Finno 2005), then it is clear that a 3D analysis would be required for most days as a result of the uneven excavated surface and erratic anchor prestressing.

The soil model used to characterize the clays in the PLAXIS simulation of the excavation is the Hardening-Soil (H-S) model (Schanz et al., 1999). This effective stress model is formulated within the framework of elasto-plasticity. Plastic strains are calculated assuming multi-surface yield criteria. Isotropic hardening is assumed for both shear and volumetric strains. The flow rule is non-associative for frictional shear hardening and associative for the volumetric cap.

Table 2 shows the initial values of the 6 basic H-S input parameters for the five clay layers that are calibrated by inverse analysis. These parameters are the friction angle, ϕ, cohesion, c, dilation angle, ψ, the reference secant Young's modulus at the 50% stress level, E_{50}^{ref}, the reference oedometer tangent modulus, E_{oed}^{ref}, and the exponent m which relates reference moduli to the stress level dependent moduli (E representing E_{50}, E_{oed}, and E_{ur}):

$$E = E^{ref} \left(\frac{c \cot \phi - \sigma_3'}{c \cot \phi + p^{ref}} \right)^m \tag{4}$$

where p^{ref} is a reference pressure equal to 100 stress units and σ_3' is the minor principal effective stress. The initial estimates of the input parameters for layers 1 to 4 were based on triaxial test results using the approach described in Calvello and Finno (2002). Because few laboratory data existed for the very stiff layer 5 soil and

41

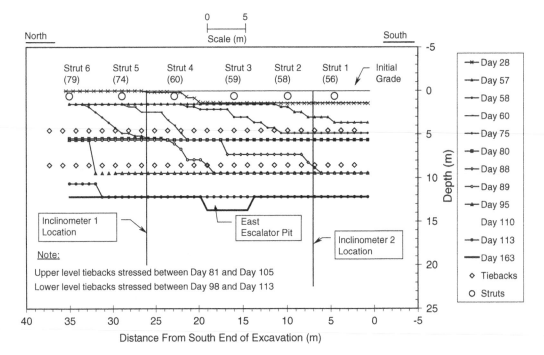

Figure 8. Excavation and support installation sequence at Chicago-State.

Table 2. H-S soil parameters for Chicago-State excavation.

Layer	1	2	3	4	5
$\Phi(°)$	23.4	23.4	25.6	32.8	32.8
c (kPa)	0.05	0.05	0.05	0.05	0.05
$\psi(°)$	0	0	0	0	0
m	0.8	0.8	0.85	0.85	0.85
E_{50}^{ref} (kPa) (final)	226 (362)	288 (362)	288 (750)	413 (2,567)	619 (3,850)
E_{oed}^{ref} (kPa) (final)	158 (253)	202 (253)	202 (525)	289 (1,797)	433 (2,695)
E_{ur}^{ref} (kPa) (final)	678 (1,086)	864 (1,086)	864 (2,250)	1,239 (7,701)	1,857 (11,550)

very small movements were observed in that stratum, the initial values of the parameters for layer 5 were selected to minimize movements in that stratum.

A sensitivity analysis indicated that the model's relevant and uncorrelated parameters are E_{50}^{ref} and ϕ (Calvello and Finno 2004). Results were also sensitive to changes in values of parameter m. However, parameter m was not included in the regression because the values of the correlation coefficients between parameters m and E_{50}^{ref} were very close to 1.0 at every layer, indicating that the two parameters were not likely to be simultaneously and uniquely optimized. When values

of ϕ were kept constant at their initial estimates, and only the stiffness parameters, E_{50}^{ref}, were optimized, the calibration of the simulation was successful. Finno and Calvello (2005) showed that shear stress levels in the soil around the excavation were much less than those corresponding to failure for the great majority of the soil. This is indeed expected for excavation support systems that are designed to restrict adjacent ground movements to acceptably small levels, and hence one would expect the stiffness parameters to have a greater effect on the simulated results than failure parameters.

See Finno and Calvello (2005) for a description of the parameters used to model the sand and fill layer and the structural elements of the excavation support system. The struts and tendons for the ground anchors were modeled as two-node elastic spring elements with constant spring stiffness, EA. The grouted portion of the anchor the tiebacks was modeled as a line element with an axial stiffness EA and no bending stiffness. Interface elements were placed between the soil and structural elements, except for the cross-lot brace.

Visual examination of the horizontal displacement distributions at the inclinometer locations provides the simplest way to evaluate the fit between computed and measured field response. When computations were made based on parameters derived from results of drained triaxial tests, the finite element model computed significantly larger displacements at every construction stage (Finno and Calvello 2005). The maximum computed horizontal displacements are about two times the measured ones and the computed displacement profiles result in significant and unrealistic movements in the lower clay layers. As one would expect, these results indicated that the stiffness properties for the clay layers based on conventional laboratory data were less than field values.

Figure 9 shows the comparison between the measured field data from the east side of the excavation and the computed horizontal displacements when parameters are optimized based on stage 1 observations. The improvement of the fit between the computed and measured response is significant. Despite the fact that the optimized set of parameters is calculated using only stage 1 observations, the positive influence on the calculated response is substantial for all construction stages. At the end of the construction (i.e. stage 5) the maximum computed displacement exceeds the measured data by only about 15%. These results are significant in that a successful recalibration of the model at an early construction stage positively affects subsequent "predictions" of the soil behavior throughout construction. Note that similar agreement was noted for the inclinometer on the other side of the excavation using the same parameters (Finno and Calvello 2005).

Analyses were also made wherein parameters were recalibrated at every stage until the final construction stage (stage 5). At every new construction stage, the inclinometer data relative to that stage were added to the observations already available. Results indicated that difference between the fit shown in Figure 9 and with those calibrated after every increment was not significant. In essence, the inverse analysis performed after the first construction stage "recalibrated" the model parameters in such a way that the main behavior of the soil layers could be accurately predicted throughout construction.

3.2 Applicability of optimized parameters at other locations in same deposit

To show the applicability of the optimized parameters that formed the basis of the good agreement in Figure 9 to other excavation sites in these soil deposits, the results of numerical simulations are presented in Figure 10 based on these optimized parameters for the Lurie and the Ford Design Center excavations. The geologic origin of the most compressible material is similar for all three cases, but the Lurie Center is located about 2 km from the Chicago-State site and the Ford Center is located about 15 km from the site. Consequently one should expect some variability in the actual parameters at each site.

For the Lurie site, examining the comparisons in the clay layers below −15 ft CCD, reasonable agreement is observed at stages 4 and 6, with significant differences seen at the intermediate stage 5. While the reasons for this are not entirely clear, the difference between the excavated levels for stages 5 and 6 was only 2 m. The observed lateral movements from stage 5 might have been impacted by the excavation process not being completely uniform. This emphasizes the need to carefully select stages for analysis that are compatible with the assumed numerical model, in this case a plane strain representation of the problem. While care was taken to do so, some simplification of the excavation process was necessary in order to obtain a complete record of the responses.

Furthermore, the computed results indicated much larger cantilever type movements than were observed. This difference is likely due to the oversimplification of the tieback supports in the plane strain simulation. These anchors were post-grouted during installation to provide adequate support. The processing of smearing the anchor stiffness into a per length value for the plane strain analysis makes it difficult to account for the benefits of the post-grouting. The larger ground anchor interface resistances were not explicitly considered in the analysis, again emphasizing the need to carefully represent the actual construction process when numerically simulating an excavation.

At the Ford Center, the numerical results shown in Figure 10 followed similar trends as the observed data, but with larger magnitudes. This is likely caused by the fact that the H-S model used herein does not include provisions to represent the large stiffness degradation with small strains. One must select moduli that represent the average strains within the soil mass, and when the movements are small, the average modulus should be higher in a model that does not consider the small strain modulus degradation. The parameters used in the analysis were based on the larger deformations that were present at the Chicago-State site, and hence resulted in larger deformations than were observed at the Ford Center. In any case, the application of the Chicago-State based optimized parameters to both the Lurie and Ford sites resulted in reasonable agreement with the observed lateral movements, within the limitations of the analyses. Application of the inverse techniques to these data would result in improved fit with minor changes to the parameters.

3.3 Effects of assumed initial conditions

The importance of accurately considering the proper effective stress conditions at the start of excavation is illustrated by a parametric study based on the conditions at the Chicago-State excavation. Table 3 summarizes the five simulations made to evaluate the effects of different initial conditions that arise from previous construction activities at the site.

All pre-excavation activities, including the construction of the tunnel tubes, the construction of the

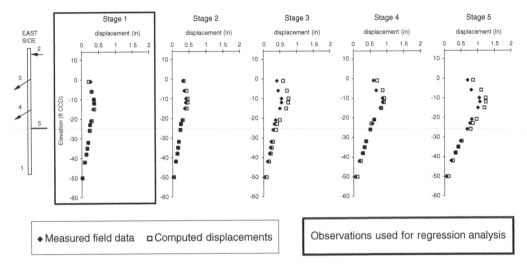

Figure 9.   Comparison of observed and computed horizontal displacements for Chicago-State.

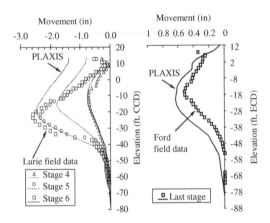

Figure 10. Computed and observed lateral movements based on parameters optimized at the Chicago-State excavation.

school and installation of the secant-pile wall, were explicitly considered in the *base-case* finite element simulation of the supported excavation system. In the *no-tunnel* simulation the tunnel was "wished-in-place" while the construction of the school and installation of the secant-pile walls were explicitly modeled. When the tunnel was "wished-in-place," it is assumed the tunnel was installed with no change in stress in the surrounding soil. Hence the stiffness of the tunnel considered in the subsequent analyses, but the effects of the stress changes associated with its installation are not. In the *no-school* simulation both the tunnel and school were wished-in-place at the beginning of the analysis, and only the installation of the secant-pile walls was explicitly modeled. In the *no-wall*

simulation, the tunnel, school and secant-pile wall were all wished-in-place at the beginning of the analysis and the wall was simulated using a beam element, in contrast to the previous cases where it was modeled with solid elements. In the *free-field* simulation the tunnel and the school were not included in the simulation, while the wall, wished-in-place at the beginning of the analysis, again was simulated using a beam element. Note that, when a construction activity was not modeled and the corresponding structure was either wished in place or not included in the finite element mesh, k_0-conditions (i.e. gravity loads) were assumed to be representative of the "initial" stresses in the soil.

The free-field analysis is included herein because that type of analysis forms the basis of semi-empirical methods used to estimate ground deformations for supported excavations (Mana and Clough, 1981; Clough et al., 1989). Predictions based on these methods do not explicitly include the effects of past construction activities that are common in urban environments, and the effects of neglecting these activities are evaluated herein.

Figure 11 shows such a comparison for the computed horizontal displacements of all simulations when the base-case best-fit values are used to define the H-S parameters of the clay layers. These results clearly show the importance of accounting for all past activities that have affected the soil stresses at an excavation site. The range of maximum movements varies 300%, depending on the assumption made. Note that the free-field case yields the largest computed movements, which are clearly unrealistic. Hence, estimates of ground movements made based on semi-empirical methods that do not explicitly account for the effects of past construction activities must be used with caution.

44

Table 3. Summary of parametric studies.

Simulation name	Structure			
	Tunnel	School	Wall	Excavation
Base-case	construction modeled	construction modeled	construction modeled	Stage-by-stage modeling
No-tunnel	wished-in-place	construction modeled	construction modeled	Stage-by-stage modeling
No-school	wished-in-place	wished-in-place	construction modeled	Stage-by-stage modeling
No-wall	wished-in-place	wished-in-place	wished-in-place	Stage-by-stage modeling
Free-field	not considered	not considered	wished-in-place	Stage-by-stage modeling

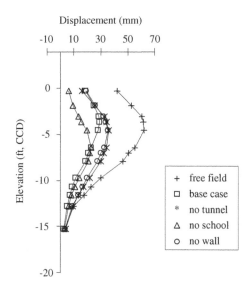

Figure 11. Effects of assumed initial conditions.

4 COMMENTS ON CALIBRATION OF SOIL MODELS BY INVERSE ANALYSIS

The calibration by inverse analysis of the various simulations presented herein indicated that the numerical methodology developed to optimize a finite element model of the excavation can be very effective in minimizing the errors between the measured and computed results. However, the convergence of an inverse analysis to an "optimal solution" (i.e. best-fit between computed results and observations) does not necessarily mean that the simulation is satisfactorily calibrated (Finno and Calvello 2005). A geotechnical evaluation of the optimized parameters is always necessary to verify the reliability of the solution. For a model to be considered "reliably" calibrated both the fit between computed and observed results must be satisfactory (i.e. errors are within desired and/or accepted accuracy) and the best-fit values of the model parameters must be reasonable. Some of the simulations discussed

in this section satisfy one of the above-specified criteria, but only the base-case simulation satisfies both of them.

The key to the successful calibration of the base-case model of the excavation lies in defining a "well posed" inverse analysis problem to calibrate the simulation. The parameters optimized by inverse analysis are few compared to the total number of parameters defining the behavior of the simulation. Indeed, the majority of the input parameters is estimated by conventional means and never "re-calibrated" (Table 2). Yet, the optimization of the base-case simulation is extremely effective because the finite element simulation of the excavation adequately reproduced the stress history of the soil on site, the soil model (H-S) adequately represented the behavior of the clays, at least in its effects on the magnitude and distribution of horizontal wall displacement, the parameters (stiffness parameters) that were most relevant to the problem under study were calibrated, and the initial estimates of the other input parameters were reasonable.

ACKNOWLEDGEMENTS

Financial support for this work was provided by National Science Foundation grant CMS-0219123 and the Infrastructure Technology Institute (ITI) of Northwestern University. The support of Dr. Richard Fragaszy, program director at NSF, and Mr. David Schulz, ITI's director, is greatly appreciated.

REFERENCES

Blackburn J.T. (2005). *Automated remote sensing and three-dimensional analysis of internally braced excavations* PhD dissertation, Northwestern University, Evanston, IL.2005.

Brinkgreve R.B.J. and Vermeer P.A. (1998). Finite Element Code for Soil and Rock Analysis. PLAXIS 7.0 manual. Balkema.

Calvello M. (2002). Inverse analysis of supported excavations through Chicago glacial clays. PhD Thesis, Northwestern University, Evanston, IL.

Calvello M. and Finno R.J. (2002). Calibration of soil models by inverse analysis. Proc. International Symposium on Numerical Models in Geomechanics, NUMOG VIII, Balkema, p. 107–116.

Calvello M. and Finno R.J. (2003). Modeling excavations in urban areas: effects of past activities. Italian Geotechnical Journal, 37(4), 9–23.

Calvello M. and Finno R.J. (2004). Selecting parameters to optimize in model calibration by inverse analysis. Computers and Geotechnics, Elsevier, in press.

Chung C.K. and Finno R.J. (1992). Influence of Depositional Processes on the Geotechnical Parameters of Chicago Glacial Clays. Engineering Geology, Vol. 32, p. 225–242.

Clough G.W., Smith E.M. and Sweeney B.P. (1989). Movement control of excavation support systems by iterative design. Current Principles and Practices, Foundation Engineering Congress, Vol. 2, ASCE, 869–884.

Finno R.J. and Blackburn J.T. (2005). *Automated monitoring of supported excavations* Proceedings, 13th Great Lakes Geotechnical and Geoenvironmental Conference, Geotechnical Applications for Transportation Infrastructure, GPP 3, ASCE, Milwaukee, WI., May, 1–12.

Hashash Y.M.A. and Finno R.J. (2005). *Development of new integrated tools for predicting, monitoring and controlling ground movements due to excavations* Proceedings, Underground Construction in Urban Environments, ASCE Metropolitan Section Geotechnical Group, New York, NY.

Finno R.J. and Calvello M. (2005). Supported excavation: The observational method and Inverse modeling. Journal of Geotechnical and Geoenvironmental Engineering. ASCE, 131 (7).

Finno R.J., Bryson L.S. and Calvello M. (2002). Performance of a stiff support system in soft clay. Journal of Geotechnical and Geoenvironmental Engineering, ASCE, Vol. 128, No. 8, p. 660–671.

Finno R.J. and Roboski J.F. (2005). *Three-dimensional Responses of a Tied-back Excavation through Clay* Journal of Geotechnical and Geoenvironmental Engineering, ASCE, Vol. 131, No. 3, March, 273–282.

Hill M.C. (1998). Methods and guidelines for effective model calibration. U.S. Geological Survey Water-Resources investigations report 98-4005, 90 pp.

Hsieh P.G. and Ou C.Y. (1998). Shape of ground surface settlement profiles caused by excavation. Canadian Geotechnical Journal, Vol. 35, p. 1004–1017.

Keidser A. and Rosjberg D. (1991). A comparison of four inverse approaches to groundwater flow and transport parameter identification, Water Resources Research, v. 27, no. 9, p. 2219–2232.

Mana A.I. and Cough G.W. (1981). Prediction of movements for braced cut in clay. Journal of Geotechnical Engineering, ASCE, New York, Vol. 107, No. 8, pp. 759–777.

Marquardt D.W. (1963). An algorithm for least-squares estimation of nonlinear parameters, Journal of the Society of Industrial and Applied Mathematics, Vol. 11 (8), p. 431–441.

Ou C.Y. and Tang Y.G. (1994). Soil parameter determination for deep excavation analysis by optimization. Journal of the Chinese Institute of Engineers, Vol. 17, No.5, p. 671–688.

Poeter E.P. and Hill M.C. (1997). Inverse Methods: A Necessary Next Step in Groundwater Modeling, Ground Water, v. 35, no. 2, p. 250–260.

Poeter E.P. and Hill M.C. (1998). Documentation of UCODE, a computer code for universal inverse modeling. U.S. Geological Survey Water-Resources investigations report 98-4080, 116 pp.

Roboski J. (2004) Three-dimensional ground movements caused by deep excavations. PhD Thesis, Northwestern University, Evanston, IL.

Schanz T., Vermeer P.A. and Bonnier P.G. (1999). The Hardening Soil model – formulation and verification. Proceedings Plaxis Symposium "Beyond 2000 in Computational Geotechnics," Amsterdam, Balkema, p. 281–296.

Numerical Modelling of Construction Processes in Geotechnical Engineering for
Urban Environment – Triantafyllidis (ed)
© 2006 Taylor & Francis Group, London, ISBN 0 415 39748 0

Influence of the construction procedure on the displacements of excavation walls

A. Hettler, T. Keiter & B. Mumme

Lehrstuhl Baugrund-Grundbau, Fakultät Bauwesen, Universität Dortmund

ABSTRACT: The realistic computation of stresses and deformations for excavation walls is a challenging task. Sophisticated stress-strain relations and the realistic modeling of construction procedures are required. This is illustrated by several examples. The first deals with extraordinary displacements which occurred while producing a jet grouting girder below the bottom of the excavation. It is shown that when using a simple elastic-plastic stress strain relation, the measured deformations can only partly be explained.

Further examples show that it is necessary to take the preloading caused by the weight of the excavation into account for a realistic modeling of the resistance of the foot of the wall. Two approaches are proposed: One by a nonlinear subgrade reaction model and another by a finite element analysis based an hypoplasticity and hypoplasticity with intergranular strain, respectively.

1 INTRODUCTION

The design of constructions in ground engineering often depends on the installation procedure. A typical example is the creation of piles: the magnitude of the skin friction can easily differ by more than a factor of three, comparing bored piles, post grouted bored piles, driven piles or micro piles. As in standard cases often a global factor $\eta = 2$ is used, it is apparent how important it is to take the installation procedure into account.

Also for excavation walls the installation process may have a big influence on the displacements and the sectional forces and therefore on the safety and on the economy of the construction. For example sheet pile walls with flexible anchor walls may be loaded by an active earth pressure with a triangular distribution, while walls with prestressed anchors show rather a rectangular load figure. In Germany, following the recommendations for excavation walls (EAB), earth pressure distributions as proposed in Figure 1 may be used for design [EAB, Weißenbach et al.].

The distribution depends mainly on the type of the wall and the amount and the position of the supports.

All distributions were mainly derived on the basis of measurements. In other countries, for example in Great Britain, distributions declining from the triangular shape, may be derived by finite-element-calculations [Weißenbach, Weißenbach et al.].

The finite-element-method requires a lot of experience and parameters have to be thoroughly determined. Although much progress has been gained by

Soldier pile walls

Sheet pile walls and concrete walls

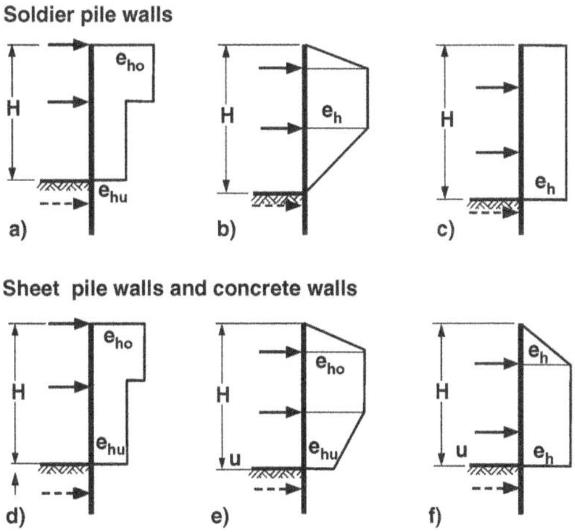

Figure 1. Typical examples for the earth pressure distributions recommended by the EAB for walls propped or anchored twice: a), b), c) For soldier pile walls, d), e), f) For sheet pile walls and concrete walls.

this method during the last years, many problems still have to be solved. In the following, different aspects are discussed:

– The first example deals with unexpected heave and wall displacements when constructing a stiffening girder system at the bottom of the wall by jet grouting and the simulation by FEM with a simple elasto-plasto stress-strain relationship.

- In the third section, the influence of props and pre-stressed anchors is studied by a numerical analysis based on hypoplasticity.
- In the forth section it is investigated whether the active or the increased active earth pressure has to be used for design when comparing sheet pile and trench slurry walls.
- Finally, the influence of the preloading by the weight of the excavation on the displacements of the foot of the wall is studied.

2 UNEXPECTED HEAVE AND DISPLACEMENTS WHEN CONSTRUCTING A SLURRY TRENCH WALL

2.1 Basic data

While constructing a jet grouting girder system at the base of a slurry trench excavation wall, an unexpected heave at the surface and displacements of the wall against the soil side were observed.

Figure 2 shows a typical cross section of the wall with the stratification. The wall was supported four times with braces and anchors. To reduce the displacements at the bottom of the wall a girder system was installed to stiffen the whole system (Figure 3).

Deformations and sectional forces were measured during the construction process to control the system.

To the big surprise, while jetting the girder system from the ground level, the soil heaved up to 25 cm in the center of the excavation and the wall moved to the soil side with a maximum at the top of about 24 mm. While excavating, the wall moved back to the inner side (Figure 4).

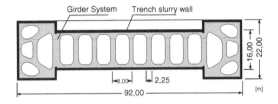

Figure 3. Plan of building pit.

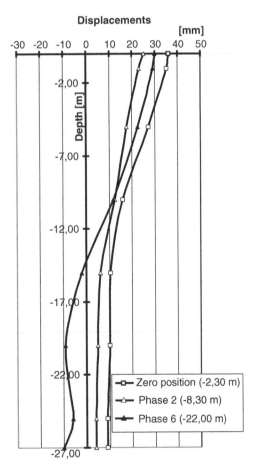

Figure 4. Measured deformations of diaphragm wall in measuring cross section M Q1 during excavation.

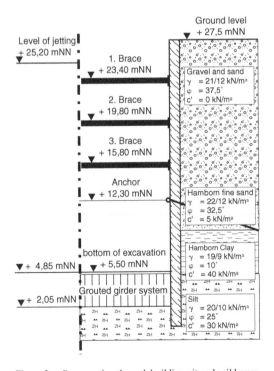

Figure 2. Cross section through building pit and soil layers.

2.2 Numerical analysis

After these unexpected deformations occurred, numerical analysis were used to try to explain this phenomenon. The calculations based on the finite-element-program PLAXIS use a linear elastic-ideal plastic model with Mohr-Coulomb's limit condition. The soil parameters were derived from the subsoil report in the sense of a real prediction.

The wall was modeled as a linear elastic stress strain relation. Interface elements with full friction angles were used at the inner side of the wall. This may be justified by the following arguments. While producing the jet grout girder system, the bentonit layer at the inner side was nearly completely replaced by a soil-concrete mixture with high friction properties. This was confirmed when excavating the soil. At the outside, the grout suspension did not move up so that the original bentonit layer was still there. Therefore, a reduction factor R = 0.5 at the soil side was chosen. FE parameter studies showed that this friction model may very well be used to explain the mechanism.

In the following, only the main results are presented. For details see [Hettler & Besler & Gutjahr].

Figure 5 shows the comparison of the deformations during the production process of the jet grout girder system. Both, the general mechanism and the horizontal displacements, agree well between measurements and calculation. The attempt to predict the deformations during the excavation of the soil by FE analysis was unfortunate. Comparing the numerical results in Figure 6 with the measurements in Figure 4

reveal some differences and specially the brace forces differed completely (Table 1).

From these results it was concluded that the simple linear elastic-ideal plastic model was sufficient to describe the more or less monotonically increasing stress-strain paths during the first loading within the grouting process. But it comes to its limitations when describing the unloading and reloading paths during the excavation of the soil.

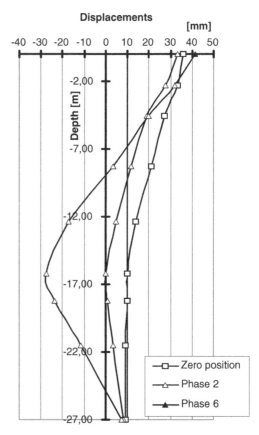

Figure 6. Calculated deformations during excavation.

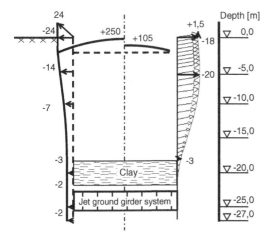

Figure 5. Measured and calculated displacements prior to excavation.

Table 1. Comparison of strut forces.

	Expected forces obtained in classical calculation	Measured forces	Forces by FEM
Strut 1.4	979 kN	2066 kN	129 kN
Strut 2.4	1030 kN	2509 kN	5022 kN
Strut 3.4	4892 kN	4002 kN	4470 kN
Anchor 4.4	455 kN	354 kN	321 kN
Sum	7356 kN	8931 kN	9942 kN

3 INFLUENCE OF BRACES AND PRESTRESSED ANCHORS ON WALL DEFORMATIONS

3.1 Basic data

Excavation walls are often designed assuming a girder model with supports which do not move. In most cases this is sufficient to determine the embedment length or the sectional forces. During the last years though the investigation of service states has become more important.

To study the influence of the boundary conditions at the supports, finite-element calculations were performed using the single propped system in Figure 7. Two types of walls were investigated: a sheet pile wall and a slurry trench wall. Referring to the results of a classical analysis two embedment lengths were used. The length $t_0 = 2,12$ m corresponds to a free earth support in the classical analysis and $t_1 = 3,65$ m corresponds to a fixed earth support according to Blum's model. The soil was assumed to be a medium dense sand with no ground water. For the FE-model hypoplasticity was implemented in ABAQUS using von Wolffersdorff's version [Kolymbas, Maier, von Wolffersdorff]. For details see [Hettler & Vega Ortiz & Mumme].

The following three cases were investigated:

- Case 1: The prop is horizontally fixed from the beginning
- Case 2: Preexcavation up to three meters, afterwards the prop is fixed horizontally
- Case 3: Preexcavation up to three meters, afterwards prestressing with 80% of the forces obtained from the classical girder calculation, after prestressing the prop is fixed.

Case 1 corresponds approximately to the classical girder model with supports which do not yield, case 2 approximately to a wall propped by braces, which

do not yield and which are not prestressed, and case 3 to a wall with prestressed anchors.

3.2 Some typical results

In the following only some typical results are presented for the sheet pile and the slurry trench wall with $t_g = 2,12$ m and an infinite surcharge $p_k = 0.0$. For further results see [Hettler & Vega Ortiz & Mumme].

Figure 8 shows the earth pressure distribution, the displacements and the bending moments for the sheet pile wall, Figure 9 for the slurry trench wall.

As expected, the displacements are quite different depending on the type of support and on the stiffness of the wall. Caused by the backward movement at the toe of the sheet pile wall and the bending deflections, the wall is partly unloaded in the middle and the earth pressure concentrates at the upper support and in the part below the excavation level. The higher stiffness of the concrete wall is connected with a mechanism similar to a rotation around the top of the wall and the displacements are smaller. In total, the magnitude of the movements is not sufficient to mobilize the active earth pressure. Therefore the bending moment in the field of the concrete wall is about twice as high as the value of the pile wall. Further calculations show

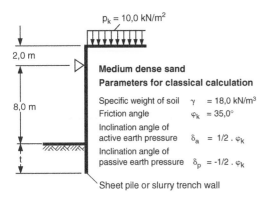

Figure 7. Statical system.

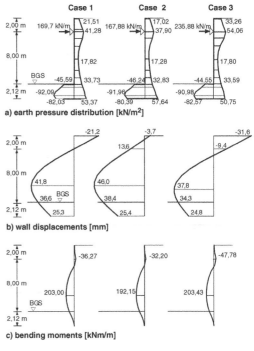

Figure 8. Results for sheet pile wall and FE-calculation for embedment depth $t_0 = 2,12$ m: a) earth pressure distribution, b) deflection line, c) bending moments.

similar results. For details see [Hettler & Vega Ortiz & Mumme].

3.3 Conclusions

The calculations show the important influence of the boundary conditions on the earth pressure distribution, the displacements and the bending moments of the wall. This may be very important when investigating service states. For example to obtain reliable predictions of building settlements adjacent to excavation walls, the construction procedure must be modeled as realistic as possible. Further it is supposed that also the installation process like vibrating the sheet piles may be of great importance. This should be subject to further research work.

4 INFLUENCE OF PRELOADING BY THE WEIGHT OF THE EXCAVATION ON THE WALL DISPLACEMENTS

In the following, the influence of the preloading by the weight of the excavated soil on the wall displacements is investigated. First, a nonlinear subgrade reaction model is used in section 4.2 and checked by field measurements. In section 4.3, the results are compared

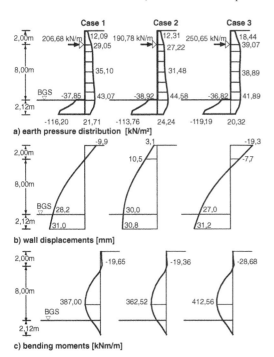

Figure 9. Results for a slurry trench wall and FE-computation for an embedment depth of $t_0 = 2{,}12$ m: a) earth pressure distribution, b) deflection line, c) bending moments.

with finite-element calculations based on hypoplasticity and hypoplasticity with intergranular strain.

4.1 Nonlinear subgrade reaction model

On the basis of model tests, Besler proposes the following nonlinear subgrade reaction model to describe the mobilized horizontal earth pressure e'_{ph} in front of the wall as a function of the displacements

$$e'_{ph} = \gamma \cdot z \left[A + \frac{B}{C + s/s_B} \right] \qquad (1)$$

with

γ : specific weight of the soil,
z : depth beneath the bottom of the soil,
A, B, C : Parameters depending on the coefficient of the earth pressure at rest K_0, the coefficient of the passive earth pressure K_p and on the displacements s_g in service states, where $e'_{ph} = 1/2\, e_{ph}$, and s_B in the limit state,
s_B : displacement in the limit state when e'_{ph} reaches the passive earth pressure e_{ph}.

For $s = 0$ the earth pressure at rest e_0 is assumed. The model was also modified to take into account a preloading vertical stress p_v. Then for $s = 0$ $e'_{ph} = p_v$ is assumed and the mobilization curve is shifted horizontally by $\Delta\xi$, where $\Delta\xi$ is the dimensionless displacement for $e'_{ph} = p_v$ in the original mobilization curve. For sands all parameters where determined in small scale model tests and also checked by comparison with field tests. For details see [Hettler & Vega Ortiz & Gutjahr].

A program was established for practical applications and parameter studies. The program and the subgrade reaction model were tested out and refined. It could be shown, that only the model which takes the preloading into account, gives realistic results for excavation walls.

As an example, Figures 10 and 11 show the deflection lines for sheet pile walls according to the statical system in Figure 7 varying the embedment length t_g. Comparing Figures 10 and 11, it is obvious that the response of the model without preloading is too soft. Following the classical calculation, a fixed earth support with a backward movement to zero deformations is already expected for $t_1 = 3{,}65$ m, whereas the model without preloading predicts an embedment length of about 7 m. It can also be shown that the model with preloading gives very good predictions for real constructions [Hettler & Vega Ortiz & Gutjahr].

4.2 Comparison with finite-element calculations

Figure 12 shows the comparison of the results for different calculation models for a sheet pile wall with an

51

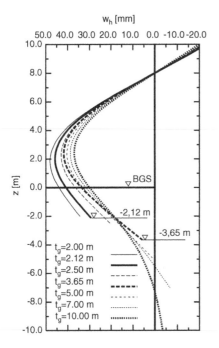

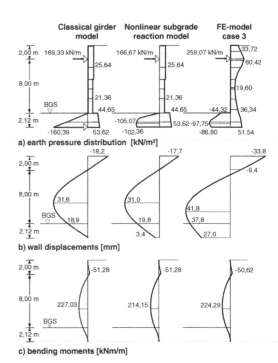

a) earth pressure distribution [kN/m²]

b) wall displacements [mm]

c) bending moments [kNm/m]

Figure 10. Subgrade reaction model without preloading: Deflection line for sheet pile walls with various embedment length.

Figure 12. Comparison of different calculation models for sheet pile wall with embedment depth $t = 2,12$ m a) earth pressure distribution, b) deflection line, c) bending moments.

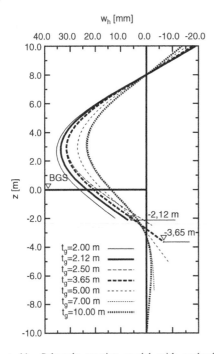

Figure 11. Subgrade reaction model with preloading: Deflection line for sheet pile walls with various embedment length.

embedment length of $t = 2,12$ m, which corresponds to the length, which is necessary for a free earth support in the classical girder calculation.

In the finite-element calculation von Wolffersdorff's hypoplasticity model was used. In all calculations a infinite surcharge $p_k = 10$ kN/m² was taken into account. The earth pressure distribution from the FE-model agrees well to the classical assumptions according to the EAB, also the bending moments coincide very well in all models. But the displacements in the FE-model are much higher. This was also observed in other parametric-studies. As the nonlinear subgrade-reaction model was checked and proved to fit well model and field tests, it can be concluded, that the stiffness in von Wolffersdorff's model is underestimated in calculations of excavation walls.

To improve the results, hypoplasticity was used with intergranular strain following Niemunis' and Herle's proposal [Niemunis, Niemunis & Herle]. As follows from Figure 13, the deflection lines of the nonlinear subgrade reaction model and the FE-model with intergranular strain approach. It is believed that although using intergranular strain the stiffness is still underestimated.

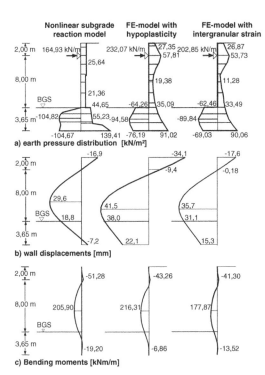

Nonlinear subgrade reaction model	FE-model with hypoplasticity	FE-model with intergranular strain

a) earth pressure distribution [kN/m²]

b) wall displacements [mm]

c) Bending moments [kNm/m]

Figure 13. Comparison of different calculation models for sheet pile wall with embedment depth t = 3,65 m a) earth pressure distribution, b) deflection line, c) bending moments.

4.3 Conclusions

To obtain realistic wall displacements in the nonlinear subgrade reaction model, the preloading by the weight of the excavated soil has to be taken into account. If not, the displacements are overestimated and the fixing effect in the soil support of the wall is underestimated.

When using the finite-element-method with von Wolffersdorff's hypoplasticity model it may be concluded that the deformations are highly overestimated.

Even stiffening the model with intergranular strain seems not to be sufficient.

It is supposed that in the hypoplasticity models used in the computations, the effects of unloading and reloading are not sufficiently taken into account. Therefore further research work is necessary to better integrate memory effects into the model.

REFERENCES

Empfehlungen des Arbeitskreis "Baugruben" EAB. 1994. 3. Auflage, Ernst und Sohn.

Hettler, A., Besler, D., Gutjahr, St.: *FE-Simulation von aufgetretenen Spannungen und Verformungen an der Baugrube Gleiswechsel in Duisburg-Meiderich*. 1998. Bautechnik 74, P. 25–32.

Hettler, A., Vega Ortiz, S., Mumme, B.: *Berechnung von Baugrubenwänden mit verschiedenen Methoden: Trägermodell*, nichtlineare *Bettung, Finite-Elemente-Methode*. 2005. Bautechnik in Vorbereitung.

Hettler, A., Vega Ortiz, S., Gutjahr, St.: *Nichtlinearer Bettungsansatz von Besler bei Baugrubenwänden*. 2005, Bautechnik 82, P. 593–604.

Kolymbas, D.: *Eine konstitutive Theorie für Böden und andere körnige Stoffe*. 1988. Universität Karlsruhe, Veröffentlichung des Institutes für Bodenmechanik und Felsmechanik der Universität Fridericana zu Karlsruhe, Heft 109.

Maier, Th.: *Numerische Modellierung der Entfestigung im Rahmen der Hypoplastizität*. 2002. Schriftenreihe des Lehrstuhls Baugrund-Grundbau der Universität Dortmund, Heft 24.

Niemunis, A.: *Extended hypoplastic models for soils*. 2003. Schriftenreihe des Institutes für Grundbau und Bodenmechanik der Ruhr-Universität Bochum, Heft 24.

Niemunis, A., Herle, I.: *Hypoplastic model for cohesionless soils with elastic strain range*. 1997. Mechanics of Cohesive-Frictional Materials 2, P. 279–299.

Weißenbach A.: *Empfehlungen des Arbeitskreises "Baugruben" der DGGT zur Anwendung des Bettungsmodulverfahrens und der Finite-Elemente-Methode*. 2003. Bautechnik 80, Heft 9, P. 569–598.

Weißenbach, A., Hettler, A., Simpson, B.: *Stability of excavations. Geotechnical Handbook*. 2003. Wiley von Wolffersdorff, P.-A.: *A hypoplastic relation for granular materials with a predefined limit state surface*. 1996. Mechanics of Cohesive-Frictional Materials 1, P. 251–271.

2. Diaphragm walls

Numerical Modelling of Construction Processes in Geotechnical Engineering for
Urban Environment – Triantafyllidis (ed)
© 2006 Taylor & Francis Group, London, ISBN 0 415 39748 0

Finite element analysis of the excavation of the new Garibaldi station of Napoli underground

L. de Sanctis & A. Mandolini
Dipartimento di Ingegneria Civile, Seconda Università di Napoli, Italy

G.M.B. Viggiani
Dipartimento di Ingegneria Civile, Università di Roma Tor Vergata, Italy

ABSTRACT: This paper describes the excavation for the new Garibaldi Station of Napoli Underground, completed in October 2004. The $44 \times 20\,m^2$ rectangular excavation, with a maximum depth of 45 m bgl., was protected by 1 m thick concrete diaphragm walls excavated using a hydromill, in 2.5 m wide panels. The excavation was carried out top down, support being provided by the floors of the station and by several levels of anchors. The paper describes a finite element analysis of the construction of the structure and a comparison of the numerical predictions with the observed behaviour. Before modelling the main sequence of excavation and propping and the changes of pore water pressures due to the excavation works, the sequence of panel excavation under bentonite slurry and concreting was examined. A significant effect of modelling wall installation versus wished in place wall analyses was found in terms of displacements of the wall and ground surface at the back of the wall, prop forces and bending moments within the wall. The available measurements were used both to assess the numerical predictions, in terms of diaphragm displacements, and to guide in the way in which some support elements, e.g. anchors, should be treated in the analyses.

1 INTRODUCTION

An increasing number of deep excavations are constructed in urban areas to meet the demands of transport and use of underground space: basements of buildings, underpasses for roads, underground car parks, underground stations, and similar.

The design of a deep excavation must address a number of issues such as, for instance, the choice of an adequate support system, the definition of the sequence of construction phases, the computation of the stresses in the structural members, and the evaluation of the magnitude and distribution of ground displacements connected to the construction phases and to possible changes of groundwater pressures connected with the excavation. In the urban environment, the ability to control and correctly predict ground movements around excavations is a crucial aspect of successful design because of their potential for causing damage to adjacent structures and services.

The increased availability of computer programs has made the prediction of ground movements using numerical analyses relatively common. However, many factors affect the numerical prediction of ground movements associated with excavations: the choice of an adequate constitutive model for the soil, the correct definition of the soil profile, the selection of representative mechanical properties for each layer, the definition of appropriate ground water conditions and initial stress states, the representation of the structural geometry and boundary conditions, and the modelling of the sequence of construction phases. Each stage of the process requires approximations and simplifications of the physical problem under examination, which will affect the results of the analyses to a greater or lesser extent.

Despite the fact that finite element programs permit to model some of the operations that occur during the construction of deep excavations, the installation of diaphragm walls is often ignored in the analyses, although it is now widely accepted that the effects associated with the installation of retaining structures can have a significant influence on the final displacements of the surrounding ground. Published surface displacements observed during construction of diaphragm walls in granular soil, soft to medium clay, and stiff to very hard clay show that the maximum settlement connected to slurry trenching can reach values of up to 0.15% of the depth of the wall and extend to a distance of up to about two times the maximum depth of the wall (Clough & O'Rourke, 1990).

Most work on installation effects, including full scale monitoring of structures (Symons & Carder, 1992; Lings *et al.*, 1991) centrifuge testing

(Powrie & Kantartzi, 1996; Richards *et al.*, 1998) and numerical modelling (Gunn *et al.*, 1992; Ng, 1994; Rampello *et al.*, 1998; Gourvenec & Powrie, 1999), has concentrated on the case of heavily overconsolidated deposits, where the *in situ* horizontal effective stress is generally larger than the vertical effective stress. In this case, wall installation tends to reduce the lateral effective stress in the soil near the wall, causing substantial ground movements towards the trench that add up with those associated with the main excavation stage. If installation is not modelled, high values of prop forces and bending moments are computed by FE analyses of diaphragm walls in overconsolidated deposits. This is because high lateral pressures on the retained side of the wall persist after excavation in front of the wall, i.e. the pressure behind the wall does not reach its active value (Potts & Fourie, 1984). In reality, in addition to relative flexibility of the wall and type of support, the service structural loads are likely to be influenced both by the initial *in situ* stress state in the ground and by the changes caused by construction. As installation reduces the applied stresses on the retained side it will also reduce prop or anchor forces, and bending moments (Gunn & Clayton, 1992).

On the other hand, in the case of a coarse grained deposit with a relatively low value of the initial coefficient of earth pressure at rest, the variation of the state of stress induced in the ground during wall installation by slurry trenching and concreting might correspond to an actual increase of the initial horizontal stress. In this case, the total horizontal stress at the boundary of the trench is increased from its *in situ* value to a value that is close to the hydrostatic pressure of the bentonite slurry, and then increased even more to a value approximating to the hydrostatic pressure of the wet concrete, at least in the upper 10 m of the wall (Gunn & Clayton, 1992). During this process, the walls of the trench will deform outwards and it is likely that some heave will be induced in the surrounding ground, before the main excavation stage.

The increase of the state of horizontal stress on the retained side, towards a value that is closer to the passive condition than its initial *in situ* state should also cause an increase in the final computed values of prop forces and bending moments in the wall relative to those obtained from finite element analyses in which wall installation is ignored.

Finally, while for a high K_0 environment a realistic constitutive model for the soil must be able to capture non linearity of behaviour at relatively small strains and features such as dependency on recent stress history, in a low K_0 environment enough non-linearity is induced by plasticity on major yield and even relatively simple constitutive models can produce realistic patterns of deformation, provided the construction processes are correctly represented. This is a relatively general finding applying not only to the prediction of ground displacements behind retaining structures but also to bored tunnels, see e.g. Addenbroke (1996) or Viggiani & Soccodato, (2004).

The example reported in this paper demonstrates that even relatively simple constitutive models and ways of modelling wall installation can improve substantially the prediction of the distribution and magnitude of ground movements around the excavation.

2 THE NEW GARIBALDI STATION

Once completed, the "Linea 1" of Napoli Underground, currently under construction, will form a closed ring connecting the North suburbs of the city, the area of the hills, the historical centre, the administrative district, and the airport for a total length of about 30 km and 25 stations. At present, about one half of the line, the so called *Tratta Alta*, between Piscinola and Dante stations, has been constructed and has been operating since 1998, with a very positive impact on the traffic at surface.

The construction of the *Tratta Bassa* (Figure 1), about 6 km long and currently under way, appears to be more problematic. All five stations included in this part of the route, namely Toledo, Municipio (Town Hall), Università, Duomo, and Garibaldi (Main Railway Station), are located in a densely built environment and have to be excavated in coarse grained deposits well below groundwater table.

Furthermore, significant direct interferences between the line and important buried archaeological remnants have arisen: at Dante Station the works were interrupted when they hit the city walls of the Greek Neapolis, the headworks for Municipio Station exposed part of the structures of the Roman port of the city, with the recovery of at least three roman ships, and at Università Station a Roman temple was uncovered by excavation. For these last two stations, in fact, it is likely that the original design will have to be changed to accommodate for the archaeological findings.

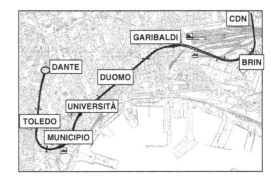

Figure 1. The Tratta Bassa of Linea 1 of Napoli Underground.

2.1 Garibaldi Station box

The new station in Piazza Garibaldi has been constructed within a $44 \times 20\,\text{m}^2$ rectangular excavation, with a maximum depth of 45 m bgl., protected by 1 m thick concrete diaphragm walls, excavated using a hydromill, in 2.5 m wide panels, see Figures 2 and 3. Four floor slabs, constructed top-down as excavation proceeded, provided support in the upper part of the excavation, while in the lower part six levels of anchors with lengths between 12 and 31 m were installed. At the top of the retaining walls one level of tubular steel props further restricted wall movements.

The sequence of main construction phases for Garibaldi Station box is summarised in Table 1. The presence of adjacent structures is unavoidable in such a dense urban environment as Napoli; in this case, two relatively big residential masonry buildings, $37 \times 74\,\text{m}^2$ size in plan, of 5 and 7 storeys, are located very close to the station box, at a minimum distance of about 3.6 m, see Figure 2.

2.2 Ground conditions

Ground conditions along the route of the *Tratta Bassa* are very variable, both horizontally and vertically.

Starting from ground level and moving downwards, the typical subsoil profile consists of made ground and remoulded ash underlain by remoulded ash and alluvial and/or *in situ* pyroclastic sand (Pozzolana) over a base layer of yellow Neapolitan tuff (Mandolini *et al.*, 2004).

Figure 4 details the soil profile and the main physical (voids ratio, e_0, dry density, γ_d, and density at the natural water content, γ) and mechanical properties (SPT blowcounts, CPT profiles, and friction angles from laboratory tests) of the ground in the area of the new Garibaldi station, as obtained from site and laboratory investigations carried out since 1997.

This included tens of boreholes formed in the vicinity of the station with SPT tests about every three meters and retrieval of disturbed samples for testing in the laboratory and CPT tests close to the excavation. The groundwater table is hydrostatic at about 9.5 m bgl.

2.3 Monitoring

An intense programme of monitoring was set up to control construction and mitigate and/or prevent the effects induced by excavation on the surrounding buildings. The layout of instruments is represented

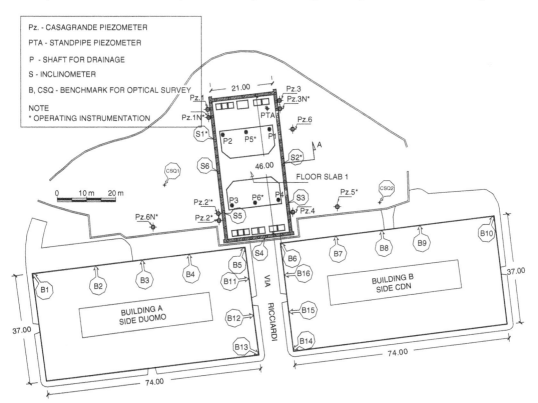

Figure 2. Plan and cross section of Garibaldi Station box.

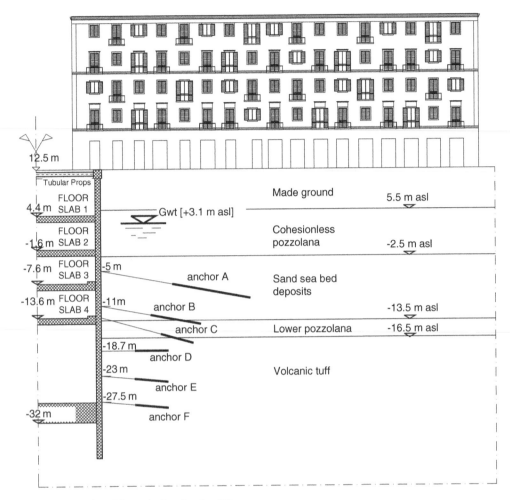

Figure 3. Cross section of the station box [section S2].

schematically in Figure 2. Monitoring activities included the measurement of: horizontal displacements of the diaphragm walls delimiting the station box, by means of 6 inclinometers; hydraulic head, by means of 10 Casagrande piezometers installed along the perimeter of the excavation and of one standpipe within the excavated area; anchor forces at installation and in working conditions, by means of 15 load cells; settlements of the ground and buildings adjacent to the excavation works, by precision levelling of 18 reference points.

Figure 5 shows the deflected wall profiles measured by inclinometers S1 and S2 at different dates during excavation. As a consequence of some faulty operation, inclinometer S2 was damaged at the end of construction phase n. 16. Therefore, the total deflections of the wall at the end of the excavation (construction

phase n. 22) are only available at the position of inclinometer S1.

From construction phase n. 4 onwards, the magnitude of the horizontal displacements measured by both inclinometers S1 and S2 increases substantially and the classical bulging shape characterizes the deflected wall profiles. The maximum measured wall deflection was about 35 mm at a depth of about 20 m bgl., and was measured at inclinometer S2 during construction stage n. 14. After reaching a depth of excavation of about 24 m bgl the deflections of the wall did not increase any more. Because inclinometers S1 and S2 were installed within the panels, the measured deflections only represent ground movements after wall installation.

The settlements of the buildings in a number of reference points installed close to the station box are

Table 1. Construction phases.

#	Description	Date
1	install diaphragm wall	19.10.02
2	inst. 1st lev. of support + exc. to 3.3 m asl	10.12.02
3	install first floor	19.02.03
4	excavate to −2.7 m asl	17.06.03
5	install second floor	18.07.03
6	excavate to −6 m asl + drill anchor level A	22.08.03
7	pre-stress anchor level A	11.09.03
8	excavate to −8.7 m asl	30.09.03
9	install 3rd floor	04.11.03
10	exc. to −12 m asl + drill anchor level B	18.11.03
11	excavate to −14 m asl	17.02.04
12	pre-stress anchor level B + drill anchor level C	28.02.04
13	pre-stress anchor level C	08.03.04
14	excavate to −14.7 m asl	30.03.04
15	install fourth floor	05.04
16	excavate to −19.7 m asl & drill anchor level D	19.06.04
17	pre-stress anchor level D	10.07.04
18	excavate to −24 m asl & drilling anchor level E	31.08.04
19	pre-stress anchor level E	14.09.04
20	excavate to −28.5 m asl + drill anchor level F	28.09.04
21	pre-stress anchor level F	
22	complete excavation at −32.2 m asl	30.10.04

plotted in Figure 6 as a function of time. The base reading for surface settlements was taken after completion of wall installation, so the measurements cannot shed any light on the magnitude of ground movements due to wall installation. The maximum absolute settlement connected to excavation ($w_{max} \cong 31$ mm) was recorded at reference point B6, installed near the corner of building B. At the early stages of construction, the rate of settlement was in the range 1–2 mm/month. During construction stage n. 10 (excavation at −12 m asl), carried out in mid November 2003, the observed rate of settlement suddenly increased to about 25 mm/month. This was connected to significant water and soil flow into the boreholes drilled to install anchor level B. After changing the technique to install the anchors, the rate of settlements decreased back to about 2.5 mm/month. The observed settlements ($w/H \sim 0.1\%$, $H =$ maximum excavation depth) are in good agreement with the experimental data reported by Clough and O'Rourke (1990) for similar retaining structures in coarse grained ground ($w/H = 0.1$–0.2%); the related angular distortions ($\beta_{max} = 1.23 \times 10^{-3}$) are smaller than those corresponding to damage in the buildings.

Figure 7 illustrates the settlement profiles of alignments B1-B5, B6-B10, B5-B13, and B6-B15 at various reference dates. An estimate of the settlements of the buildings induced only by excavation can be obtained correcting the time displacements plots of

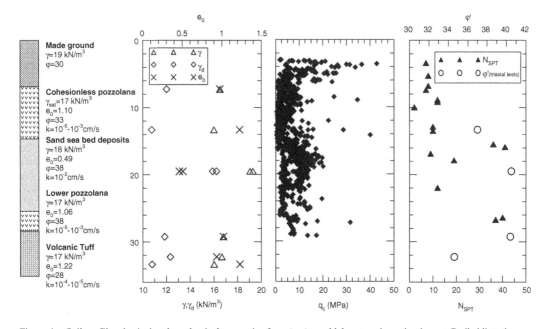

Figure 4. Soil profile, physical and mechanical properties from *in situ* and laboratory investigations at Garibaldi station.

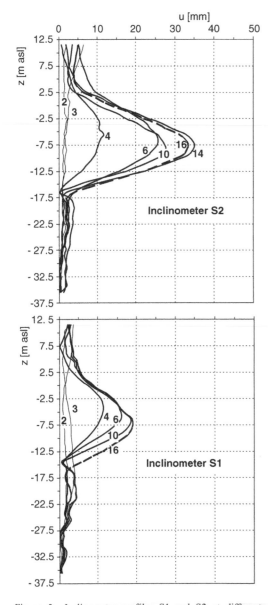

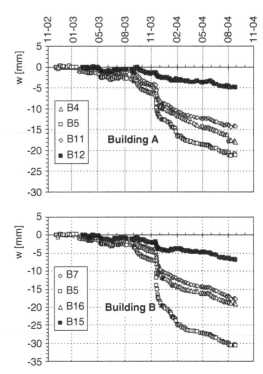

Figure 6. Foundation settlements measured after completion of wall installation.

Figure 5. Inclinometer profiles S1 and S2 at different construction stages [number refer to phases listed in Table 1].

Figure 6, to subtract the sudden increase of settlements observed at construction stage n. 10, as in Figures 8 and 9.

3   NUMERICAL ANALYSES

All numerical analyses were carried out in plane strain condition using version 6.4 of the commercial finite element code ABAQUS. The analysed section

of the station box corresponds to the position of inclinometer S2.

Figure 9 shows the finite element mesh adopted in the analyses. Four-noded isoparametric elements were used to represent both the soil and the diaphragm wall. The vertical sides of the mesh were restrained in the horizontal direction, while the base was not allowed to move in either the vertical or the horizontal directions. Interface elements had a thickness of $0.1 \times d$, where d is the diaphragm wall thickness ($=1$ m in the example at hand).

The ground was taken to behave fully drained both during installation and subsequent excavation; for construction phases in which the bottom of the excavation was below groundwater table, pore fluid pressures were calculated for steady state seepage with the following hydraulic boundary conditions: impervious wall, pervious boundary at the bottom of the excavation with zero pore water pressure, impervious boundaries at the sides and at the base of the mesh, constant hydraulic head at the back of the wall, corresponding to the hydrostatic condition before excavation.

The soil was assumed to behave like a linear elastic – perfectly plastic material with Mohr-Coulomb failure criterion while the diaphragm wall was modelled

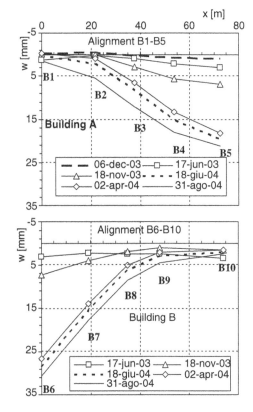

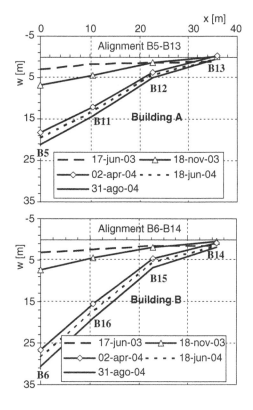

Figure 7. Foundation settlements across different alignments.

as a linear elastic solid with a Young's modulus $E = 31.2$ GPa and a Poisson's ratio $\nu = 0.15$.

Anchors were modelled as constant nodal forces, see Table 2; this is because the measurements of anchor forces at installation and in exercise showed that there was less than 6% variation from the initial value as excavation proceeded below each anchor level. Floor slabs were modelled as linear springs whose stiffness was obtained by finite element analyses of the floor slabs to take into account their geometry with irregular openings for mucking, see Table 3.

The friction angles of cohesionless layers were obtained from *in situ* and laboratory tests. The values of cohesion and friction angle for the yellow tuff were derived from a few available results of unconfined compression tests and published data. Interface elements within the tuff were given cohesion of 20 kPa and no tractions were allowed between the wall and the tuff.

Elastic modules of the different layers were obtained by back analyses of wall deflections according to the following procedure: the relative stiffness of the different layers was evaluated first, based on the results of site cone and standard penetration tests (CPT, SPT) and then the absolute value of the stiffness

of each layer was fixed by fitting the wall deflection profile measured by inclinometer S2 at stage n. 16. The above procedure was applied first in conventional analyses were wall installation was not modelled (WIP, or Wished In Place analyses) and then again in analyses were the process of slurry trenching and concreting was modelled (WIM, or Wall Installation Modelled analyses). The stiffness parameters obtained from WIM analyses were then used to re-analyse the main construction sequence for a wished in place wall.

In the first stage of WIM analyses, the soil elements within the wall were removed and simultaneously replaced by nodal forces corresponding to the hydrostatic pressure of bentonite slurry ($\gamma_f = 11$ kN/m^3). Nodal forces in the trench were then increased to the value corresponding to wet concrete ($\gamma_c = 25$ kN/m^3). Finally, removing horizontal pressures and simultaneously activating diaphragm wall elements simulated concreting of the panels.

Recent experiences of diaphragm walls built in Singapore (Poh *et al.*, 2001) showed that there is a significant effect of the width of the panels on the magnitude of the horizontal displacements of the ground during installation. In particular the maximum horizontal displacement increases as the area of the lateral

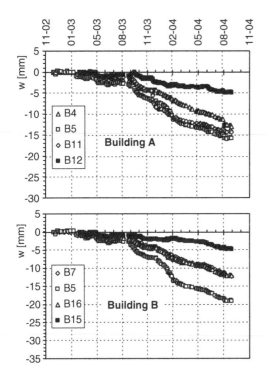

Figure 8. Corrected time displacements plots.

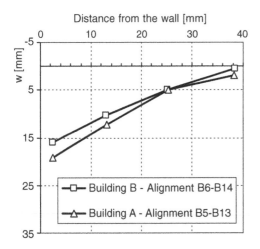

Figure 9. Corrected ground surface profiles at the back of the wall.

Table 2. Anchor forces adopted in the FE analyses.

Anchor level	Force (kN/m)
A	537
B	386
C	286
D	868
E	929
F	806

Table 3. Stiffness of floor slabs.

Slab	EA (kN/m)	K (kN/m/m)
1	$1.68 \cdot 10^6$	$1.58 \cdot 10^5$
2	$1.86 \cdot 10^5$	$1.75 \cdot 10^4$
3	$1.68 \cdot 10^6$	$1.58 \cdot 10^5$
4	$1.86 \cdot 10^5$	$1.75 \cdot 10^4$

is provided by WIP analyses. To model wall installation realistically, a fully three dimensional approach would be required. Still, the comparison between the two bounds provided by the plane strain conditions (WIP and WIM) highlights a number of fundamental issues on installation effects.

Figure 11 shows the deflected wall profile obtained from the first set of WIP analyses. The pattern of predicted deflections of the wall can be made to match the experimental observations quite closely, but for the upper part of the diaphragm wall ($z = 0$–10 m bgl), the measured displacements are nearly constant ($w \sim 5$ mm) while the computed values increase almost linearly from a minimum value of about 4 mm at surface to about 20 mm at a depth of 10 m bgl.

Figure 12 shows the deflected wall profile obtained from WIM analyses. In this case, in order to compare with the measured values, the computed horizontal displacements are increments relative to the end of the installation stage. As before, the match between the predicted and expected deflections is satisfactory, with the exception of the upper part of the diaphragm wall.

The surface displacements predicted at the back of the wall are in good agreement with the observed values, at least for distances from the wall greater than 10 m, see Figure 13. When the loads transmitted by the adjacent buildings to the ground (about 60 kPa, in plane strain conditions) were included in the WIM analyses, the pattern of computed surface settlements at the back of the wall did not change substantially.

However, a very substantial improvement of the pattern of the predicted deflections of the wall was obtained when thermal effects were explicitly considered in the analyses. Significant changes in temperature were recorded during construction as shown in

surface of each excavated panel increases. Because they are carried out in plane strain conditions, WIM analyses model wall installation as the excavation of an infinitely long slot, and, therefore, installation effects are likely to be overestimated, providing one bound of the problem. Another bound to the solution

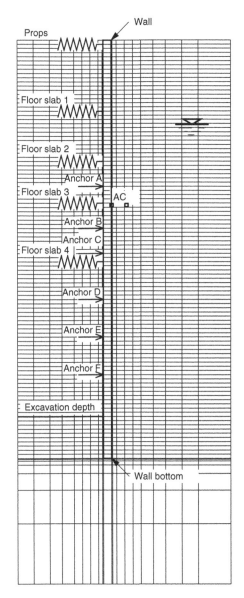

Figure 10. Finite element mesh around the wall.

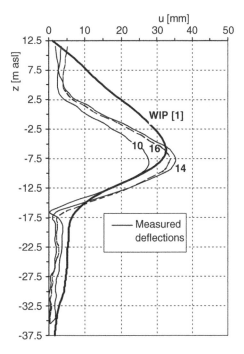

Figure 11. Deflection wall profile obtained from the first set of WIP analyses.

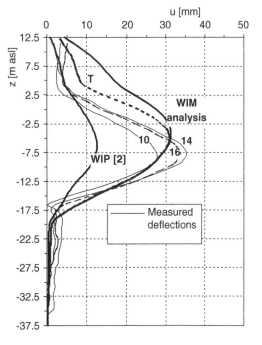

Figure 12. Deflection wall profile obtained from different types of analyses.

Figure 14. These temperature changes affect primarily the tubular steel props at ground level and floor slab 1 (see Figure 3), i.e. those support elements directly exposed to sun radiation. WIM analyses were carried out in which 3 uniform thermal distortions corresponding to the changes of temperature summarised in Table 4 were applied to the two upmost levels of support. The results of these analyses correspond to the deflection profile labelled T in Figure 12 and are in very good agreement with the experimental observations.

65

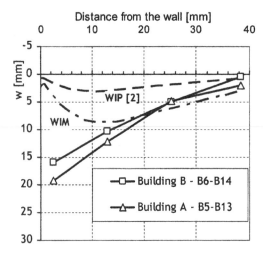

Figure 13. Surface settlements at the back of the wall.

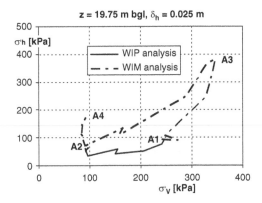

Figure 15. Stress path followed by stress state at point A [δ_h = distance from the wall].

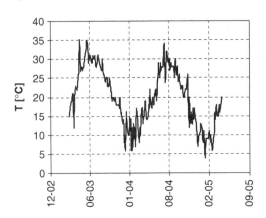

Figure 14. Temperature changes recorded during construction.

Table 4. Temperature changes corresponding to thermal distortions applied in the analysis.

Period	ΔT [°C]
January–July 2003	~20
September–December 2003	~−20
January–August 2004	~20

The deflected wall profiles obtained from WIP analyses carried out using the same soil stiffness obtained by back analysis of the experimental data in WIM analyses, about four times larger than those obtained from WIP analyses, are also reported in Figure 12 for comparison. As expected, adopting the same set of soil parameters, the computed horizontal displacements induced by excavation in front of the wall are much larger for WIM analyses than for WIP analyses. The

same conclusion can be arrived at when comparing surface settlements, see Figure 13.

The differences between predicted wall deflections in the two types of analyses can be explained inspecting the changes of stress state for soil elements behind the wall. As an example, Figure 15 illustrates the stress paths followed by the effective stress state at point A located immediately behind the wall at a depth of 19.75 m bgl (see Figure 10). For WIP analyses, the stress state moves directly from point A1 (*in situ* stress condition) to point A2 (end of excavation) with a total horizontal stress relief of about 26 kPa. For WIM analysis, the stress state moves first from A1 to A3 (end of installation, including slurry trenching and concreting) and then to A4 corresponding to the end of the main excavation sequence. The total horizontal stress relief along path A3-A4 is 210 kPa or about eight times that corresponding to path A1-A2. The reduction of effective vertical stress in path A3-A4 is due to the increase of shear stress at the soil-wall interface as the excavation depth increases, and the wall deforms outwards.

Also prop forces and bending moments in the wall computed by WIM analyses are larger than those obtained from the corresponding WIP analyses, for a given set of soil stiffnesses. The computed bending moments in the wall are shown in Figure 16; these were obtained integrating the stress in the Gauss points of the elements representing the wall.

4 CONCLUSIONS

This paper detailed a finite element analysis of the excavation of the new Garibaldi Station of Napoli Underground, completed in October 2004. A 44 × 20 m² rectangular excavation, with a maximum depth of 45 m bgl., was created to accommodate the new station box under the protection of 1 m thick concrete diaphragm.

66

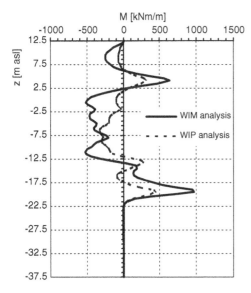

Figure 16. Bending moments predicted by the WIP and WIM analyses.

In the design of such deep excavations, the control and correct prediction of ground movements is a crucial aspect of design, because of their potential for causing damage to adjacent buildings and services. This is a complex soil-structure interaction problem because the ground movements that are of interest are generally at the surface or near surface while the driving unloading processes take place away from the surface at a relatively distant boundary. In fact, lateral wall movements are relatively easy to predict and not very sensitive to the soil model adopted, while observed surface ground movements behind the wall are much more difficult to match and very sensitive to the details of the analysis.

In this case, before modelling the main sequence of excavation and propping and the changes of pore water pressures due to the excavation works, the sequence of panel excavation under bentonite slurry and concreting was examined in plane strain conditions. The available measurements were used both to assess the numerical predictions, in terms of diaphragm displacements, and to guide the way in which some support elements, e.g. anchors, should be treated in the analyses.

A significant effect of modelling wall installation versus wished in place wall analyses was found in terms of displacements of the wall and ground surface at the back of the wall, prop forces and bending moments within the wall.

Finally, the temperature changes due to exposure of direct sun radiation on the upper levels of support affect significantly the computed profile of horizontal displacements of the wall, making it closer to the experimental observations.

ACKNOWLEDGEMENTS

The Authors wish to thank Ms Alessandra Belà for many valuable contributions at the early stages of this work.

REFERENCES

Addenbroke T.J. 1996. Numerical analysis of tunneling in stiff clay. *PhD Thesis*, Imperial College of Science, Technology and Medicine, University of London.
Clough G.W. & O'Rourke T.D. 1990. Construction induced movements of *in situ* wall. *Design and Performance of Earth Retaining Structures*, Lambe & Hansen, Ithaca (NY). ASCE GSP 25: 430–470.
Gourvenec S.M. & Powrie W. 1999. Three-dimensional finite element analysis of diaphragm wall installation. *Géotechnique* 49, No. 6, 801–823.
Gunn M.J. & Clayton C.R.I. 1992. Installation effects and their importance in the design of earth-retaining structures. Tech. Note, *Géotechnique* 42, No. 1, 137–141.
Lings M.L., Nash D.F.T., Ng C.W.W. & Boyce M.D. 1991. Observed behaviour of a deep excavation in Gault clay: a preliminary appraisal. *Proc. X ECSMFE*, Florence 2, 467–470.
Long M. 2001. Database for Retaining Wall and Ground Movements due to Deep Excavations. *Journal of Geotechnical and Geoenvironmental Engineering, ASCE*, 127(3): 203–224.
Mandolini A., Viggiani C. & e Viggiani G.M.B. 2004. Interazione fra Linee Metropolitane e il tessuto urbano storico e monumentale della città di Napoli. Incontro Annuale dei Ricercatori di Geotecnica, IARG 2004, Trento.
Ng C.W.W. 1994. Effects of modelling wall installation on multi-staged excavations in stiff clays. *Proc. 1st Int. Conf. Pre-failure Deformation of Geomaterials*, Sapporo 1, 595–600.
Poh T.Y., Chee-Goh A.T. & Wong I.H. 2001. Ground Movements Associated with Wall Construction: Case Histories. *Journal of Geotechnical and Geoenvironmental Engineering*, ASCE, 127(12): 1061–1069.
Potts D.M. & Fourie A.B. 1984. The behaviour of a propped retaining wall: results of a numerical experiment. *Géotechnique* 34, No. 3, 347–352.
Powrie W. & Kantartzi C. 1996. Ground response during diaphragm wall installation in clay: centrifuge model tests. *Géotechnique* 46, No. 4, 725–739.
Rampello S., Stallebrass S.E. & Viggiani G.M.B. 1998. Ground movements associated with excavations in stiff clays: current prediction capability. *Proc. II Int Symp. On Hard Soils/Soft Rocks*, Napoli, vol. III, 1527–1540
Richards D.J., Powrie W. & Page J.R.T. Investigations of retaining wall installation and performance using centrifuge modeling techniques. *Proc. ICE, Geotechnical Engineering* 131, No. 3, 163–170
Symons I.F. & Carder D.R. 1992. Stress changes in stiff clay caused by the installation of embedded retaining walls. *Proc. ICE, Retaining Structures*: 227–236.
Viggiani G.M.B. & Soccodato F.M. 2004. Predicting tunnelling-induced displacements and associated damage to structures. *Rivista Italiana di Geotecnica*, No. 4, 11–25.

Numerical Modelling of Construction Processes in Geotechnical Engineering for
Urban Environment – Triantafyllidis (ed)
© 2006 Taylor & Francis Group, London, ISBN 0 415 39748 0

The impact of diaphragm wall construction on the serviceability of adjacent strip foundations in soft ground

R. Schäfer & Th. Triantafyllidis
Institute of Soil Mechanics and Foundation Engineering, Ruhr-University Bochum, Germany

ABSTRACT: To avoid excessive settlements of buildings in the vicinity of deep excavation pits, diaphragm walls are frequently used as retaining structures. However, in conventional design the effect of the diaphragm wall installation on the deformation of the ground and its impact on the serviceability state of nearby structures is usually neglected. In this article we discuss the deformation behaviour of the wall itself as well as of a strip footing resting on soft clay. A three dimensional finite element model with a fully coupled consolidation analysis is used. The stepwise excavation and pouring process of a continuous wall section are simulated. The stress-strain behaviour of the soft clay is described by a visco-hypoplastic constitutive model, which includes the rheological properties of the ground and increased stiffness for small strains. Based on the numerical results, several recommendations are formulated for simplified estimation of the settlements. A suitable construction technique of diaphragm walls in soft ground is proposed in order to reduce deformations during the installation.

1 INTRODUCTION

For deep excavation pits in urban areas, diaphragm walls are frequently chosen as supporting structures. Walls with high bending stiffness in combination with sufficiently designed anchors or struts can reduce wall and ground movements towards the pit. One of the requirements in urban environment is maintaining of the serviceability of nearby structures during the entire excavation process and afterwards. In the EC7 §8.7.2 (2), (3) the following issue is formulated:

(2) A cautious estimate of the distortion and displacement of retaining walls, and the effects on supported structures and services, shall always be made on the basis of comparable experience. This estimate shall include the effects of construction of the wall. It shall be verified that the estimated displacements do not exceed the constrains. (3) If the initial cautious estimate of displacement exceeds the constrains, the design may be justified by a more detailed investigation including displacement calculations.

Therefore appropriate models and procedures should be developed and established. The construction process of the diaphragm wall is usually neglected and thus, ground movements resulting from the installation of the wall panels are disregarded. However, the construction process of the diaphragm wall can create substantial settlements of adjacent buildings. They may be even larger than the settlements caused by the pit excavation. The higher the magnitudes of the settlements during the construction process are, the lower

have to be those allowed for during the pit excavation in order to satisfy the given constrain criteria.

For a conventional diaphragm wall design in the practice it is usual to restrict the contruction induced settlements by increasing the global safety factor η of the open trench stability. The German code of practice DIN 4126 requires a global safety factor of the open trench of $\eta = 1,3$ (instead of $\eta = 1,1$), if the panel is excavated within a critical distance to an adjacent structure. The higher global safety factor is meant to consider influences of the panel geometry, the shear resistance mobilisation of the ground, the loading of the footings etc. These influences cannot be exactly quantified and therefore the resulting deformation behaviour or settlements of the ground and the nearby structures cannot be predicted.

Moreover, it is difficult to calculate the stability of the open trench at presence of adjacent structures. For strip footings, for example, the safety factor η can be reasonably evaluated if the stiffness of the superstructure is taken into account (Kilchert and Karstedt, 1984). The general increase of the global safety factor does not cover the effect of the installation process of the diaphragm wall.

In this paper, we calculate the deformations of strip footings resulting from the construction process of a continuous diaphragm wall section in soft clayey ground. For this purpose, a three dimensional finite element model is generated which simulates the excavation process and the subsequent pouring process of wall panels. A few calculations are done in order to quantify the influence of the panel geometry, the loading of the footings as well as the position of the footings with reference to the diaphragm wall on the resulting settlements of the building. Finally, several conclusions are drawn pertaining the construction sequence of the trenches or the division of the diaphragm wall section into different panels for minimising the construction-induced settlements.

2 NUMERICAL MODEL

2.1 Finite-element-model

For the numerical studies, a three-dimensional finite-element-model of a plain diaphragm wall section was developed by the use of the programme code ABAQUS 6.3. The model comprises the stepwise construction procedure of three adjacent wall panels including the trench excavation under slurry support and the subsequent pouring process. Figure 1 presents the finite-element model for the example of an adjacent strip footing. The finite elements in front of the model are removed for the purpose of a better illustration. The model with the dimensions of 110 m × 57,6 m × 45 m consists of about 24000 tri-linear elements for a coupled consolidation analysis.

Each of the considered diaphragm wall panels has a depth of 35 m and a width of 0,90 m. In the numerical calculations, the panel length L varies in the range of L = 3,60 m, 5,40 m and 7,20 m. The adjacent strip foundation is located at a depth of 5,6 m below the surface ground level. The strip footing has a width of b = 1,80 m.

In the first step of the calculation the geostatic stress conditions of the soil and the hydrostatic pore water pressure distribution are adopted. Thereafter the loading of the strip footing is applied, followed by the dissipation of the excess pore water pressure. The magnitude of the strip loading is chosen under the condition that the respective settlement is

$s_{cal} = 1$ cm. For the numerical calculations, the settlement values $s_{cal} = 1$ cm and $s_{cal} = 2$ cm have been adopted corresponding to the loadings $p'_{1cm} = 153$ kPa and $p'_{2cm} = 213$ kPa resp. Finally, the installation process of the diaphragm wall section is modelled. For each panel, the instalation under slurry support is modelled by removing the finite elements inside the trench and applying disributed load equivalent to the hydrostatic slurry pressure. The unit weight of the bentonite slurry is $\gamma_b = 10,3$ kN/m³. Subsequently, in order to simulate the pouring of concrete into the trench, the pressure is increased. In contrast to the slurry pressure, the distribution of the fresh concrete pressure is not hydrostatic over the entire panel depth. Due to a beginning of the hydration process and the aggregates intergranular friction the pressure gradient is lower than γ_c (specific weight of the fresh concrete) at greater depth (Lings et al., 1991). In the presented simulation model, the approximation of the fresh concrete pressure versus depth follows a bi-linear relation as proposal by Lings et al. (1991):

$$p_c = \begin{cases} \gamma_c \cdot z & z \leq h_{crit} \\ \gamma_b \cdot z + (\gamma_c - \gamma_b) \cdot h_{crit} & z > h_{crit} \end{cases} \quad (1)$$

where:
p_c = fresh concrete pressure
γ_c = bulk unit weight of the concrete (23,5 kN/m³)
γ_b = bulk unit weight of the slurry (10,3 kN/m³)
h_{crit} = critical depth (m)
z = depth below surface ground level (m)

The critical depth h_{crit} for the model example is assumed to be 20% of the panel depth.

Finally, the distributed loads are removed and additional finite elements are placed into the trench, which represent the concrete. The hardening of the fresh concrete due to aging is expressed by the Young's Modulus E and the Poisson ratio v development as functions of time:

$$E_c(t) = E_c \cdot \left[\exp \left\{ s \left[1 - \frac{28}{(t/t_1)} \right]^{\frac{1}{2}} \right\} \right]^{\frac{1}{2}} \quad (2)$$

$$v(t) = 0,5 - (0,5 - v_{28}) \cdot E_c(t) / E_c$$

where $E_c = 30500$ MPa and $v_{28} = 0.2$ are the values of $E_c(t)$ and Poisson's ratio at the age of concrete of 28 days, t_1 is 1 day and s is a parameter depending on the cement type used (Schäfer, 2004).

2.2 Soil conditions and constitutive equations

Figure 2 represents the chosen subsurface soil conditions, which correspond to the Sungshan Formation of the Taipeh Basin. The geotechnical properties of the Sungshan Formation have been accurately determined

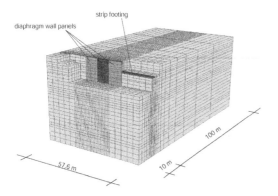

Figure 1. 3-D FEM-Model of a D-wall segment consisting of three panels adjacent to one strip footing.

in the course of the construction of the Taipeh National Enterprise Center (TNEC) in 1991.

Generally, the subsoil is mainly characterised by a soft, normally- to slightly over-consolidated silty clay deposit of medium to low plasticity with a thickness of 25 m. The natural water content w is close to the liquid limit w_l of the clay and ranges between 30 and 38%. The over-consolidation-ratio (OCR) decreases with the depth of the layer from 1.8 to 1.05. The approximate ratio of the undrained shear strength and the effective overburden pressure c_u/σ'_v is 0.34 and 0.21 in case of triaxial compression and extension, respectively. The coefficient of permeability is $k = 4 \cdot 10^{-8}$ m/s. This clay layer is overlain by a loose silty sand and an over-consolidated clay deposit at the surface ground level. Below, the soft clay alternating medium dense silty sand and normally consolidated clay layers can be found. The natural bedrock stratum is located at a depth of approx 46 m below ground level. All the soil parameters have been determined by triaxial compression and extension, field vane shear and cone penetration tests with pore-water pressure measurement. In the following numerical calculations, the required soil parameters and state variables are either directly taken from the literature. Ou et al. (1998), (2000a), (2000b), or correlated with given quantities or estimated. Table 1 gives an overview over the chosen soil parameters and state variables.

The homogenous slightly overconsolidated soil deposit is described by an incremental non-linear visco-hypoplastic constitutive equation according to Niemunis (2003), where:

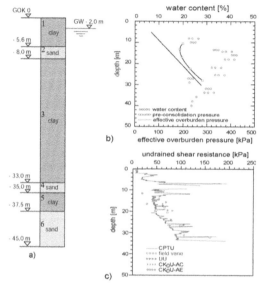

$$\dot{T} = L\,(T,e) : (D - D^v) \tag{3}$$

$$D^v = -\frac{B}{\|B\|}\dot{\gamma}\left(\frac{1}{OCR}\right)^{\frac{1}{I_v}} \tag{4}$$

with $\dot{T}$ being the stress rate, D the deformation rate and $\dot{\gamma}$ the reference creep rate.

The visco-hypoplastic model incorporates the rate dependence as well as the relaxation and creep

Figure 2. Subsurface soil conditions of Taipeh-Basin: a) idealised soil layers, b) water content, effective overburden pressure and c) undrained shear resistance c_u versus depth.

Table 1. Soil parameters and state variables.

Layer		1	2	3	4	5	6
γ_r	(kN/m³)	18,3	18,9	18,2	19,3	18,2	19,6
w	(%)	32	25	35	24	28	30
w_l	(%)	34	–	35	–	33	–
I_p	(%)	23	–	15	–	21	–
φ'	(°)	29	32	29	32	31	32
c'	(kPa)	–	–	–	–	–	–
C_c	(–)	0,098		0,098		0,098	
C_s	(–)	0,019	0,002	0,019	0,0005	0,019	0,0005
I_v	(%)	2	–	2	–	1,9	–
k	(m/s)	4E–8	1E–4	4E–8	1E–4	4E–8	1E–4
e_{eo}	(–)	0,81	–	0,98	–	0,81	–
OCR	(–)	5	–	1,05 –1,9	–	1,05	–
K_0	(–)	1,1	0,47	0,52 –0,68	0,47	0,5	0,47

where: γ_r = bulk density of the ground, w = water content, w_l = liquid limit, I_p = plasticity index, φ' = friction angle (in case of the clay layers it is the critical friction angle φ_c), c' = cohesion, C_c/C_s = compression/swelling index (e–ln p-diagram), I_v = viscosity index, k = water permeability of the soil, e_{eo} = void ratio at p_{eo} = 100 kPa, OCR = over-consolidation-ratio, K_0 = earth pressure coefficient at rest.

behaviour of the clayey ground. The total strain rate D is decomposed into an elastic $D^e = D - D^v$ and a viscous part D^v, which includes plastic strain increments. The magnitude of the viscous strain increment is proportional to 1/OCR with a power of $1/I_v$ (the viscosity index I_v (Leinenkugel, 1976) is given in Table 1). The direction of D^v is given by the 2nd rank tensor **B**. The stiffness tensor L depends on the actual effective stress T and the void ratio e. Both tensors **B** and L result from the theory of hypoplastic soil behaviour (Niemunis, 2003).

The increased small-strain stiffness of the soil is accounted by an additional state variable called intergranular strain h (Niemunis, 2003). Directly after a 180° strain path reversal, soils exhibit an increased elastic stiffness within a small strain range. During further straining, the plastic proportion of the strains increases, which causes a gradual decrease of the stiffness. At the end of a strain path with the length ε_{som} (som = swept-out of memory, Gudehus et al., 1977) measured from the point of strain reversal, the soil recovers the original stiffness during monotonic straining. The intergranular strain h leads to a stiffness increase via the scalar multpliers m_R and m_T. The stress-strain relation of the clay is generally proposed to be:

$$\dot{T} = M : D - L : D^{vis} \qquad (5)$$

The stiffness tensor M takes into account the strain history and reads as:

$$M = \left[\rho^\chi m_T + (1 - \rho^\chi) m_R \right] L + \begin{cases} \rho^\chi (1 - m_T) L : \hat{h}\hat{h}, & \hat{h} : D \leq 0 \\ \rho^\chi (m_R - m_T) L : \hat{h}\hat{h}, & \hat{h} : D > 0 \end{cases} \quad (6)$$

where: $\rho = \|\hat{h}\|/R$ normalized amphitude
$\hat{h}^* = \hat{h}/\|\hat{h}\|$ direction of intergranular strain
with: m_T, m_R magnification factor
χ interpolation factor
The evolution equation of the intergranular strain rate, which is slightly smaller than D is proposed as:

$$\dot{h} = \left(I - \hat{h}^* \hat{h}^* \rho^{\beta_R} \right) : D, \hat{h}^* : D > 0$$
$$\dot{h} = D \qquad , \hat{h}^* : D \leq 0 \qquad (7)$$

The evolution equation for **h** is objective and requires the material parameters β_R, R. Further material parameters in eq. (6) are m_T, m_R, and χ. All those five intergranular parameters used in the calculation for the soft clay are given in Table 2.

Table 2. Intergranular parameters for the soft clay layer.

R	m_R	m_T	β_R	χ
10^{-4}	5	2	0,5	6

The intergranular parameters influence only the elastic part of the model and the relaxation of the deformation is not affected. For a detailed description of the visco-hypoplastic model the interested reader is referred to Niemunis (2003).

The mechanical behaviour of the sand layers is modelled by the use of the elasto-plastic Drucker-Prager model implemented in the ABAQUS code. The yield surface is modified in such a way, that different strengths of the soil in case of triaxial compression and extension stress states can be taken into account. The elastic stiffness of the model is proportional to the actual mean effective stress level $p' = \text{tr}T$. For further details, the interested reader is referred to the paper of Schäfer and Triantafyllidis (2004).

3 NUMERICAL RESULTS

3.1 Single wall panel excavation

At first, the construction process of a single diaphragm wall panel is simulated and the arising settlements of a strip footing are analysed. The clear distance between panel and strip footing is 1,0 m.

Figure 3 illustrates the settlement troughs of the strip footing caused by the excavation of a single trench with lengths L = 3,6 m, 5,4 m and 7,2 m resp. The black lines represent the settlements of the footing with $s_{cal} = 2$ cm and the gray lines those with $s_{cal} = 1$ cm. Due to the symmetry of the model, the greatest deformation can be observed at the centerline of the trench. The magnitudes of the settlement increase with the length of the open trench: f.E. doubling of the panel length from 3,6 m to 7,2 m results in a triplication of the deformation magnitudes. Similarly, a higher loading of the footing (increase of s_{cal} from 1 cm to 2 cm) leads to greater settlements during the excavation process for the same panel length.

For a panel length of L = 7,2 m, Figure 4 compares the settlements troughs of the strip footing caused by the the trench excavation and the pouring process of the

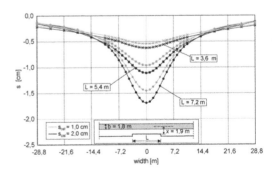

Figure 3. Dimensionless settlement s/L of a single footing due to the excavation of a single D-wall panel of the length L.

panel. The placing of fresh concrete leads to a distinct reduction of the settlements resulting from the excavation process of the trench. In the considered cases of the footing loading the settlement decrease due to pouring is calculated to be 30–35% of the maximum deformation magnitude.

The influence of the distance between the wall panel and the strip footing is shown in Figure 5, where the settlements of the footing due to trench excavation versus the normalized distance x/L is presented. The panel length was L = 7,2 m and the adopted effective loading p' = 213 kPa. As expected, the deformations decrease with increasing distance x/L. Nevertheless, the deformation behaviour of the footing is affected by the construction procedure of the trench, even if the distance is x/L = 1,0.

With respect to the settlement behaviour, the bending stiffness of the strip footing is of great importance. For the stiffness variation of the footing different thicknesses d = 0,25 m, 0,5 m and 0,75 m in the FE–simulations have been adopted. In Figure 6, the numerical results for the settlement curves of these variations together with the limit bending stiffnesses EI → 0

and EI → ∞ are presented. The rigid strip footing (EI → ∞) shows the lowest deformation magnitudes during the excavation process. In the case of older buildings with partly damaged flexible strip footings, only a negligible bending stiffness can be assumed. Figure 6 shows, that in the case of EI → 0 the settlement magnitudes due to trench excavation process considerably increase up to 6,0 cm because no transfer of the strip load to areas beside the open trench takes place.

3.2 Settlement of strip footing during the excavation of D-wall panels

In order to analyse the deformation behaviour of strip footings during the construction process of an adjacent diaphragm wall, the installation of a single panel is not sufficient. For the construction of a diaphragm wall several adjacent panels are excavated and poured. In order to analyse the impact of the construction process on the serviceability state of a nearby footing, it is necessary to consider a representative wall section. Schäfer and Triantafyllidis (2004) have already shown that the stress condition of the soil at the centerline of a diaphragm wall panel is affected by the installation of the panel itself and the construction process of the two adjacent panels (one on each side of the panel under consideration). Accordingly, a diaphragm wall section consisting on three continuous wall panels (Figure 1) has been generated in the FE-simulations for the determination of the deformation behaviour of the strip footing.

In Figures 7 and 8, the development of the settlement behaviour of the strip foundation during the construction process of three adjacent D-wall panels for the oscillating (Fig. 7) and continuous (Fig. 8) installation sequence are presented. The panel length is L = 7,2 m, the clear distance between panel and strip footing is 1,0 m (the distance between the centerlines of the trench and footing is x = 1,9 m, i.e.: x/L = 0,26)

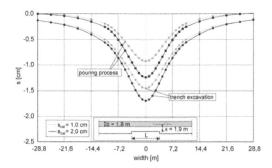

Figure 4. Development of the footing settlements during the sequence of construction of a D-wall segment in oscillating order.

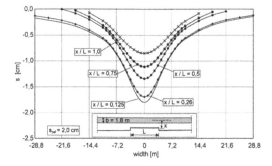

Figure 5. Comparison between the maximum settlements of a foundation located in the centerline (position 1) of a D-wall panel (L = 7,2 m) and at the transition line between two panels (position 2) for the oscillating construction sequence.

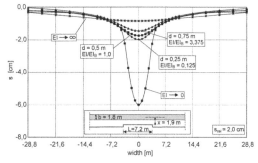

Figure 6. Comparison of the settlement development for foundations located in the centerline (position 1) of a D-wall panel (L = 7,2 m) and at the transition line between two panels (position 2) for the continuous construction sequence.

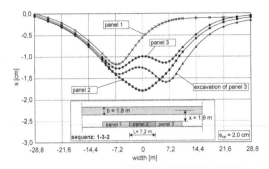

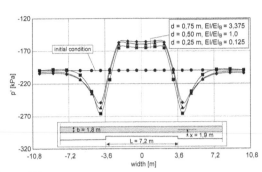

Figure 7. Settlement of a strip footing due to the excavation of an adjacent single panel of the length $L = 3,6$ m $5,4$ m and $7,2$ m. The loading of the foundation is derived for a maximum settlement (without trench) of $s_{zul} = 1,0$ cm (gray) and $s_{zul} = 2,0$ cm (black).

Figure 9. Settlement behaviour of a strip footing with allowable pressure of $p' = 213$ kPa ($s_{zul} = 2$ cm) due to the excavation of a single panel ($L = 7,2$ m) located in x/L distance to the footing (x: distance between face and centre of strip foundation).

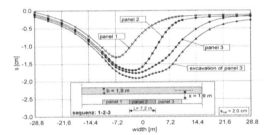

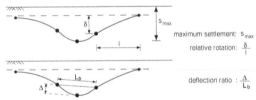

Figure 8. Settlement of strip footing with different bending stiffness EI due to the excavation of a $L = 7,2$ m long panel. EI_B is the reference bending stiffness for a $d = 0.5$ m high footing.

Figure 10. Development of the settlement behaviour of a strip footing due to construction of three D-wall panels in oscillating construction order (sequence 1-3-2). The panel number indicates the settlement of the strip footing after the completion of that respective panel.

and the strip load is $p' = 213$ kPa. In the two Figures (7 and 8), the settlement behaviour of the strip footing is shown after the completion of each wall panel.

In addition, the temporary deformations during the excavation under slurry support of panel 3 are illustrated. Although the total settlement is in the same order for both installation sequences, it is interesting to notice how the settlement distribution changes during the progress of the construction activities. In the case of continuous sequence of construction the settlement curve is of the trough-shape and increases continuously with the work progress. The development of the settlement curve is not associated with changes between the trough-shape and the saddle-shape. This seems to be unavoidable in the case of an oscillating diaphragm wall installation sequence.

In addition, the temporary deformations during the excavation under slurry support of panel 3 are illustrated.

To ensure the serviceability of the building structure near the excavated diaphragm wall, adequate criteria for the damage risk resulting from the settlements have to be established. Generally, two frequent used

criteria exist as shown in Figure 10: the relative rotation δ/L and the deflection ratio Δ/L. If $\delta/L < 1/500$ (0,2%) the settlements are considered to be noncritical. A structure is more sensitive to a settlement development of the saddle-shape (hogging) rather than of the trough-shape (sagging). Thus, limiting values of Δ/L are defined to be 1/4000 (0,025%) for the former and 1/2000 (0,05%) for the latter.

Figure 11 illustrates the calculated relative rotations and deflection ratios of the strip footing versus the normalized length of the trench L/L_{elas}, L_{elas} being the elastic length of the footing. The panel length is adopted to $L = 7,2$ m and the effective strip loading reads $p' = 153$ kPa ($s_{cal} = 1$ cm) and $p' = 213$ kPa ($s_{cal} = 2$ cm). The limit magnitudes of $\delta/L_{lim} = 0,2\%$ and $\Delta/L_{lim} = 0,05\%$ are also given in Figure 11. With respect to the relative rotation of the strip footing, a thickness of $d = 0,2$ m in case of $s_{cal} = 2$ cm is necessary in order to keep the limit of δ/L_{lim}. On the contrary, the deflection ratio requires a respective footing thickness of $d = 0,56$ cm to exclude possible damages of the super structure. Therefore, the deflection ratio Δ/L defines a severe requirement for the serviceability

74

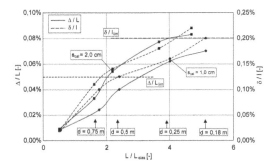

Figure 11. Development of the settlement behaviour of a strip footing due to construction of three D-wall panels in a continuous construction order (sequence 1-2-3). The panel number indicates the settlement of the strip footing after the completion of that respective panel.

state. Moreover, Figure 11 shows that the excavation process of a panel inevitably causes damages to the nearby structure if the bending stiffness is very low, because relative rotations of >0,2% and deflection ratios of >0,08% occur. In such cases, a limitation of the panel length would be obligatory.

In the case of a continuous construction sequence of a diaphragm wall like the one in Figure 9 (panel number 1-2-3), the settlement trough develops parallel to the progress of the installation procedure. Due to the interaction of the settlement troughs of each panel the deformation magnitudes continuously increase. The results of the excavation step of panel 3 illustrate, that the maximum deformations correspond approximately to those, which could be observed during the installation of a single panel. However, this simplification is not true as far as the serviceabilty of the strip footing is concerned. While the relative rotation δ/L remains more or less constant during the installation steps of the wall section, the deflection ratio Δ/L decreases due to an increasing width of the trough. Therefore, the excavation step of a single trench (f.E. the starting panel) is decisive with respect to the serviceability state of the nearby structure. In contrast to a continuous construction sequence, an oscillating installation of the panels (1-3-2, see Figure 7) temporarily causes a saddle shape (hogging) for the deformations of the strip footing, which can cause damages in the super structure. In order to avoid the possible damage, the acceptable deflection ratio in the case of hogging is reduced by a factor of 2 leading to a limit value of $\Delta/L = 1/4000$. In the numerical example the value of the deflection ratio of the saddle shaped deformation of the strip footing obtained within the course of the oscillating construction procedure (see Fig. 7) is 1/5000, which is not far away from the critical limit value of 1/4000. It can be easily seen that a small reduction in the bending stiffness will lead to this limiting value of the deflection ratio and therefore in cases

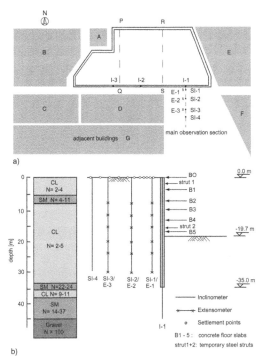

Figure 12. a) Plan view and b) Vertical section of the excavation pit for the TNEC-building.

where the strip foundation stiffness is low (f.E. old buildings with partly damaged foundation elements) it is recommended to construct the D-wall with the continuous sequence.

3.3 Numerical results for the supporting system

In this section, the influence of the construction method on the deformation of the supporting system itself is analysed. In order to make comparisons between numerical results and measurements, the well documented excavation pit for the TNEC (Taipeh National Enterprise Center)-Building has been chosen, Ou et al. (1998), (2000a, b).

The TNEC constructed in 1991 as an 18-story building with five basement levels. Figure 12 a, b illustrates a plan view of the site and a vertical section view of the excavation. The excavation pit is 19,7 m deep showing in plan the dimensions 106 × 43 m and was supported by a 35 m deep and 0,9 m thick diaphragm wall. The length of the D-wall panels varied between 4 m to 6 m. The building was constructed according to the top-down construction method and the retaining wall was supported by five permanent concrete floor slabs and two prestressed temporary struts (see Fig. 12b). Inside the excavation pit, piles have been

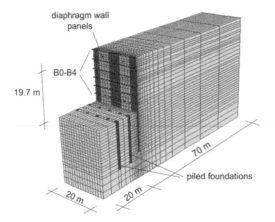

diaphragm wall panels

B0-B4

19.7 m

70 m

piled foundations

20 m   20 m

Figure 13.   3D FEM model of the TNEC excavation pit.

Table 3.   Effective moduli $E_{c,eff}$ and stiffness of the slabs B0 to B4.

Concrete slab	$E_{c,eff}$ [MPa]	S_i [kN/m]
B0	9114	85440
B1	7418	69540
B2	9114	85440
B3	6510	61031
B4	8395	78703

placed in order to support the steel staunchions carrying the loads of the super-structure. With a monitoring programme the ground surface settlements and wall deflections have been observed during the entire excavation process. The soil conditions, the material model for soil and concrete have already been described in section 2.2.

In Figure 13, the 3-dimensional FEM model of the TNEC-excavation pit is presented. The model consists of 2400 tri-linear brick elements for a coupled consolidation analysis and represents a plane section of the wall which comprises 5 D-wall panels. The pore pressure is calculated at one point per element (see ABAQUS, reference manual). Two different types of FEM calculation have been run (*wip, wim*). The conventional *wip-* (*wished in place*) calculation totally neglects the installation process of the retaining wall, assumes the earth pressure at rest and the hydrostatic pore water pressure as initial conditions at the beginning of the excavation. For the simulation of the pit excavation process, the finite elements in front of the wall are gradually removed in accordance with the scheduled construction activities on the site. The temporary struts and permanent floor slabs are represented by spring-elements with a corresponding stiffness and are implemented into the model in dependence of the current excavation depth.

The increased stiffness of the concrete due to aging is considered in the *wim* – model by the evolution of the Young's Modulus E and the Poisson ratio v as given in the previous example (see eq. 2). In addition to this, for the permanent concrete floor slabs creep and shrinkage of the concrete has been considered.

For the calculation of the slab stiffness the Young's Modulus of concrete $E_{cm} = 29000$ Mpa, the slab thickness of 15 cm and the pit span of 40 m is taken into account and therefore the stiffness value of 217,5 MN/m is obtained. For the consideration of the

creep deformation of the concrete slab, the modification of the Young's Modules to a "creep effective" one is taken into account (Reinhardt, 2001),

$$E_{c,eff} = \frac{1,1 E_{cm}}{1,1 + \varphi(t,t_0)}$$ (8)

where the creep number φ depends on the age of the concrete at the time t_0 of load application, the relative humidity, concrete strength, the thickness of the slab etc. For the specific project, the construction time and the rest of the required parameters lead to values of φ between 2,4 and 3,8. In Table 3, the effective Young's Modulus $E_{c,eff}$ of the slab and the resulting spring stiffness for the FEM-simulation are presented (for more details reference is made to Schäfer, 2004).

The total slab concrete shrinkage strain due to drying is expected to be of the order of 0,03 % (Schäfer, 2004), which in case of a length of 20 m (half of the slab span) would lead to approx. 6 mm of additional slab deformation. Due to the slab openings as well as different expansion joints, it is expected that the deformations will be much lower than the calculated 6 mm, and therefore the impact of shrinkage is neglected in the present analysis. For the simulation of the construction method according to the *wim*-model the bentonite pressure and concrete placement follows the same procedure of element removal and placement as outlined in the previous numerical example. The completion of a single panel construction is assumed to take one day, and the construction of the piles took in total 60 days after the completion of the D-wall, where in the *wim*-model the simulation of the construction procedure for the piles is not taken into account. The piles, having a cross-section of $1,25 \times 1,25$ m, act as bending and stiffening elements inside the excavation reducing the deformations of the soil on the passive site of the retaining wall. The piles together with the steel staunchions have a lumped stiffness of $E_{piles} = 1,9 E7$ kPa. The time for the installation of the slabs following the documented site records (Ou et al., 1998, 2000a,2000b) is summarized in a table in Schäfer (2004). During excavation, the soil elements in front of the wall are removed and the respective

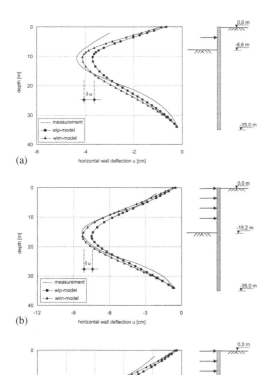

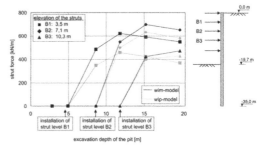

Figure 15. Strut loadings of the levels B1, B2 and B3 versus the excavation depth of the pit.

Figure 14. Horizontal wall movement due to the pit excavation of the TNEC building up to a a depth: a) $z_a = 8,6$ m b) $z_a = 15,2$ m and c) $z_a = 19,7$ m.

spring elements are activated at the respective time of the slab construction.

For the evaluation of the impact of the construction process on the serviceability of the retaining structure, the horizontal wall deflections calculated by the *wim*- and the *wip*-model are compared for different excavation stages of the pit. Furthermore, these results are additionally compared with the in-situ measurements of the inclinometer I1 (Fig. 12b) in order to check the quality of the numerical model and the constitutive relation.

Figure 14a–c presents the horizontal wall movements due to a pit excavation up to a depth of $z_a = 8.6$ m, 15.2 m and 19.7 m resp. The results of the *wim*- and the *wip*-model are characterised by black-filled squares and triangles, respectively, and the in-situ measurements are shown as a continuous line.

The greatest deflections can always be observed close to the bottom of excavation and the diaphragm wall shows a typical deformation behaviour of a strutted retaining structure.

For the final excavation stage with $z_a = 19.7$ m, the greatest measured movement is about $u_{max} = 10.5$ cm and corresponds to approximately 0.5% of the excavation depth. On the contrary, the *wim*- and the *wip*-model predict deflections of $u_{wim} = 10$ cm and $u_{wip} = 9$ cm respectively. Accordingly, the in-situ deflection behaviour of the structure can be very well back-calculated. The consideration of the construction process leads to a slight improvement of the prediction, although the difference Δu in the deformation results between the two models is only 10–15%. Considering an excavation depth of the pit of $z_a = 15.2$ m (Fig. 14b), the measurements of the inclinometer I1 is reproduced almost exactly by the *wim*-model. The deflections especially at the top of the wall are evidently underestimated for lower excavation depths of the pit ($z_a = 8.6$ m, see Fig. 14a). Nevertheless, the *wim*-calculated wall movements exceed those predicted by a conventional *wip* model adopting the earth pressure at rest at the beginning of simulation, throughout the whole excavation process.

Figure 15 presents the forces of the strut layers B1 ($z_s = 3.5$ m), B2 ($z_s = 7.1$ m) and B3 ($z_s = 10.3$ m) versus the excavation depth of the pit as results from the *wim*- and *wip* calculations. Unfortunately, there are no measurements of the corresponding strut forces, which can be compared with the numerical results.

Anyway, the strut loadings increase with an increasing excavation depth and with respect to the construction process of the diaphragm wall, greater magnitudes are calculated. The greatest difference between the results of the *wim*- and the *wip*-model can be observed in case of the strut level B1 and an excavation depth of $z_a = 19.7$ m: in consideration of the construction process up to 50% higher strut forces are predicted. However, the impact of the diaphragm wall installation on the development of the strut force decreases, the lower the strut level is arranged below the surface ground level. In case of the level B3 at a depth of $z_s = 10.3$ m, only a negligible influence can finally be shown and both FE-models predict a maximum loading of approximately 430 kN/m. A higher stress level prior to the pit excavation not only affects the deformation behaviour of the system, but also causes a substantial increase of the strut force (Fig. 15). The higher the arrangement of the strut level, the greater is the increase of the strut loading during the excavation of the pit. This result can be attributed to the fresh concrete pressure distribution versus the depth of the trench. Due to the hydrostatic pressure increase down to the critical depth h_{crit}, the passive earth resistance is partially mobilized, especially in the upper half of the trench associated with a respective increase of the initial stress level (Schäfer and Triantafyllidis, 2004). Considering the strut level B1, an up to 50% higher loading can be expected accordingly. However, this effect is probably of minor importance in case of the considered top-down construction, because the concrete floor slabs can easily sustain the higher loadings. But considering conventional bottom-up excavation pits, which are frequently braced by temporary steel struts with a safety factor of $\eta = 1.5$ for example, loadings close to the ultimate limit state are supposable or can be expected.

The presented numerical results of the TNEC-excavation pit show, that the detailed consideration of the stepwise construction process of the diaphragm wall predicts greater wall deflections and surface ground settlements and higher strut loadings during the excavation process of the pit. These results apparently contradict the *wip* ones, which have been presented for highly over-consolidated soil layers in Great Britain (Ng and Yan 1999, Gourvenec and Powrie, 1999). Simpson (1992) and Lings et al. (1991) show, that a conventional FE-prediction adopting the earth pressure at rest at the beginning of the calculation distinctly overestimates the wall deflections and strut loadings. This development can primarily be attributed to the construction process of the diaphragm wall. The excavation and the pouring of wall panels in over-consolidated soil deposits with a high earth pressure coefficient at rest K_0 result in a decreasing lateral effective stress in the adjacent ground. Due to the lower stress level prior to the pit excavation process, smaller wall movements and lower strut loadings can subsequently be expected.

Schäfer and Triantafyllidis (2004) have already shown, that the pouring process of wall panels may lead to a partial mobilisation of the passive earth pressure in soft soil deposits, especially in the upper half of the wall. Figure 16 illustrates the lateral effective stress σ'_{hm} at $z_s = 9.75$ m (point A) averaged by the width of the model versus the excavation depth of the pit. Considering the curve calculated by the *wip*-model,

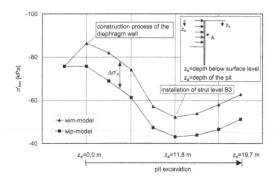

Figure 16. Development of the in width averaged effective horizontal earth pressure σ'_{hm} at a depth of $z_s = 9.75$ m (Point A) below the surface ground level during different construction steps.

the stress decreases from the condition at rest with an increasing depth of the pit. The minimum magnitude of $\sigma'_{hm} = 43$ kPa, which still exceeds those of the active limit state ($\varphi_s = 29°$ and $\delta_a = 2/3\varphi \rightarrow \sigma'_a = 32.4$ kPa), occurs during an excavation depth of $z_a = 11.8$ m. However, the subsequent installation of the strut level B3 ($z_s = 10.3$ m) initiates a rearrangement of the earth pressure behind the wall resulting in a stress increase in point A during the final construction steps of the excavation process. By comparison, the *wim*-model calculates a stress development similar to the conventional one.

However, the construction process of the diaphragm wall leads to a stress increase of $\Delta\sigma'_{hm} \approx 13$ kPa (about 17% of the initial magnitude at rest) at the beginning. Due to this stress increase, higher earth pressure magnitudes arise throughout the whole excavation process and greater wall deflections take place. Nevertheless, both finite-element-models reproduce well the in-situ deformation behaviour of the soil-structure-system, which speaks in favour of the visco-hypoplastic constitutive model for soft clayey soil deposits.

Figure 17 represents the over 5 panels averaged horizontal effective stress for a depth of $z_a = 23,2$ m at the passive side of the wall. At the beginning of the excavation, the difference in horizontal effective stress between *wip*- and *wim*-model was approx. 10%, which represents the reduction of the K_0-situation due to panel excavation. In an excavation depth of approx. 5 metres the difference in the stress level between *wip*- and *wim*-model approaches very small values and the stresses of the *wim*-model are approaching the K_0-situation with increasing excavation depth of the pit. For both models, as it can be seen in Figure 17, the passive resistance is reached at the excavation depth of $z_a = 17,3$ m. Any excavation below that depth is associated with a reduction of the vertical stress in the toe of the wall and therefore the horizontal stresses

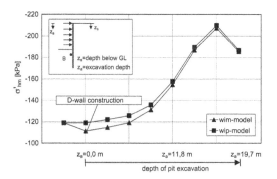

Figure 17. Development of the in width averaged effective horizontal effective earth pressure on the passive wall side at a depth of $z_a = 23.2$ m (Point B) during different construction steps.

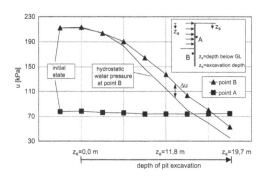

Figure 18. Development of the pore water pressure at the depth $z_a = 9.75$ m (Point A) and $z_a = 23.2$ m (Point B) versus the pit excavation depth.

also reduces until the final depth of excavation at $z_a = 19.7$ m has been reached.

Figure 18 shows for both positions A and B the development of the pore water pressure during the excavation steps of the pit. Especially for the position B the hydrostatic water pressure is presented if one assumes that the water level inside the excavation area coincides with the excavation level. From the results it can be seen that under the excavation level still a pore water pressure during the excavation process even for a long construction period is present. Parallel to the excavation work a dissipation of the pore water pressure appears, so that excess of pore water pressure is limited only to 20 kPa and the soil gains in strength due to the consolidation effects. This postulate is supported by the comparison of the yield curves at the beginning and the end of the construction process as presented in Figure 20 for the *wim*- and *wip*-model, where the stress paths at the end of the excavation process ($z_a = 19.7$) are located very close to the critical

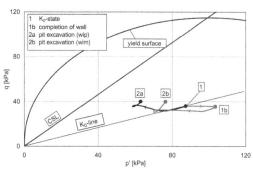

Figure 19. Effective stress path for point A without (*wip*-model) and with (*wim*-model) the simulation of the wall construction process.

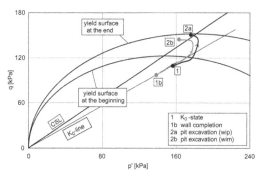

Figure 20. Effective stress path for point B without (*wip*-model) and with (*wim*-model) the simulation of the wall construction process.

state line for triaxial extension and therefore the soil at the position B is on the passive limit state.

In Figure 20 further the Roscoe-invariants p' and q are presented for the considered positions a and B

$$\left(p' = -tr \ T/3 \ \text{and} \ q = \sqrt{\frac{3}{2}} \left\| T^* \right\|, \right.$$

T^* being the deviator stress), where in the *wim*-model starting from the K_0-state prior to the D-wall construction, a reduction of the p' and q-stresses (Point $1 \rightarrow 1b$ in Fig. 20) takes place and the excavation inside the pit causes a reloading of the soil with a stress path starting below the K_0-line (Point $1 \rightarrow 2a$ or $1b \rightarrow 2b$ in Fig. 20) and approaching the CSL (critical state line).

On the contrary, the stress path development at Point A (s. Fig. 16) leads for the entire construction process to levels which are far below the CSL for triaxial compression. The construction of the D-wall results to an increase of the isotropic stress p' (Point $1 \rightarrow 1b$, in Fig. 19) so that for the *wim*-model the excavation starts from a higher isotropic level ($\Delta p' = 15$ KPa). At the

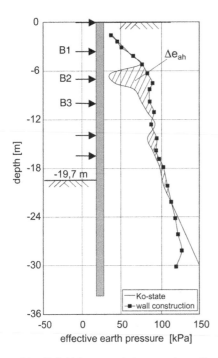

Figure 21. K_o-initial state and the over the wall width averaged earth pressure after the completion of the wall.

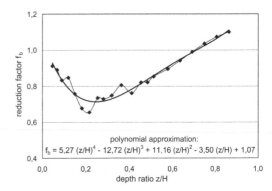

polynomial approximation:
$f_b = 5,27\ (z/H)^4 - 12,72\ (z/H)^3 + 11,16\ (z/H)^2 - 3,50\ (z/H) + 1,07$

Figure 22. Reduction factor f_b for the calculation of the modified earth pressure coefficient at rest K_0^* versus the depth ratio z/H (H: trench depth).

end of the excavation (Points 2a or 2b, in Fig. 19), the stress levels are sufficiently away from the critical state so that the active soil pressure is not fully mobilized.

It can be concluded, that the difference δu of the displacements between the *wim*- and *wip*-model results from the higher earth pressure in the *wim*-model. In Figure 21, the K_o-stress situation and the respective one after completion of the D-wall over an average wall width are presented. The area with shaded lines represents the increased earth pressure Δe_{ah} in comparison to the pressure at rest. This difference reaches a maximum at a depth $z = 6,35$ m, where the sand layer shows a small earth pressure coefficient, and the fresh concrete pressure is increasing linearly with the depth until the depth $h_{crit} = 7,0$ m has been reached. Both curves for the stresses show smaller differences at greater depths, because the fresh concrete pressure below the depth h_{crit} increases with a smaller gradient and the earth pressure at rest for the deeper located clay layers is higher than the respective one of the sand layer. The observed increase of the earth pressure causes also the difference in the strut forces between *wim*- and *wip* model as given in Figure 15, where the *wim*-model shows higher force levels in general.

The simulation of the entire construction process in real projects is at present very time consuming. In order to be able the Eurocode requirement with the existing commercial software to satisfy, that we also have to estimate the effects of the wall construction on the deformations, one of the possible ways would be to use a modified initial earth pressure coefficient at rest K_0^*, which might vary with the depth and includes the effects of the construction process.

Due to the presented results for soft clay, it is prudent to assume that the modification of K_0^* can be correlated to the fresh concrete pressure in the trench (Mayer, 2000), so that the influence of the sequence of the D-wall installation and the influence of the trench geometry can be neglected in a first stage. Therefore the following relation for the K_0^* value may be adopted:

$$K_o^* = \frac{f_b}{\gamma \cdot z}\left(\left\{\begin{matrix} \gamma_c^z & : z \le h_{crit} \\ (\gamma_c \cdot h_{crit}) + \gamma_b(z - h_{crit}) & : z > h_{crit} \end{matrix}\right\} - \gamma_w(z - z_w)\right) \quad (9)$$

with:
γ_b: unit weight of the bentonite slurry
γ_c: unit weight of the fresh concrete
γ_w: unit weight of water
z: depth below ground surface
z_w: height of the ground water table
h_{crit}: critical depth of the concrete pressure
f_b: reduction factor according to Figure 22

The reduction factor f_b takes into account that the stress level versus depth after the completion of the diaphragm wall differs from the respective fresh concrete pressure versus depth and can be calculated for every level from the ratio between them (see Fig. 21 and equation (1)).

The resulting value of f_b is related to the dimensionless depth z/H, H being the total depth of the diaphragm wall (see Fig. 22). For the calculation of K_o^*, the readings for f_b can be either directly taken from Figure 22 versus the ratio z/H or calculated with the forth order polynomial approximation given also in Fig. 22.

It has to be pointed out, that this approximation has been proven for the given example only. For a more general approximation, additional examples should calculated and compared.

4 CONCLUSIONS

A diaphragm wall construction process in soft clays produces considerable settlements on nearby structures. The maximum settlement magnitudes, the relative distortions as well as the deflection ratio of strip footings increase with the panel length, the strip load and decreasing bending stiffness of the footing. With respect to the serviceability of the nearby structure and for the prevention of possible damages, the deflection ratio as the most decisive criterion has been established. The largest settlements occur during the excavation of a single wall panel, especially if the bending stiffness of the footing is low. The subsequent pouring of the trench leads to a heave of the footing due to the fresh concrete pressure in the upper third height of the wall panel. In the case of a continuous diaphragm wall installation, the excavation of the starting panel is the critical construction step as far as the maximum deformation magnitude is concerned. During the subsequent construction steps of the adjacent panels the settlement trough width increases, leading to a decreasing deflection ratio. The continuous installation sequence of the panels never leads to a saddle shaped deformation of the strip footing. The starting panel in the vicinity of a building should be limited in length in order to reduce the sagging ratios of the footing. In the course of the subsequent installation steps, the panel length can be increased up to the size resulting from the stability analysis of the open trench.

In normally to slightly over-consolidated clay layers, conventional finite element calculations neglecting the construction process of the diaphragm wall and adopting the earth pressure at rest as initial condition underestimate the wall deflections and settlements. The increase of deformation, which varies between 10–15% compared to the *wip* calculated magnitudes, can be attributed to larger effective stresses in the adjacent ground resulting from the pouring of the individual wall panels one by one. Although the additional movements are usually relatively small (as in case of the analysed TNEC-pit), under some conditions, like normally consolidated clay layers up to the surface ground level, the installation process of the retaining structure becomes of importance for the resulting wall deformations.

Moreover, the modified earth pressure additionally cause higher strut loadings. The closer the arrangement of the struts below the surface ground level, the greater is the impact of the installation on the strut force. Considering, for example, a strut level at a depth of 3.5 m, up to 50% higher strut loadings are calculated in comparison to a conventional model. Especially in the case of a bottom-up excavation pit with temporary steel struts and a safety factor of $\eta = 1.5$, loadings close to the ultimate limit state cannot totally be excluded any longer.

Considering the serviceability of nearby structures, the settlements due to the excavation process of the pit itself have to be superimposed with the construction-induced movements of the diaphragm wall. Thereby it is important to emphasise, that the excavation of a trench under slurry support can already produce substantial settlements, if the footing is located close to the panel. For a panel length of $L = 7.2$ m, settlements of a single footing up to 2 cm have been calculated during the installation of the wall.

Further, we have demonstrated that the visco-hypoplastic model gives very good results for engineering applications in soft clay deposits and the construction procedure of diaphragm walls can be very well simulated enabling accurate predictions of deformations. The promising results may serve as the basis of a guidance for practical use for the design of retaining structures in urban environment.

ACKNOWLEDGEMENT

The authors gratefully acknowledge the financial support from the German Research Council (DFG, AZ 218/4-1) and the Federal Ministry of Research and Technology for the grant of the scientific program "structure prevention oriented geotechnics". Furthermore they like to express their gratitude to Dr. A. Niemunis, who developed the FORTRAN-routine of the constitutive model, which has been used to obtain the presented results.

REFERENCES

ABAQUS 6.3, Theory Manual, Version 6.3.

August 1986. DIN 4126 – *Ortbeton-Schlitzwände; Konstruktion und Ausführung*

Franke, E. 1980.*Überlegungen zu Bewertungskriterien für zulässige Setzungsdifferenzen. Geotechnik 3*: S. 53–59.

Kilchert M., Karstedt J.: Standsicherheitsberechnungen von Schlitzwänden nach DIN 4126. Beuth-Kommentare, 1984.

Leinenkugel, H.J.: 1976 *Deformations- und Festigkeitsverhalten bindiger Erdstoffe*. Schriftenreihe des Institutes für Bodenmechanik und Felsmechanik, Universität Fridericiana, Karlsruhe. Heft Nr. 66.

Lings, M.L., Ng C.W.W., Nash D.F.T.: 1994. *The lateral pressure of wet concrete in diaphragm wall panels cast under bentonite*. Proc. of Institution of Civil Engineers, Geotechnical Engineering 107. S. 163–172.

Mayer, P.-M.: 2000. *Verformungen und Spannungsänderungen im Boden durch Schlitzwandherstellung und Baugrubenaushub*. Institut für Bodenmechanik und Felsmechanik der Universität Fridericiana Karlsruhe. Heft 151.

Niemunis, A.: 2003. *Extended hypoplastic models for soils. Habilitation.* Schriftenreihe des Instituts für Grundbau und Bodenmechanik, Ruhr-Universität Bochum. Heft Nr. 34.

Ou, C.-Y., Liao J.-T. Cheng W.-L.: 2000a. *Building response and ground movements induced by a deep excavation.* Géotechnique 50, Nr. 3. S. 209–220.

Ou, C.-Y., Liao J.-T., Lin H.-D.: 1998. *Performance of a diaphragm wall constructed using top-down method.* Journal of Geotechnical and Geoenvironmental Engineering 124. Nr. 9. S. 798–808.

Ou, C.-Y., Shiau B.-Y., Wang I.-W. : 2000b. *Three-dimensional deformation behavior of the Taipei National Enterprise Center (TNEC) excavation case history.* Canadian Geotechnical Journal 37. S. 438–448.

Reinhardt, H.-W.: *Beton.* Beton-Kalender 2001, Ernst & Sohn Verlag.

RUBSchlitz 2001. *Programm zu Standsicherheitsnachweisen suspensionsgestützter Schlitze.* Aufbauend und erweitert auf Grundlage der Veröffentlichung von Triantafyllidis et al. Verfügbar am Lehrstuhl für Grundbau und Bodenmechanik der Ruhr-Universität Bochum.

Schäfer, R.: 2004. *Einfluss der Herstellungsmethode auf das Verformungsverhalten von Schlitzwänden in weichen bindigen Böden.* Dissertation, Schriftenreihe des Instituts für Grundbau und Bodenmechanik, Ruhr-Universität Bochum. Heft Nr. 36.

Schäfer, R., Triantafyllidis T. 2004. *Modelling of earth- and pore water pressure development during diaphragm wall construction in soft clay.* International Journal for Numerical and Analytical Methods in Geomechanics. Nr. 28, pp. 1305–1326.

Triantafyllidis, Th., König D., Sonntag M.: 2001. *Zur äußeren Standsicherheit von nicht-ebenen suspensionsgestützten Erdschlitzen.* Bautechnik 78. Heft 2. S. 77–88.

Simpson, B., 1992. *Retaining structures – displacement and design. Thirty-second Rankine Lecture.* Géotechnique 42. No. 4, pp. 541–576.

Gudehus, G., Goldscheider M., Winter H.: 1977 *Mechanical properties of sand and numerical integration methods: some source of errors and bounds of accuracy.* In *Finite Elements for Geomechanics.* (Ed. Gudehus). John Wiley. New York

Lings, M.L., Nash D.F.T., Ng C.W.W., Boyce M.D. 1991 *Observed behaviour of a deep excavation in Gault Clay: a preliminary appraisal.* Proc. 10th European Conference on Soil Mechanics and Foundation Engineering, pp. 467–470.

Ng, C.W.W., Yan R.W.M. 1999. *Three-dimensional modelling of a diaphragm wall construction sequence.* Géotechnique, Vol. 49. Nr. 6. pp. 825–834.

Gourvenec, S.M., Powrie W. 1999. *Three-dimensional finite-element analysis of diaphragm wall installation.* Géotechnique, Vol. 49. Nr. 6. pp. 801–823.

Numerical Modelling of Construction Processes in Geotechnical Engineering for Urban Environment – Triantafyllidis (ed)
© 2006 Taylor & Francis Group, London, ISBN 0 415 39748 0

A comparison between monitoring data and numerical calculation of a diaphragm wall construction in Rotterdam

A. Lächler, H.P. Neher & G. Gebeyehu
Züblin AG, Zentrale Technik, Technisches Büro Tiefbau, Stuttgart

ABSTRACT: Constructions of diaphragm walls cause displacements and state changes in the adjacent ground. To study the underlying mechanisms, the installation of a diaphragm wall in Rotterdam, the Netherlands, is accompanied by an extensive monitoring program. This program includes data records during the construction of the diaphragm wall panels and the excavation of the pit. This contribution focuses on records produced during the installation of the diaphragm wall. The construction is modelled with the Finite Element Method (FEM) using an elastic-ideal-plastic model with a Mohr-Coulomb yield criterion (the Mohr-Coulomb model). Finally, the numerical results are discussed and compared with the gained monitoring data.

1 INTRODUCTION

In the framework of the RandstadRail rapid transit project, a new railway track is being constructed in the province of South Holland, the Netherlands. The new route directly connects the city centers of Rotterdam, Den Haag and Zoetermeer. As part of the project, the existing "Erasmuslijn" (metroline in Rotterdam) is connected with the existing "Hofpleinlijn" (track between Den Haag and Rotterdam) with a new link called "Statenwegtracé" in Rotterdam.

A plan view of the passage "Statenwegtracé" in Rotterdam is illustrated in Figure 1. This passage is realised with two single track tunnels, each with a total length of about 2.94 km. Around 80% of the total length is constructed as a hydro shield driven tunnel with segmental lining, while the rest 20% is a cut-and-cover tunnel. Each tunnel has an outside diameter of 6.5 m and the thickness of the tunnel lining segments is 0.35 m. They are the first bored tunnels below Rotterdam's urban area. The tunnels begin at "Centraal Station", crossing under the tracks of the "Nederlandse Spoorwegen" and continuing along the street "Statenweg". The tunnels extend northwards passing under the "Noorderkanaal" and the motor-way "Rijksweg A20". The passage ends at "Sint Franciscus Driehoek", traversing before under the "Schieplein" and the "Goudselijn". In the middle of the street Statenweg, the new station "Station Blijdorp" is being built. The project contractor SATURN (Samenwerking Tunnelrealisatie Nederland) is a joint venture of the Ed. Züblin AG, Germany and the Dura Vermeer Group N.V., the Netherlands. The client is the public transport company RET (Rotterdamse Electrische Tram).

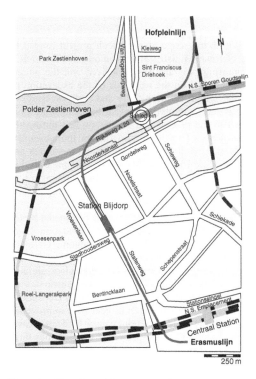

Figure 1. Plan view of Statenwegtracé.

The design of the geotechnical works and the site supervision is run by the communal engineering office of Rotterdam (Ingenieursbureau Gemeentewerke Rotterdam).

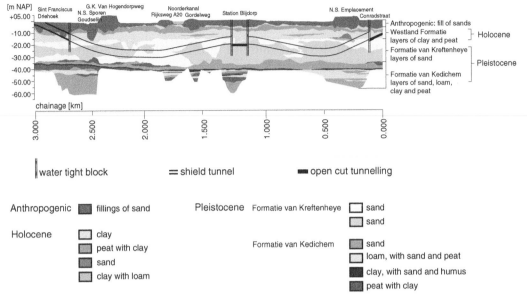

Figure 2. Geology of Statenwegtracé.

Within the scope of a research project funded by the German Ministry of Education and Research, a monitoring program was conceived and the gauges were installed on the west side of the excavation pit for the station Blijdorp. Pore and earth pressures as well as deformations (vertical and horizontal) are monitored in the ground and on the surface during the construction of the diaphragm wall and the subsequent excavation of the pit. In addition, temperatures and pressures are logged inside the diaphragm wall during concreting. The deflection of the diaphragm wall is recorded during the whole excavation process. Strut forces are logged, as well.

The title of the research project, conducted by the Institute of Soil Mechanics and Foundation Engineering of the Ruhr-University Bochum, Germany (RUB) and the Geotechnical Department of the Ed. Züblin AG, is "Bauwerksschonende Geotechnik – Optimierung der Herstellung von Verbauwänden im Hinblick auf Verformungen benachbarter Gebäude".

The aim of this research project is to improve the predictability of deformations caused by deep excavations in soft soils. The project is divided into three parts (A, B and C). Part A (RUB) covers element and model tests, as well as laboratory tests on soil samples to obtain material parameters for the numerical modelling. The soil samples are retrieved at the location of the station Blijdorp. The numerical simulations of diaphragm wall installations and the respective pit excavations as well as the modelling of bored diaphragms are covered in Part B (RUB). Ed. Züblin AG deals with the sampling, the monitoring program

at Station Blijdorp (Part C) and modelling of the diaphragm wall at station Blijdorp (Part B). The above mentioned geotechnical monitoring program is used to verify the numerical models. The simulation, covers the stages of the construction process, e.g. trenching with slurry support and concreting.

2 GEOLOGY

The underground of Rotterdam can be subdivided into four sub-horizontal layers. The anthropogenic fill generally consists of sand. This soil layer, with a thickness of about 1.5 m to 14.0 m, was placed roughly 80 years ago to create a constructible underground. It peters out north of the motor-way "Rijksweg A20", where the Zestienhoven Polder is. The filling is under-lain by a 6.0 m to 13.0 m thick layer of peat and clay ("Westland Formatie") from the Holocene. These peaty and clayey deposits are of fluvial origin and very heterogenous, their transition is blurred. Below the Holocene formation the Pleistocene formations "Formatie van Kreftenheye" and "Formatie van Kedichem" are located. The "Formatie van Kreftenheye" is a sand layer, about 19.0 m thick. The "Formatie van Kedichem" is made up of sand, peat, clay and loam layers. The phreatic groundwater level averages about 2.5 m below NAP (Normaal Amsterdams Peil) along the Statenwegtracé. The most of the designated tunnel lies in the sand, "Formatie van Kreftenheye". In some parts, the ground below the tunnel is improved by jet-grout bodies

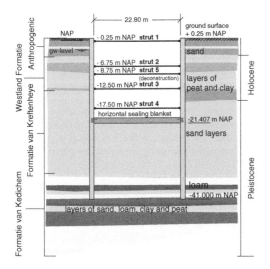

Figure 3. Geological cross section along station Blijdorp.

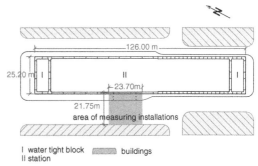

I water tight block
II station

buildings

Figure 4. Plan view of the station Blijdorp excavation.

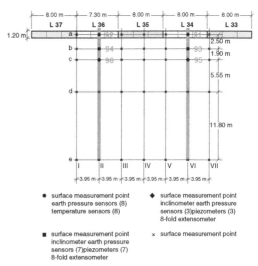

● surface measurement point
earth pressure sensors (8)
temperature sensors (8)

◆ surface measurement point
inclinometer earth pressure
sensors (3)piezometers (3)
8-fold extensometer

■ surface measurement point
inclinometer earth pressure
sensors (7)piezometers (7)
8-fold extensometer

× surface measurement point

Figure 5. Location of the monitoring instruments.

or cement columns, as well as injecting gel or soil replacement.

Figure 3 shows a geological cross section of the Station Blijdorp. Near the station, the anthropogenic sand fill is 4.5 to 6.0 m thick, underlain by soft soils from the Holocene. These soft soils, consisting of peat and clay, overlie a sand layer and reach in depths beyond 15.0 to 18.0 m NAP. The sand layer has a thickness of about 17.0 to 22.5 m at this location. Below the sand layer lies the "Formatie van Kedichem", where the diaphragm wall panels are footed. As shown in Figure 3, the diaphragm wall panels have a depth of 41.0 m. The width of the panels varies between 1.2 m and 1.5 m. In order to stabilize the excavation of the 22.80 m wide pit, four sets of struts are placed at different depths. Three of the struts are located "in the Holocene". A fifth set of struts is installed in a depth of 8.75 m. This is done during dismantling the earlier placed struts.

3 MONITORING PROGRAM

Figure 4 shows a plan view of the station Blijdorp excavation. A detailed location plan of the monitoring gauges is shown in Figure 5. The devices are installed in five rows (axes a, b, c, d and e) parallel to the diaphragm wall. The distance between these rows varies from 1.90 m near the diaphragm wall to 11.80 m between axes d and e. There are seven axes (axes I, II,...,VI and VII) in the orthogonal direction having a constant spacing of 3.95 m. The deformation of the ground surface is recorded at the points where the axes intersect each other.

Along the two main axes (axis II and axis VI) additional information is recorded over the depth. At the locations marked with 93 to 96, the vertical and horizontal displacements, the pore pressures and the lateral earth pressures are recorded in the ground. In axis a, located in the middle of the diaphragm wall, the slurry and the wet concrete pressures as well as the respective temperatures are recorded before and during concreting the panels L34 and L36. The depth of the gauges are shown in Figures 6 and 7. In axis a the sensors are placed at a vertical spacing of 5 m down to the depth of 40 m NAP (see Figure 6). Thus, there are eight points in each panel to record the changes in pressure and temperature. The temperature gauges and the pressure cells are fixed on a steel pipe with a diameter of 200 mm. This pipe is placed inside the reinforcement cage, which is already lowered into the trench. After concreting the panels, the inclinometer tube is run inside the steel pipe to record the deflection of the diaphragm wall during the excavation of the pit.

Figure 7 illustrates the cross section along axis b. The position of the different monitoring instruments

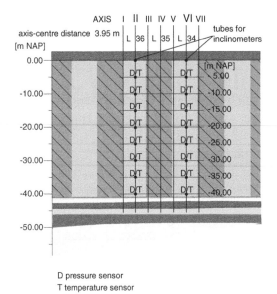

D pressure sensor
T temperature sensor

Figure 6. Depth of monitoring instruments, axis a.

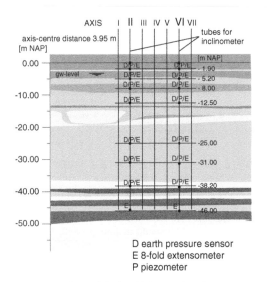

D earth pressure sensor
E 8-fold extensometer
P piezometer

Figure 7. Depth of monitoring instruments, axis b.

in the ground is shown. The monitoring data from the piezometers and ground pressure cells are recorded at seven different depths, see Figure 7. Moreover, vertical displacements are obtained from an 8-fold extensometer, located at the same depths. Horizontal displacements are recorded with an inclinometer. The locations of the measuring instruments in axis *c* are similar to those in axis *b*. However, piezometers and pressure cells are installed only at depths of 1.9 m, 12.5 m and 25.0 m.

4 MONITORING RESULTS

In this section the monitoring results of wet concrete pressures, horizontal and vertical displacements and pore pressures are presented. The records from the earth pressure sensors are unfortunately useless. All of them show more or less similar results like the piezometers. The records of wet concrete pressures are shown for the panels L34 and L36 and serve as input data for modelling the wet concrete pressure. Displacements and pore pressures are presented for axis VI (panel L34). An overview of the construction process is shown in Figure 8.

4.1 Wet concrete pressure

Figures 9 and 10 show the development of the wet concrete pressure during the first 20 hours. The first recorded value reflects the initial hydrostatic pressure of the slurry, in the respective depths before concreting. As the concrete level passes a sensor, the wet concrete pressure firstly increases. This increase corresponds to the rising concrete level. The maximum pressure is recorded in the deepest pressure cell. The maximum pressure of the other pressure sensors, located higher, show a time lag. One to two hours after concreting, the pressure starts to decrease (see Figures 9 and 10). This is due to the setting of the concrete (hydration process). All pressure sensors re-reach the level of the initial hydrostatic slurry pressure after 15 hours.

The pouring of concrete is accidently interrupted in panel L34. This interruption is noticeable at the pressure cells located in depths of 5 m and 10 m below NAP and is illustrated with their two plateaus, as seen in Figure 9. After this break, a loss of power causes a lack of data for approximately two and a half hours. Therefore, no data are plotted for this period.

The averaged maximum change in pressure during the first 20 hours is 102 kN/m^2 in panel L34 and 69 kN/m^2 in panel L36. The reason for this different pressure change is the varying velocity of concreting. Due to the fact, that the wet concrete pressure is reduced by its setting behaviour and increases with a rising concrete level, the final wet concrete pressure depends on the velocity of pouring. A high velocity of concreting results in a high maximum wet concrete pressure. The reason for this behaviour is a rapidly increasing concrete level and a small influence of the concrete setting in the beginning. On the other hand, a slow velocity of concreting influences the setting behaviour more significantly as the concrete level increases much slower. Thus, the maximum wet concrete pressure is smaller.

The development of the wet concrete pressure over the depth during concreting is exemplarily shown for panel L34 in Figure 11. The hydrostatic slurry pressure and the hydrostatic wet concrete pressure are

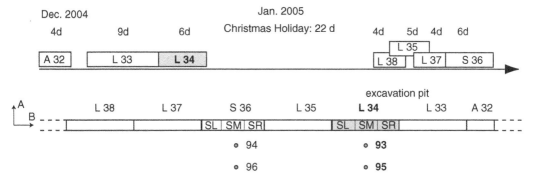

Figure 8. Production stages of the diaphragm wall.

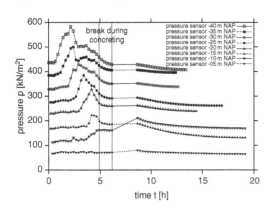

Figure 9. Pressure-time-history of the wet concrete pressures in panel L34 at location 91.

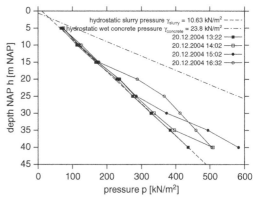

Figure 11. Wet concrete pressures versus depth for panel L34 at location 91.

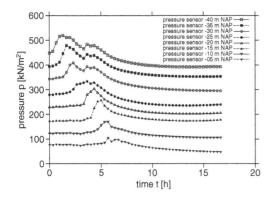

Figure 10. Pressure-time-history of the wet concrete pressures in panel L33 at location 92.

plotted as reference lines, respectively. The line for the hydrostatic slurry pressure intersects the abscissa because the slurry level is over NAP. Similarly, the line for the hydrostatic wet concrete pressure subtends the ordinate because the panel is only poured with concrete up to 0.6 m below NAP. The first reading is

taken at 13:22, just before the pouring process started and is thus equal to the hydrostatic slurry pressure. The first two envelopes of the wet concrete pressure (14:02 and 15:02) follow first the hydrostatic slurry pressure line and they are subsequently parallel to the hydrostatic wet concrete pressure line. This implies that the wet concrete pressure increases in the first phase linearly with the concrete level. At that time the setting behaviour of the concrete has not started yet. The envelope of the data at 16:32 shows a trilinear shape. At first, the envelope follows the hydrostatic slurry pressure line, then it is parallel to the hydrostatic wet concrete pressure line, and lastly again parallel to the hydrostatic slurry pressure line. The pressures at −35 m NAP and −40 m NAP are at that time smaller than at 15:02. Thus, the setting behaviour of the concrete has already started. A higher concrete level does not longer increase the wet concrete pressure in greater depths, rather the setting behaviour reduces the wet concrete pressure.

Regarding the entire pouring process, it is observed that the wet concrete pressure at a given level behaves

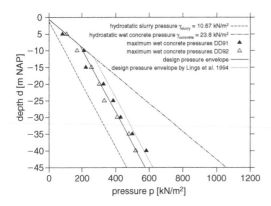

Figure 12. Compilation of the maximum wet concrete pressures versus depth for panel L34.

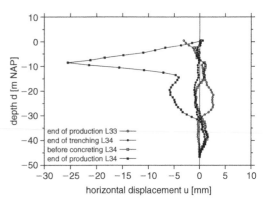

Figure 13. Horizontal displacements versus depth at location 93 perpendicular to panel L34.

at first hydrostatically (approximately one to two hours, see Figures 9 and 10) and then the pressure decreases due to the concrete setting. Thus, the slope of the wet concrete pressure envelope shows for recently poured concrete a hydrostatic performance, that gradually changes back to the hydrostatic slurry pressure line and finally vanishes.

In Figure 12, the maximum pressures in panel L36 (location 91) and panel L36 (location 92) are plotted versus depth. Analogous to Figures 9 and 10 the hydrostatic slurry pressure and the hydrostatic wet concrete pressure are drawn in as reference lines. Moreover, the design pressure envelope is added. This line is defined by the hydrostatic wet concrete pressure up to a critical depth and thereafter by the inclination of the hydrostatic slurry pressure. The chosen design pressure envelope has a similar shape as given by Lings and co-workers (Lings et al. 1994), based on the CIRIA design method (Clear and Harrison 1985). The only difference is the selected critical depth. The recorded critical depth of the diaphragm wall at Station Blijdorp is approximately 8.5 m (see Figure 12). Compared to Lings and co-workers (Lings et al. 1994), who assume a critical depth of less than one third of the trench depth. Which in our case would yield approximately 12 m.

4.2 Displacements

In this section the horizontal and vertical displacements, observed during the installation of panel L34, are presented. The construction of panel L34 begins shortly after the completion of panel L33 and takes 6 days (see Figure 8). The trenching is performed in three phases, beginning with the right stitch, then the left and at last the middle. It takes two days to cut the trench for the whole panel. Pouring the concrete followed 4 days later and is completed within 8 hours.

4.2.1 Horizontal displacements

In Figure 13, the horizontal displacements are illustrated over the depth at location 93 (row b, 1.9 m behind the wall) perpendicular to panel L34. The four curves show the horizontal displacements before and after trenching as well as before and after concreting panel L34. Positive displacements denote a movement into the trench, whereas negative displacements away from it. During the excavation of panel L34, the horizontal displacements point away from the trench in the depth range between 6 and 31 m below NAP. However, above and below the soil moves into the trench. The maximum horizontal displacements during trenching are less than 3.5 mm. Only very small displacement changes are observed for the time period of the slurry supported open trench. They sum up to 1 to 2 mm in the peat and the clay (8 to 15 m below NAP) with soil moving into the trench. After pouring, minor displacements are recorded due to the high wet concrete pressure near the foot (beyond the depth of 30 m below NAP). The maximum horizontal displacement of about 27 mm is recorded in the peat at 9 m below NAP.

As shown in Figure 14, the displacements in row c (location 95, 3.8 m behind the wall) differ from those in row b in magnitude and extent. Amazingly, these horizontal displacements (pointing away from the trench) increase to 4.5 mm, despite of the distance from the trench during the excavation. While the trench is supported only by the slurry, displacements go back more or less. So the displacements before the concreting of the panel L34 are almost similar to those before the excavation started. The displacements in the adjacent ground increases after concreting due to the wet concrete pressure. As a matter of fact, the horizontal displacements decrease with increasing distance from the trench. The maximum horizontal displacement is again located in the soft soil stratum with about 13 mm.

In general, trenching does not cause considerable horizontal displacements. They are all less than

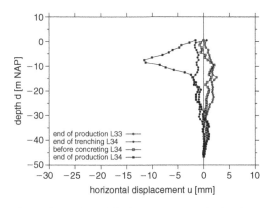

Figure 14. Horizontal displacements versus depth at location 95 perpendicular to panel L34.

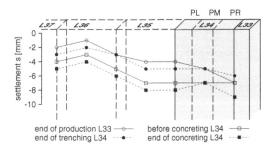

Figure 15. Surface settlements in axis b.

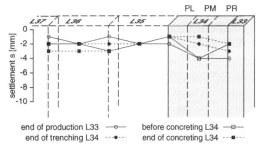

Figure 16. Surface settlements in axis c.

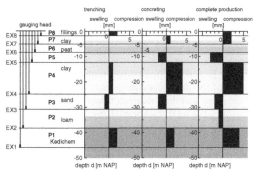

Figure 17. Thickness change of different ground packages at location 93.

3.5 mm. Moreover, the horizontal movements during the time period of the open slurry supported trench are negligible. This is apparently shown by the two almost overlapping curves of "end of trenching L34" and "before concreting L34". The monitoring data taken from the inclinometers at the locations 93 and 95 show clearly that the process of concreting produces the maximum horizontal displacements.

4.2.2 *Surface settlements*

The surface settlements along axis b (1.9 m behind the wall) and axis c (3.8 m behind the wall) are illustrated in Figures 15 and 16, respectively.

Movements at the surface hardly occur along axis b in consequence of trenching. The surface heaves 1 mm opposite to the right edge of panel L34 (PR) while the points opposite to the center and between panel L36 and L35 stay at their initial level. All other points settle 1 mm. The upward movement of the surface opposite to the right edge of panel L34 is reversed during the time period of the open slurry supported trench. All other points settle between 1 mm and 2 mm in that time period. This downward directed movement of the surface is increased further as the trench is concreted. However, the installation of panel L34 causes settlements between 2 mm and 4 mm along axis b. The

maximum settlement in axis b at the end of concreting panel L34 add up to 9 mm.

In axis c the surface points opposite to the right and the left edges of panel L34 heave 1 mm during the excavation of the trench. A heave of 2 mm is observed opposite to the center of the panel L34. Between panel L37 and L36 as well as between panel L36 and L35 a settlement of 1 mm is recorded. During the time period of the open slurry supported trench 1 mm settlement is logged between panel L36 and L35. Moreover, the surface point opposite to the center of panel L34 settles 2 mm and the surface point opposite to the right edge of panel L34 heaves 1 mm. A heave of 3 mm is observed in axis c in opposite to the center of panel L34 during concreting. Between panel L36 and L37 and in the middle of panel L36, the surface settles about 1 mm. However, along axis c there are only small settlement changes up to 3 mm caused by the installation of panel L34. The maximum settlement in axis c at the end of concreting panel L34 add up to 4 mm.

4.2.3 *Subsurface deformations*

Figures 17 and 18 show relative vertical displacements caused by the installation of panel L34, between two extensometer reading points. These points denote the soil packages (P1 to P8) for which thickness

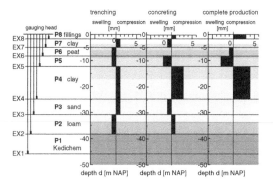

Figure 18. Thickness change of different ground packages at location 95.

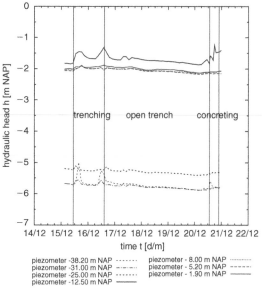

Figure 19. Piezometer level during construction of panel L34 for different depths at location 93.

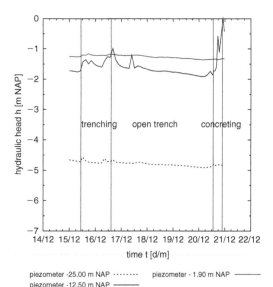

Figure 20. Piezometer level during construction of panel L34 for different depths at location 95.

changes are recorded. Compression is defined positive, whereas swelling is set to be negative. In both, Figures 17 and 18, the first two plots show the distribution of the thickness change during the excavation of the panel and the subsequent pouring. The last plot illustrates the total thickness change recorded during the whole construction process of panel L34.

In Figure 17, the thickness change of the corresponding ground packages at a distance of 1.9 m behind the wall are presented. The process of trenching causes a compression of 1 mm in the clay-sand package (P4). During the subsequent concreting the compression is 4 mm. For the total construction process the maximum compression of 5 mm is recorded in the ground package P4. Furthermore, the upper two packages P7 and P8 as well as the layers of the "Formatie van Kedichem" (package P1) are compressed during the construction process. No deformation is identified in ground package P6. In the packages P2, P3 and P5 swelling is recorded.

At a distance of 3.8 m behind the diaphragm wall (see Figure 18), similar behaviour for the ground packages P4 to P8 is observed. The maximum compression is again recorded in package P4 with 4 mm for the whole construction process. The recorded deformations for the lower packages (P1 to P3) are negligible.

4.3 Pore pressure

In Figures 19 and 20, the development of the pore pressures (plotted as hydraulic head) at locations 93 and 95, respectively, is given in different depths for the construction of panel L34. One can easily recognize two different groundwater aquifers in both plots. The phreatic aquifer corresponds to the higher hydraulic head. In the sand layer down from -16 m NAP and the deeper layers a second aquifer with lower piezometric level is present.

Significant changes are only observed for the piezometers located in the clay layer at a depth of 12.5 m NAP. And even these are only within a range of 5 to 7 kN/m^2 during trenching. Concreting panel L34 surprisingly gives at location 95 the highest pore pressure increase of about 18 kN/m^2. Smaller changes are recorded during trenching and concreting in the piezometer located at a depth of 38.2 m NAP in axis b.

90

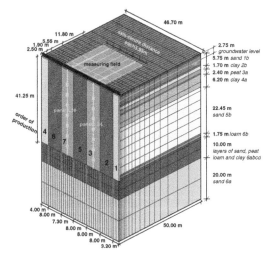

Figure 21. Part of the finite element model.

All other piezometers show more or less a constant value.

5 NUMERICAL MODELLING

The diaphragm wall installation is simulated with a 3D finite element model using the Mohr-Coulomb model (see next section). Aiming to reproduce the construction process of the wall for the station Blijdorp. The numerical values obtained will be compared with the recorded data.

5.1 Finite element model

The FE-program TOCHNOG is used for the numerical simulation. The 3D model consists of 21000 hexahedron elements (23472 nodes) with a linear interpolation function. It is 62.0 m wide, about 71.0 m deep, and 46.7 m long. The width is subdivided into three sections. 11.4 m are located in front of the diaphragm wall representing half of the excavation pit. A symmetry condition is assumed in the center of the excavation. The modelled diaphragm wall is 1.2 m wide. Behind the wall the mesh extends another 49.4 m. Figure 21 shows a section through the center of the diaphragm wall with the monitoring field behind it.

Horizontal rollers are located at the vertical edges of the model. At the bottom vertical rollers are applied. The groundwater level is assumed for all layers at −2.5 m NAP (−2.75 m below surface level). The second water level is not modelled for simplicity reasons. Moreover, drained soil behaviour is assumed. This assumption is also done for the soft layers (clay 2b, peat 3a and clay 4a), even though, they are generally calculated undrained in that case. But this is done for

Table 1. Numerical simulation stages.

Step	Description
1	generation of in situ stress state
2	guide wall installation
3	stepwise excavation of panel A32, including activation of slurry pressure
4	stepwise concreting of panel A32
5	stepwise excavation of panel L33 (left stitch), including activation of slurry pressure
6	stepwise excavation of panel L33 (right stitch), including activation of slurry pressure
7	stepwise excavation of panel L33 (middle stitch), including activation of slurry pressure
8	stepwise concreting of panel L33
9	stepwise excavation of panel L34(left stitch), including activation of slurry pressure
10	stepwise excavation of panel L34(right stitch), including activation of slurry pressure
11	stepwise excavation of panel L34(middle stitch), including activation of slurry pressure
12	stepwise concreting of panel L34

simplification and considering, that almost no pore pressure changes are recorded (see Figures 19 and 20).

5.2 Simulation of installation

After the generation of the in situ stress state, the installation of the guide wall is modelled. Thereafter, the sequential construction steps of panels A32, L33 and L34 are considered. The panels are excavated in three stages (left, right and middle stitch) except for panel L32 (one stitch). Each process of an excavation stitch is simulated with 10 segments. The same segments are used to model the concreting of the trenches. The slurry pressure is modelled as a surface load and applied when the segments are removed. The pouring of the concrete is simulated by increasing the surface load, starting from the bottom of the trench. The critical height for the hydrostatic wet concrete pressure is assumed to 8.5 m (see section 4). After the concrete has reached that level, the wet concrete pressure is modelled bilinearly (first part hydrostatic wet concrete pressure, second part slope like hydrostatic slurry pressure). The final shape of the wet concrete pressure for the filled trench is given in Figure 12. An overview on the numerical simulation stages is given in Table 1.

6 CONSTITUTIVE LAW AND MATERIAL PARAMETERS

In this section, the used constitutive law is explained. All soil layers are modelled with an elastic-ideal-plastic constitutive model with a Mohr-Coulomb

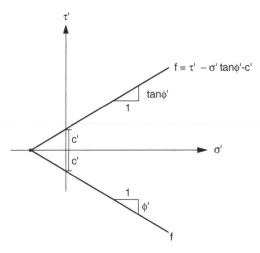

Figure 22. Mohr-Coulomb yield condition in a σ'-τ-diagram.

yield criterion. The Mohr-Coulomb model is briefly introduced and the material parameters for all soil layers are given.

6.1 Mohr-Coulomb model

The Mohr-Coulomb model is a linear-elastic ideal-plastic constitutive law. The total strain is divided into an elastic and a plastic part. The mechanical soil behaviour is described with Hooke's linear elasticity up to the yield stress f. The yield stress f is determined by Mohr-Coulomb's criterion (see Figure 22), using the effective cohesion c' and the effective friction angle φ'.

$$f = \tau - c' - \sigma' \cdot tan\varphi' \qquad (1)$$

The soil behaviour is linear-elastic below the yield surface, defining the yield condition. Stress states upon the yield surface generate ideal-plastic deformations. Stress states above the yield surface are not possible. A plot of the Mohr-Coulomb yield condition in a σ'-τ-diagram is given in Figure 22. Compression is defined positive.

The input parameters of the Mohr-Coulomb model are the effective shear parameters c' and φ' as well as the dilatancy angle ψ for the description of the plastic soil behaviour. The use of a dilatancy angle implies a non-associated flow rule for the plastic strains. The Young modulus E and the Poisson ratio ν are necessary to describe the elastic soil behaviour.

Figure 22 shows, that soils with cohesion are theoretically able to sustain tensile stresses. However, generally soils do not bear any tensile stress. Therefore, a so-called "tension cut-off" criterion is used to

Table 2. Material parameters for the Mohr-Coulomb model.

Layer	γ_r [kN/m³]	E [kN/m²]	φ' [°]	c' [kN/m²]
sand1b	19.5	60000	29.0	1.0
clay2b	15.8	9000	15.0	10.0
peat3a	10.5	6000	13.5	10.0
clay4a	15.0	12000	15.0	10.0
sand5b	20.0	80000	29.0	0.0
loam6b	21.5	18000	24.5	4.0
layers6abcd	21.5	20000	13.0	3.5
sand6a	20.0	120000	29.0	0.0

chop off the tensile area of the Mohr-Coulomb model in the calculation.

6.2 Material parameters

Table 2 shows the used material parameters of the different layers for the Mohr-Coulomb model. The Poisson ratio is assumed to be 0.33 for all layers. The dilatancy angle ψ is $0.0°$ for all soil types. The specific soil weight γ_r and the strength parameters φ' and c' are selected according to given subsoil evaluations. However, the Young's modulus E (un-/reloading) is based on the subsoil evaluations and experience.

7 DISCUSSION OF DATA AND CALCULATION RESULTS

The present calculation is the first modelling of the detailed construction process of the panels A32, L33 and L34 at Station Blijdorp. Therefore, the aim is not a perfect agreement with the monitoring data, but checking the possibility of simulating the sequential construction process with simple soil models. It should be mentioned, that there is nearly no plastic behaviour in the soft layers. The results from the calculation are illustrated in this section.

7.1 Horizontal displacements

Figures 23 and 24 show a comparison between the calculated and recorded horizontal displacements at location 93 and 95 during trenching, respectively. The trenching of panel A32 and L33 causes movements away from the trench in the calculation. The main displacements of about 4 mm occur in the soft soil layers (clay 2b, peat 3a and clay 4a). The recorded data show displacements with no major pointing direction. However, the metering precision of the 46 m long inclinometers are in a range of a few millimetres (see e.g. (Marte et al. 1998)). Only small displacements changes are recorded after the completion of trenching panel L34. These movements point mainly away

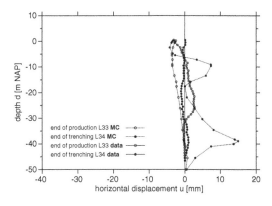

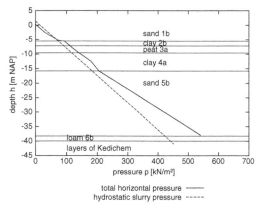

Figure 23. Comparison between inclinometer data and calculation at location 93 during trenching.

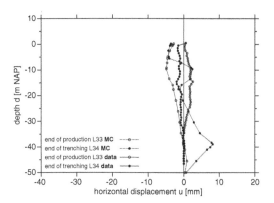

Figure 25. Comparison between total horizontal and the hydrostatic slurry pressure versus depth.

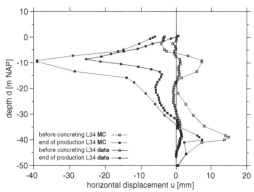

Figure 24. Comparison between inclinometer data and calculation at location 95 during trenching.

Figure 26. Comparison between inclinometer data and calculation at location 93 during concreting.

from the trench. This behaviour cannot be modelled so far. It seems, the stiffness for the loam 6a is underestimated, because of the big calculated movements (up to 16 mm at location 93 and 9 mm at location 95) into the trench. Moreover, in the soft soil layers smaller movements into the trench are calculated. Whereas, in the fill displacements into the trench are simulated. The calculated displacements can be explained with the help of Figure 25. In there, the total horizontal and the hydrostatic slurry pressure used in the calculation are plotted versus depth. The resulting horizontal pressure down to a depth of about 5 m is pointing outward of the trench. Below this depth the resulting pressure is pointing inward. For this reason, the general behaviour of the calculation result is explainable. May be the small differences between the total horizontal pressure and the hydrostatic slurry pressure are not big enough to cause big deformations in reality.

Moreover, the total horizontal pressure, at least in the sand 5b, is in reality lower, because of the lower ground water level in the second aquifer (see Figures

19 and 20). Thus, the resulting horizontal pressure is in reality smaller than in the calculation.

The small displacements changes recorded during the open trench period (see Figures 13 and 14) can not be modelled so far, because no time-dependent constitutive law is used and drained soil behaviour is assumed. Thus, the calculated displacements after trenching the panel L34 are the same as those just before the concreting of panel L34 starts.

Figures 26 and 27 show a comparison between the calculated and recorded horizontal displacements at location 93 and 95 during concreting, respectively. The effect of concreting on the horizontal displacements is qualitatively the same in the calculated as well as the recorded results. In both cases the wet concrete pressure causes movements away from the trench. Moreover, the maximum horizontal displacements occur in the soft soil layers for both, the recorded and the calculated data. But the calculated displacements are bigger then the recorded. Thus, the soft soil

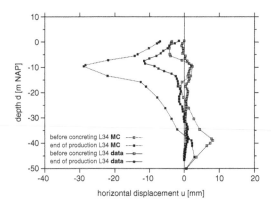

Figure 27. Comparison between inclinometer data and calculation at location 95 during concreting.

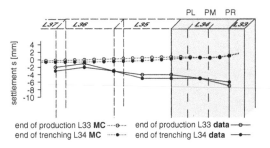

Figure 28. Comparison between settlement data and calculation in axis *b* during trenching.

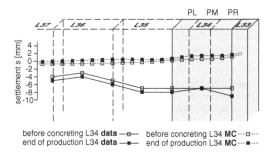

Figure 29. Comparison between settlement data and calculation in axis *b* during concreting.

layers are modelled like the loam 6a probably to soft or the strength parameters are to low. As expected the calculated and the recorded displacements at location 95 are smaller than at location 93.

7.2 Surface settlements

In the following Figures 28 to 31, the settlements on the surface are presented. Comparisons are made for the settlements in axis *b* and axis *c* during trenching and during concreting. The calculated settlement

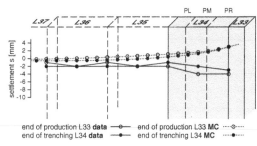

Figure 30. Comparison between settlement data and calculation in axis *c* during trenching.

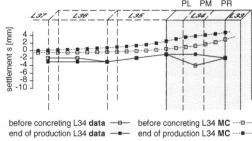

Figure 31. Comparison between settlement data and calculation in axis *c* during concreting.

trough before trenching in both axis differs from the recorded. The main difference is a calculated heave in front of panel L34, whereas settlements are recorded. An explanation for the difference may be the use of the Mohr-Coulomb model, that cover the soil behaviour not correctly. Thus, further investigation is needed here. During trenching, small settlements of 1 mm are calculated. Because no time-dependent constitutive law is used and drained soil behaviour is assumed no settlements are simulated during the open trench period. The records show settlements for this time period (see Figures 15 and 16). In the calculation results, the heave increases due to concreting, especially in front of panel L34. This effect is due to the wet concrete pressure acting on the soil. However, the recorded data show no clear direction. In axis *b* small settlements of 1 mm occur, where in axis *c* a heave is recorded in front of panel L34.

8 CONCLUSIONS

The data recorded during the construction of diaphragm wall panels in Rotterdam show only small horizontal displacements during trenching. This is probably due to the fact, that the total horizontal pressure is not much bigger than the slurry pressure (see Figure 25). Moreover, only half of the panels are located in the soft soil layers and the other half in the stiff sand layer. In case of concreting the panels, horizontal displacements pointing out of the trench mainly occur in the soft soil layers. This behaviour is caused by the wet concrete pressure. The recorded settlements show a downward direction of a few millimetres caused by the panel installation.

The results obtained from the numerical simulation with the Mohr-Coulomb model qualitatively cover the horizontal displacements during concreting. Thus, the chosen approach of modelling the fresh concrete pressure, based on the obtained records, is acceptable. But regarding the too big displacements during concreting and the mismatching of the movements during trenching the material parameters should be adjusted in a next calculation. Moreover, the Mohr-Coulomb model is not able to simulate the recorded settlements at least qualitatively. Thus, in further simulations higher constitutive laws have to be used. In addition it is planned to simulate the installation of more panels.

ACKNOWLEDGEMENT

The research is supported by the German Ministry of Education and Research grant 19W2086B. The authors are responsible for the content.

REFERENCES

Clear, C. A. and T. A. Harrison (1985). Concrete pressure on formwork. Technical Report 108, Construction Industry Research and Information Association.

Lings, M. L., C. W. W. Ng, and D. F. T. Nash (1994). The lateral pressure of wet concrete in diaphragm wall panels cast under bentonite. *Proceedings of Institution of Civil Engineers, Geotechnical Engineering 107*, 163–172.

Marte, R., S. Semprich, M. Fritz, and W. Weber (1998). Messungenauigkeiten von Inklinometer-Messergebnissen. In *Messen in der Geotechnik 1998, Braunschweig*, Mitteilungen des Institutes für Grundbau und Bodenmechanik, TU Braunschweig, Heft-Nr. 55, pp. 327–350.

van Zanten, D. C., M. de Vries, and H. M. A. Pachen (2004). Door de rotterdamse ondergrond met twee boortunnels. *Geotechniek 8*(3), 68–75.

Numerical Modelling of Construction Processes in Geotechnical Engineering for
Urban Environment – Triantafyllidis (ed)
© 2006 Taylor & Francis Group, London, ISBN 0 415 39748 0

Small scale model tests on fresh concrete pressure in diaphragm wall panels

C. Loreck, D. König & Th. Triantafyllidis
Institute for Foundation Engineering and Soil Mechanics, Ruhr University Bochum, Germany

ABSTRACT: To clarify the development of the fresh concrete pressure within diaphragm wall concreting, model scale tests were performed in 1 g and 30 g condition. As a model-concrete a 2 mm limestone based grout was used, that acts equivalent to the original 32 mm cement-based concrete without the implication of setting. This enables to study whether geometric effects, chemical-induced changes of the fresh concrete rheology or the setting are responsible for the pressure distribution that is experienced on site. Results show, that the flowable concrete acts as a heavy fluid that shows no inner friction and that reductions of the concrete flowability lead to a decrease of pressures on the open trench walls.

1 INTRODUCTION

The consideration of the entire construction process of the wall in a 3-D FE-model of a pit wall excavation leads to better prediction of the deflections. In contrary to the ordinary 2-D simulation starting with K_0-stress state ("wished-in-place" wip-simulation) a 3-D simulation offers this advantage because all relevant occurring changes in major stress states can be taken into account before the pit excavation starts ("wall-installation-modeled" wim-simulation).

Schäfer (2004) showed that for a diaphragm wall construction in soft soils the stress distribution versus depth of the fresh concrete pressure has a non-negligible effect on the forces of the upper wall struts and therefore on the horizontal wall deflections (Figure 1).

The reason for this effect is that the fresh concrete pressure at the upper third depth of the trench exceeds the K_0-stress state and – due to the reason of horizontal equilibrium – leads to an initial stress state prior to the wall-excavation that is on the passive side of the earth pressure. This stress state works as a pre-stressing of the wall, that has to be carried either by the upper supports or needs an increased deflection to be de-stressed to an active stress state.

For his simulation, Schäfer (2004) used the bi-linear stress distribution of the fresh concrete pressure proposed by Lings (1994) (Figure 2).

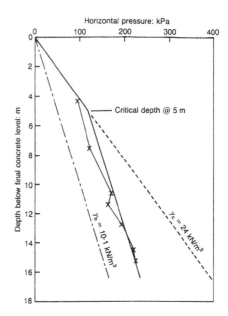

Figure 2. Bi-linear empiric distribution of fresh concrete pressure according to Lings (1994).

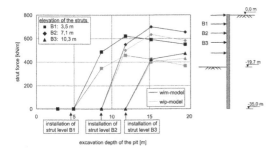

Figure 1. Development of strut forces with (wim) and without (wip) modelling of the construction process according to Schäfer (2004).

Lings analyzed different site surveys and concluded the stated empiric distribution that shows a fresh concrete-weight hydrostatic pressure for the upper third of the trench and a water-weight hydrostatic pressure below that "critical depth".

2 ACTUAL INVESTIGATION

2.1 *Motivation*

As a continuation of Schäfer (2004) and with respect to actual questions related to the serviceability of nearby structures, the Institute of foundation engineering and soil mechanics of the Ruhr-University Bochum, Germany, was assigned by the German Federal Ministry of Education and Research to undertake further investigations of the development of the fresh concrete pressure in diaphragm wall panels.

To eliminate site-related effects and to perform a precise analysis of relevant parameters it was decided to carry out model-size tests. In order to prevent scale-related shortcomings by a – possible – inappropriate scaling of geometric size and material strength the tests were performed in the geotechnical centrifuge located at the Institute. This procedure allows for an original "strength" of the fresh concrete that simplifies the identification of the model-concrete properties (Jessberger and Güttler 1988).

2.2 *Fundamentals of fresh concrete pressure*

The fresh concrete pressure is of main interest for the formwork in high-rise constructions and many investigations and design rules exist already. Out of these, the known main parameters for the fresh concrete pressure are:

- Concreting speed in relation to hydration progress and consistency
- Type of grain (rounded, angular)
- Stiffness of formwork

The first mentioned aspect covers a wide range of site- and material-related parameters that prevent a close comparison of pressure measurements between different sites and even different structural parts of one site. These parameters are:

- all aspects of concrete recipe and used materials as type of cement, water-cement ratio, mixing conditions, chemical additives, etc.
- environmental conditions, especially ambient temperature and temperature of concrete.
- detailed schedule of concreting (pouring-speed and interruptions) and concrete-age at pouring time

The specific geometry (slenderness) and concreting conditions (tremie-method/under-water concreting, duration of concreting) of the trench differs significantly from the ordinary concreting of the superstructure. With regard to the known pressure distribution (Figure 2), that shows no participation of the gravel weight to the total pressure recordings below the critical depth, the following parameters were also considered:

- consolidation of the grain structure
- arching- and/or silo-effects of the grain structure

The latter might occur by initial consolidation (before the setting starts) and/or may change with the hydration state of the concrete.

2.3 *Concept for the model-size tests*

The more-or-less uncontrollable (after initialization) progress of the hydration would need an exact timing of the concreting procedure in all laboratory tests to produce redundant and comparable results. In preliminary tests this was tested to be hardly achievable (e.g. the ambient temperature). Therefore it was decided to use a non-hardening concrete with a constant rheology.

Moreover this enables to separate the parameters in question in two major categories: change of rheology by time-dependent parameters (loss of flowability, setting) and mechanical effects by geometry-dependent parameters (height of fresh concrete column/slenderness).

The model scale was selected to be 1:20. Reasons for this scale were the maximum model-height in the centrifuge (about 1,50 m) and the width of the model trench in interaction with the maximum aggregate grain size of the concrete and model tremie pipe diameter. With a model panel size of 940 × 450 × 66 mm (depth × length × width) and an acceleration of 20 times the earths' gravity the model-panel resembles a trench size of about 19 × 9 × 1,30 m.

A main aspect of the test preparation was the design of an adequate model-size concrete that fulfils the following requirements:

- Comparable (equal) rheology to the original size concrete (workability, flowability)
- Maximum grain size of 2 mm (about 32 mm real-size diameter scaled down by about 1:20)
- Constant rheology (non-setting)
- Stable against demixing during centrifugal acceleration

3 MODEL-SIZE CONCRETE

3.1 *Reference real-size concrete*

To develop a model-size concrete a reference real-size concrete was needed. This was set up as a standard mixture following the minimum requirements of the European design code EN 1538/EN 206:

- Minimum cement-content: 350 kg (32 mm max grain size)

- Maximum allowable PFA: 30 kg
- Maximum water-cement ratio: 0,55
- Flowtable test: 55 ≤ a ≤ 60 cm
- Slumb: 180–210 mm

Using a local standard CEM II/A-LL 32.5 R the mixture was tested to a flowtable spreading of 55 to 57 cm (Figure 4). To achieve the required flowability without increasing the cement-paste volume or w/c-ratio, admixtures are used in reality. As they have a distinct time-dependent effect on the rheology they are not used here. According to Krell (1985), an increase of the effective water-cement ratio without changing the overall cement-paste volume shows an equivalent effect (see chapter 3.2) and therefore this approach was used instead of an admixture in the experiments presented here.

3.2 Reduction of aggregate grain size

The maximum grain size of the model concrete aggregate was reduced to 2 mm, so that the model-concrete is – practically spoken – a "grout". The grain size distribution was derived by the modified *Fuller-distribution* (Wesche 1993) with a constant Fuller-Exponent $a = 0,4$ for both of the 32 mm real-size and 2 mm model-size concrete.

Due to the increased accumulated aggregate surface and void volume of the granular structure the water consumption increases significantly. To maintain the original rheology of the real-size concrete, the revised water content was derived by the theory of Krell (1985).

The principle of this theory is shown in Figure 3. The flowability (measured by flowtable testing that depends mainly on the flow limit, Bonzel & Krell 1984) depends on the effective water-cement ratio and the so-called "equivalent cement-paste thickness" μ_{eq}. Changes in concrete flowability (spreading width a, Figure 3) may be induced by changes of the effective

w/c-ratio (1) or the equivalent cement-paste thickness μ_{eq} (2) or both of them (3).

μ_{eq} is defined as the theoretical thickness of the cement-paste, that is not used to fill the void volume of the granular structure, related to a characteristic "equivalent" aggregate grain size D_{eq}:

$$\mu_{eq} = \mu/D_{eq}$$
$$\mu = V_{p,ef}/A_G$$
$$D_{eq} = \sqrt{\frac{A_G}{N_G \cdot \pi}}$$

$V_{p,ef} = V_p - V_v$ = effective cement paste volume
V_p = total cement-paste volume
V_v = void volume of the dense granular structure
A_G = total surface area of the grain size distribution [m² per m³ of concrete]
N_G = total quantity of grains [nos per m³ of concrete]

The adequate water content for the model size concrete was derived by keeping the μ_{eq} and effective water-cement ratio constant for the reduced grain size (which results to a change of D_{eq}). To verify this design approach, a variety of aggregate distributions (all with different maximum grain size and thereby D_{eq}) were tested. The results of flowtable tests are shown in Figure 4, demonstrating that the model-size-aggregate concrete meet the flowtable properties of the reference real-size 32 mm concrete. With these results, the model concrete proves to be equivalent to the reference concrete by the theoretical derivation according to Krell and for practical testing.

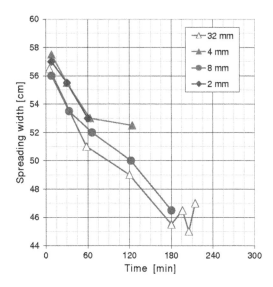

Figure 4. Flowtable spreading width for reference (32 mm) and model-size concrete (2–8 mm) with $\mu_{eq} =$ const. and eff.-w/c = const.

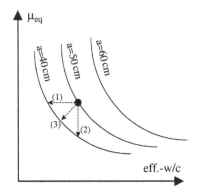

Figure 3. Dependence of flowtable test results (spreading width a) according to Krell (1985).

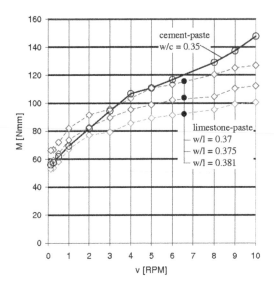

Figure 5. Rheological profiles of cement and limestone pastes.

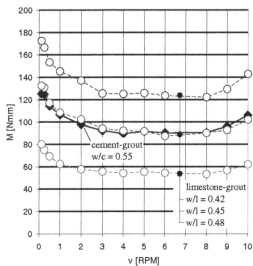

Figure 6. Rheological profiles of cement and limestone grouts.

3.3 Rheology requirements

As Figure 4 shows, the spreading width as measured in the flowtable test changes rapidly with time, due to initial reactions of the cement and the later setting. The initial flow limit of a cement paste, that prevents the aggregate grains from demixing, comes from a floc-structure of the cement particles due to different electrostatic surface charges (Spanka et al. 1995 and Keck 1999).

In order to obtain a constant flowability, the cement was replaced by a comparable particle sized inert limestone powder ($CaCO_3$). This provides a similar floc structure that is able to keep the aggregate grains in the grout mixture without separation.

Figure 5 shows rheological flow-profiles (resisting moment of a paddle that is vertically inserted in the paste-/grout-sample versus rotation velocity measured in a Schleibinger "Viskomat NT" testing device) of the original cement paste (using effective w/c-ratio = 0.35) and limestone pastes with different effective w/l-ratios. Figure 6 shows the corresponding flow-profiles for the original 2 mm cement-grout and lime-grouts with different w/l-values but constant equivalent paste thickness μ_{eq}. Both pictures show, that a limestone-paste as well as a limestone grout can be designed with equivalent properties as cement-based mixtures.

An interesting detail of Figure 6, that will be discussed later, is the increase in flow resistance with decreasing shearing speed, that stands in contrary to the behavior of the pastes. This effect must be assigned to the interaction of the aggregates in the grout.

The overall result of the model-size concrete development is an inert limestone-based 2 mm model-size concrete, that behaves similar to the model size cement-based concrete, but does have a constant flowability versus time. A visual comparison of the original 32 mm concrete to the model-size concretes is shown in Figure 7a to 7c.

3.4 Fresh concrete-stability

In normal case centrifugal acceleration of suspensions is used to separate their constituents. Exactly this has to be avoided for the testing of fresh concrete under increased gravity (ng).

Under normal gravity (1 g) the flow limit of the cement paste prevents the aggregate grains from demixing. For the model-size 2 mm concrete, the 2 mm grain has at 20-g a similar specific surface as the 32 mm grain at 1 g. Due to this, a paste with 1 g flow limit should be able to carry a 2 mm grain at 20 g as well as a 32 mm grain at 1 g. Therefore, the stabilization of the paste with an unchanged paste flow limit was the target to achieve a centrifuge- stable concrete mixture.

Referring to Stoke's law for sedimentation, the viscosity of the water and the specific surface of the particles are the parameters that could be used to design a centrifuge-stable paste.

A stringent application of Stoke's law proved to fail, what might be explained with the volumetric ratio of fluid and solid and/or the floc-structure of the cement and limestone particles.

By variation of a viscosity-heightening stabilizer (standard concrete admixture) in preliminary tests, the

Figure 7. (a) Reference 32 mm concrete (a = 54/56 cm); (b) Cement-based 2 mm concrete (a = 55/56 cm); (c) Limestone-based 2 mm concrete (a = 53/54 cm).

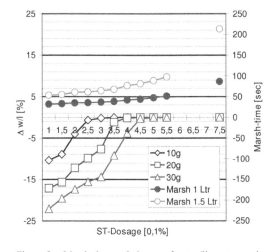

Figure 8. Marsh-time and change of water/limestone ratio as function of stabilizer-dosage after subjection to increased g-level for 20 minutes (initial w/l-ratio = 0.48 = 100%).

minimum increase of viscosity was determined that proved to provide a stable model concrete (Figure 8). As a result, the required increase in viscosity was significantly lower (0,45 weight-% for 20 g) than a calculation by Stoke's law would deliver (0,75% stabilizer would be required to achieve a theoretically required viscosity for the model scale of 1:20).

4 MODEL TESTS

4.1 *Objective*

As a first step, the basic concrete behavior under centrifugal acceleration was tested. The question was if the detected frictional effects of the granular structure in the grout (Figure 6) leads to any kind of arching or silo-effect in the trench.

For these tests, the concreting under centrifugal acceleration was skipped and the concreted model-panel was put in centrifuge after tremie-concreting at rest.

4.2 *Setup*

A sketch of the testing device and a photo are given in Figure 9.

The model panel was built as a stiff steel/acryl box with dimensions of 940 × 450 × 66 mm (depth × length × width).

The vertical stress distribution was measured by 4 horizontal oriented pressure cells for total pressures, 1 vertical oriented pressure cell for the total pressure at the bottom of the trench and 3 piezometers measuring porewater pressure at different depths of the trench.

In order to obtain a direct measurement of the expected aggregate friction in the grout a vertical "anchor" was placed in the fresh grout (before concreting), that was vertically pulled out (load-controlled) after finishing the concreting.

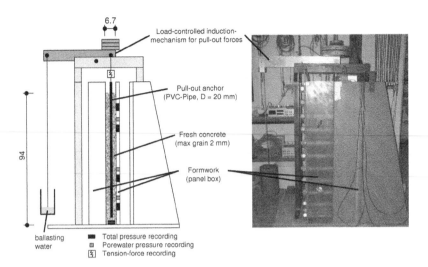

Figure 9. Test-setup.

4.3 Execution of experiments

The behavior of the limestone model-size concrete was tested in 1-, 10-, 20- and 30-g acceleration levels. The model-concrete for the 1-g condition used conventional unstabilized water. Concrete mixtures for 10- to 30-g conditions used stabilized paste as described before.

Concreting was realized by use of a model-size tremie pipe without and with water-surcharge and took about 45 minutes to complete.

In two tests on 20 g level less stabilizer was used than required for a centrifuge-stable mixture. That lead to partly de-mixing during centrifugal acceleration.

The time between finishing the concreting and before pulling the anchor was kept for a minimum of 30 minutes (for 10- to 30-g level tests subjected to centrifugal acceleration).

4.4 Results

4.4.1 Tremie-concreting

All pressure cells recorded hydrostatic conditions (with the concrete specific weight of 22 kN/m³) at all stages of concreting no matter whether water-surcharge was used or not (see Figures 10 & 11).

4.4.2 Resting time

Concrete mixtures which did not demix, showed no changes in total- and pore water pressure readings during resting time, even under ng conditions (see Figure 12).

Concrete mixtures that partly demixed showed a parallel decrease of total- and pore water pressure readings (Figure 13) that slowed down versus time and seemed to meet a final stable state before the complete separation of their constituents was reached.

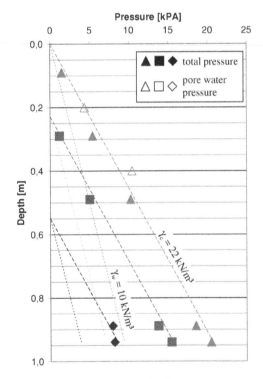

Figure 10. Vertical stress distribution during concreting without water-surcharge (three stages).

4.4.3 Bleeding during resting time

All mixtures showed a bleeding (of water) even under 1 g as well as under ng. The separated water was weighted and the final w/l-ratio backcalculated.

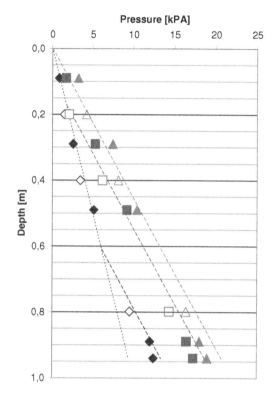

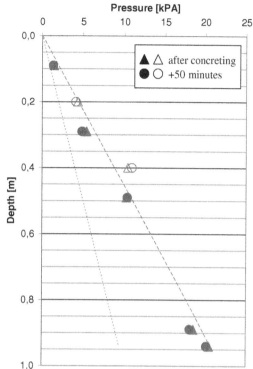

Figure 11. Vertical stress distribution during concreting with water surcharge (under-water concreting); three stages.

Figure 12. Vertical stress distribution of a non-demixing concrete during resting time at 1 g. Tests at ng showed similar distributions (with the stress levels corresponding to g-level).

For the non-demixing concrete mixtures, the separated water weight ranged from 30 to 83 g (initial water content of the concreted panel-volume 7051 g) that changed the initial w/l-ratio of 0.48 by less than 0.005 (final-w/l > 0.475).

For the demixing concrete the final-w/l-ratio achieved values between 0.45 and 0.39 (20 g-level).

4.4.4 Anchor pull-out forces

Concrete mixtures that did not demix showed pull-out forces close to the self-weight of the anchor (Table 1). An exact backcalculation is hardly achievable, because the test setup was nearby its measurement accuracy.

In general, the measured values are significantly far away from any conservative assumption regarding a minimum frictional resistance, e.g. (for 1 g)

$$R_S = A_S \cdot \frac{1}{2} \cdot \gamma \cdot L^2 \cdot K_a \cdot \mu = 26\,N$$

$A_S = 565\,cm^2$ (L × Ø = 900 × 20 mm)
$\gamma' = 8{,}5\,kN/m^3$
$K_a = 0{,}33$ ($\varphi = 30°$)
$\mu = 0{,}40$

On the other hand the concrete mixtures, partly demixed during the experiment, showed pull-out forces that are significantly far away from the measurement accuracy (Table 2).

5 DISCUSSION

For the non-demixing concrete mixtures, the vertical pressure distributions showed to be hydrostatic with the concrete-weight and stayed unchanged in the 1 g as well as in the 10-, 20- or 30-g tests. The pressure recordings correspond to the anchor pull-out forces that were nearby zero, that also verifies the hydrostatic state without the presence of any inner friction.

The increasing shear resistance for decreasing shearing speeds, which was detected with the rheologic profile (Figure 6), seems to drop when the grout is complete at rest.

Out of these observations, it can be said that a flowable concrete, as used for diaphragm walls, acts as a heavy fluid, no matter what column height or gravitation acts. No arching- or silo-effects can be detected.

For the partly demixed concrete mixtures a rapid decrease from hydrostatic concrete-weight distribution to hydrostatic water-weight distribution takes place. Even the bottom pressure cell, that measures the vertical pressure and was supposed to bear the total weight of the grout-column,

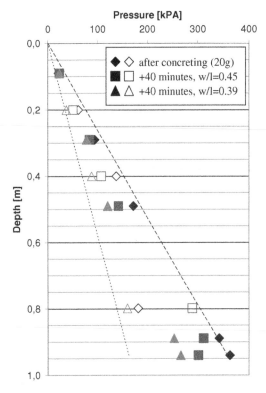

Pressure [kPA]

◆ ◇ after concreting (20g)
■ □ +40 minutes, w/l=0.45
▲ △ +40 minutes, w/l=0.39

Figure 13. Vertical stress distribution of the partly-demixing concrete-mixtures during resting time at 20 g (malfunction of lowest pore water measurement).

Table 1. Pull-out forces of the vertical anchor in non-demixing concrete mixtures.

	g-level			
	1	10	20	30
Self-weight of anchor [N]	9,2	86	169	249
Pull-out forces assigned to friction [N]	3	~0	~0	~0

Table 2. Pull-out forces of the vertical anchor in demixing concrete mixtures.

	end-w/l-ratio (20 g-level)	
	0.45	0.39
Pull-out forces assigned to friction [N]	150	385

recorded reduced hydrostatic pressures also. Similarly the pull-out forces of the anchor increased significantly and verified the action of effective pressures.

As can be backcalculated, the resisting paste volume (reduced by the separated water from bleeding) still exceeded the void volume of the granular structure (in dense state), what means that a reduced paste-thickness is statistically still maintained on the grain's surface.

As the effective w/l-rate is also affected by the partly dewatering of the mixture, a change in concrete-consistency according to path (3) shown in Figure 3 took place, which is – practically spoken – equivalent to a stiffening of the concrete as usually occurs by loss of superplasticizer effects or setting of the cement. Within this context, comparable results using path (1) and (2) as shown in Figure 3 are needed to obtain a complete view of the interrelations.

6 CONCLUSION

At this state of the investigation, the laboratory tests reveal a plausible concept to explain the site experiences: The flowable fresh concrete acts as a heavy fluid after pouring. With a change in the consistency of the fresh concrete – before the later setting starts – an immediate reduction of horizontal pressures take place. The "critical depth" suggested by Lings (1994) must therefore be related to an interaction of concreting speed and loss of flowability of the specific concrete mixture and not to a geometric constant of one third of the trench's depth.

REFERENCES

R. Schäfer. *Einfluß der Herstellungsmethode auf das Verformungsverhalten von Schlitzwänden in weichen bindigen Böden*. Schriftenreihe des Institutes für Grundbau und Bodenmechanik der Ruhr-Universität Bochum, Heft 36, 2004

M. L. Lings, C. W. W. Ng, D. F. T. Nash. *The lateral pressure of wet concrete in diaphragm wall panels cast under bentonite*. Proc. of Institution of Civil Engineers, Geotechnical Engineering, (107): 163–172, July 1994

H. L. Jessberger, U. Güttler. *Geotechnische Großzentrifuge Bochum – Modellversuche im erhöhten Schwerefeld*. Geotechnik (11): 85–98, 1988

J. Krell. *Die Konsistenz von Zementleim, Mörtel und Beton und ihre zeitliche Veränderung*. Schriftenreihe der Zementindustrie, Heft 46/1985

K. Wesche. Baustoffe für tragende Bauteile, Band 2: Beton und Mauerwerk, P. 132, Bauverlag 3. Auflage 1993

J. Bonzel, J. Krell. *Konsistenzprüfung von Frischbeton*. beton 2/1984, P. 61–66 and beton 3/1984, P. 101–104

H.-J. Keck. *Untersuchung des Fließverhaltens von Zementleim anhand rheologischer Messungen*. Mitteilung aus dem Institute für Bauphysik und Materialwissenschaft der Universität Essen. Heft 5 1999

Spanka et al. *Operative mechanism of plasticizing concrete admixtures*. beton 12: P. 876–881, 1995

Numerical Modelling of Construction Processes in Geotechnical Engineering for Urban Environment – Triantafyllidis (ed)
© 2006 Taylor & Francis Group, London, ISBN 0 415 39748 0

Metro line U5 Berlin – Deformation of diaphragm walls

L. Speier & S. Görtz
Zerna, Köpper & Partner, Bochum, Germany

U. Schran
CDM Consult, Berlin, Germany

ABSTRACT: In Berlin, the metro line station "Brandenburger Tor" is under construction. Starting in 2007, it will serve as a temporary terminus for the metro shuttle connecting the Brandenburg gate to the new central station and the Reichstag. As a first step of construction, two deep excavation pits are built using the cut-and-cover method. Between these two the ground will be stabilized by freezing and the platform building of the station will be driven by conventional mined excavation. The station is situated close to a number of existing buildings, noticeably the directly adjacent suburban railway station constructed in the nineteen thirties. Prior to the construction of the new station the effects on the existing sensitive structures had to be analysed. Herefore Finite-Element-models are used which are capable to realistically describe the complex soil-structure-interaction. However, the numerical model has to be kept simple to yield practicable results within the tight time frame of an ongoing construction project.

1 PROJECT

1.1 Station building "Brandenburger Tor"

In the centre of Berlin, the metro line station "Brandenburger Tor" is under construction. This station will serve as a temporary terminus for the metro shuttle connecting the Brandenburg gate to the new central station Lehrter Bahnhof and the Reichstag. The shuttle service is scheduled to be put into operation in 2007. On behalf of the Berliner Verkehrsbetriebe BVG (Berlin public transit authority), the design and the tender documents for the station were prepared by the joint venture "Ingenieurgemeinschaft U5" (Zerna, Köpper & Partner Ingenieurgesellschaft für Bautechnik, PSP Beratende Ingenieure, CDM Jessberger und DMT Bauconsulting). The site investigation was conducted and the geotechnical report prepared by CDM Baugrund Berlin.

The construction of the station is divided into three parts: Two entrance buildings at the eastern and western ends of the station and a platform building with a length of approx. 90 m between them. The plan view is shown in Fig. 1.

Fig. 2 shows a view of the construction site in front of the Brandenburg gate. Due to its exposed location at the tourist site "Brandenburger Tor" and directly in front of the hotel Adlon adverse effects at the surface had to be reduced to a minimum. Therefore the subterranean platform building will be driven by

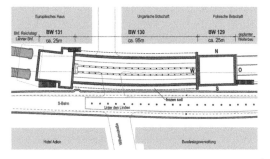

Figure 1. Plan view of metro station "Brandenburger Tor".

conventional mined excavation into a load bearing body of frozen soil.

As a first step, two excavation pits are constructed using the cut-and-cover method. They consist of reinforced concrete diaphragm walls with multiple bracing and a jet grouted base. Starting from those pits and perforating the wall, 30 micro tunnels are driven which will later accommodate the freezing lines. This requires a high degree of apertures in the reinforcement of the diaphragm walls facing the platform building.

The metro line station is situated alongside a suburban railway tunnel plus station, which were constructed in the nineteen thirties below ground.

Every step of construction of the new metro line had to be planned in such way that bearing capacity

Figure 2. View of the construction site in front of the Brandenburg Gate.

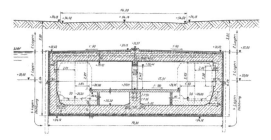

Figure 3. Cross-section of suburban railway tunnel near excavation pit BW 129.

and serviceability of the old railway tunnel were completely maintained during and after construction. The railway tunnel had been constructed in an open pit kept dry by drawdown of the groundwater. Its cross section consists of a reinforced concrete trough covered by a WIB ceiling, whose bearing conditions are assumed as hinged. It is sealed against the groundwater by several layers of tar board. Fig. 3 shows the cross section of the old tunnel near excavation pit BW 129.

In order to maintain its bearing capacity and serviceability, the following effects on the suburban railway tunnel had to be minimized in both the design phase and the construction phase of the metro line station:

- Additional loading of the reinforced concrete
- Settlements or deformations

- Reduction of contact pressure on the tar board sealing

The probable effects of construction on the railway tunnel were studied as part of the design phase by FE numerical simulations. These simulations took into account possible ranges for characteristic soil values and variations in the material properties of the historic construction.

1.2 Ground conditions and hydrology

The boulevard "Unter den Linden" and the Pariser Platz in the inner city of Berlin are located in the so-called Urstromtal, a stream bed formed by glacial runoffs during the ice age, see in Assmann (1957) oder Cepek (1973). Here, below a layer of fill and debris approximately 3.3 m thick, the ground relevant for construction consists of glacially deposited soils. The upper soil strata mostly consist of fine to medium sands with some gravel (alluvial sands and so-called Talsand) in a medium dense state. Below this, in a depth of approximately 15 m below the surface, lies the ground moraine of the Saale glaciation in form of either a heavily overconsolidated glacial till ("boulder marl") or – as a result of erosion – a layer of boulders in the former base of the moraine. Even where the marl is completely eroded this characteristic layer of boulders remains. Its base is located at depths between 16 m and 26 m below street level. Underlying the marl there are dense glaciofluviatile sands, gravels and till. In a depth of about 42 m, marls or silts and clays containing xylite were encountered. Closer to the surface lenses of peat and soft organic clays or trenches filled with sapropel can be found.

The sand and gravel strata form one coherent aquifer with a shallow gradient and slow flow velocity. The groundwater table is situated approximately 3.2 m below the surface with a range between high and low water of about 1 m.

The new metro line mainly runs through poorly graded non-cohesive soils. To simulate the ground in the deformation analysis the following soil parameters are used as given in the site report (BBI, 1996):

We assume the strata below the Saale marl and the deeper parts of the "Talsand" to be overconsolidated due to the weight of the glacier. With regard to the soil parameters assigned to the strata this is taken into account by appropriately higher K_0- and μ-values compared to the upper layers and by high stiffness moduli.

2 SIMULATION OF LOADING PHASES OF THE DIAPHRAGM WALLS

Table 1. Soil parameters, from Baugrundgutachten (1996).

Stratum	γ/γ' [kN/m^3]	ϕ' [°]	E_S [kN/m^2]	K_0 [-]	μ [-]	E [kN/m^2]
Fill type II	17.0/10.0	33	50,000	0.5	0.33	33,750
Alluvial sands	17.7/10.45	34	75,000	0.45	0.31	54,100
Sands	17.7/10.45	34	120,000	0.425	0.30	89,150
Sands, overconsolidated	17.8/10.7	34	250,000	0.65	0.39	125,300
Sands and till	17.8/10.7	36	350,000	0.65	0.39	150,000

Cohesion: $c' = 0$ kN/m^2

Table 2. Overview of the design objectives for diaphragm walls South (S) and West (W).

	Diaphragm wall S	Diaphragm wall W
Design of diaphragm wall	0 no special objectives compared to the other diaphragm walls	+ due to high degree of perforation flow of force to be studied closely
Deformation of diaphragm wall and backfill	+ close proximity of suburban railway tunnel	− no adjacent structures

+ high significance
0 normal significance
− minor significance

2.1 General description

The bearing resistance and the deformation behaviour of the diaphragm walls had to be analysed closely, taking into account the close proximity of the suburban railway tunnel and the high degree of apertures in the walls facing the platform building. To illustrate this, for this paper the eastern construction pit BW 129 was chosen as an example.

Here the evaluation of the two walls listed below is most important:

- Calculation of the behaviour of the southern diaphragm wall (Wall S) including its influence on the adjacent railway tunnel.
- Calculation of the western diaphragm wall (Wall W) with its numerous perforations to allow for micro tunnelling.

Calculations were performed numerically using the Finite-Element (FE) method.

In this calculation the formulation of the material laws is of particular importance. For the reinforced concrete structures (diaphragm walls, suburban railway tunnel), as well as for the surrounding soil, these material laws are high-grade non-linear. A number of models already exist to describe both materials on macro level. They are mainly based on one-dimensional stress-strain-relationships which are then combined to two- or three-dimensional models taking into account the anisotropic material behaviour (e.g. Drucker and Prager (1952), Ottoson (1977) and others).

2.2 Models used

The development of new material laws on the basis of the rapidly growing computer capacities leads to increasingly complex descriptions. The essential part of the engineer's work therefore is to define the relevant objectives of calculation and to abstract systems and material laws with respect to these individual core questions. This process aims at achieving realistic calculation results using an acceptable amount of effort and straightforward models. In a practical design process within tight time constrains this is of utmost importance.

With regard to their loading and their deformations, the two diaphragm walls discussed here (walls S and W) required different design objectives (Table 2), which had to be taken into account in the modelling process.

From the design objectives given in Table 2 the following model parameters were developed, as shown in Table 3. In both cases, isoparametric finite-element-approaches were used.

Behind wall W no adjacent building exists in the direct sphere of influence – therefore performing a realistic prognosis of deformations is unnecessary. The aim of the calculations is a design both economic and safe, and for this a linear-elastic model is appropriate.

Because of the adjacent railway tunnel, the effects of the construction of the diaphragm wall W itself, as well as those of the excavation of pit BW 129, have to be analysed more closely. For this realistic, non-linear material laws have to be utilised. Due to the highly non-linear behaviour of concrete, a completely non-linear analysis was deemed too complex, so the model described below was developed:

1. Soil:
   The soil is represented by an elastoplastic material law with prismatic yield surface and a non-associated yield rule according to Mohr-Coulomb (eqs 1 and 2).

   $$f \leq 0 \qquad (1)$$

   where:

   $$f = \frac{1}{3}I_1 \sin\varphi + \sqrt{I_2}\left(\cos\theta - \frac{\sin\theta\sin\varphi}{\sqrt{3}}\right) - c \cdot \cos\varphi \qquad (2)$$

   I_1, I_2, Invariants of the stress tensor
   The according soil parameters φ und c are taken out of the site investigation report (also see Table 1).
2. Reinforced concrete (suburban railway tunnel and diaphragm walls):
   The decrease in stiffness due to cracking is modelled accordingly by assuming a reduced value of stiffness, gained by an analysis of the moment-bending-relationship. This is explained more closely in section 2.3.
3. Soil-structure interaction:
   Soil-structure interaction is modelled by spring elements placed normally and tangentially to the plane of contact. Normal to the plane of contact a linear elastic behaviour is assumed, including *loss of a cut-off value for (the) tensile spring(s)*. Tangential to the plane of contact the friction force can be transmitted as a maximum.
4. Steel bracing:
   The flexibility of both the struts and the waling are taken into account. For the waling only half the value of the maximum calculated in the middle between two struts is used, because of the different deflection in the longitudinal direction. Since the struts are designed not to yield under working loads, a linear-elastic material law suffices to represent them.
5. Reinforced concrete and jet-grouted bracing:
   To represent the reinforced concrete base slab acting as a panel under compressive loading, as well as the jet-grouted sealing base, linear-elastic material laws are used.

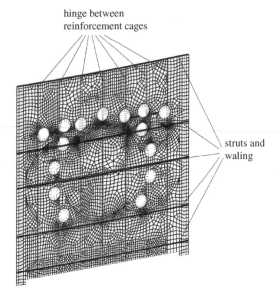

Figure 4. Shell-model of diaphragm wall W.

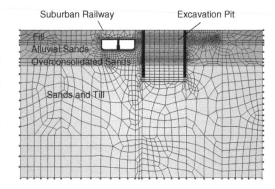

Figure 5. Panel model of wall S.

FE-models of walls W and S are shown in Figs. 4 and 5, respectively.

2.3 *Reduced stiffness of reinforced concrete*

The reduction of stiffness in reinforced concrete due to progressive cracking is represented by using one single reduced value of stiffness. To determine this value, the normal-force dependent relationship between moment and curvature is established, using the simplified method described in the DAfStb-booklet 525 (2003). As a rule, the reduction of stiffness increases with increasing loads up to the point of complete cracking. Once cracking is completed, the residual stiffness does not vary significantly anymore. It can be safely assumed that a seventy-year-old concrete tunnel structure has completed cracking, especially in its areas

Table 3. Overview of significant model parameters.

		Model parameters	
		Diaphragm wall S	Diaphragm wall W
Elementation	Diaphragm wall	Panel	Shell
	Soil	Panel	–
			(acts as loading)
	Structure	Panel	–
Material laws	Diaphragm wall	linear-elastic with reduced stiffness	linear-elastic
	Soil	Mohr-Coulomb	–
	Structure	linear-elastic with reduced stiffness	linear-elastic

of high loading. Therefore the stiffness given by the curve at the point of stabilized cracking can be utilised in the calculations, thereby approximating the highly non-linear material behaviour by using a linear relationship.

Establishing the relationship between moment and curvature after the simplified method described in the DAfStb-booklet 525 (2003):

1. Iteration of the strain field at the crack
   => strain of steel ε_{sr} at the crack and strain of concrete in the compression zone ε_c
2. Determination of mean steel strain
   Formation of single cracking state:

$$\varepsilon_{sm} = \varepsilon_{s2} - \frac{\beta_t \cdot (\sigma_s - \sigma_{sr}) + (1,3 \cdot \sigma_{sr} - \sigma_s)}{0,3 \cdot \sigma_{sr}} \cdot (\varepsilon_{sr2} - \varepsilon_{sr1}) \quad (3)$$

Stabilized cracking state:

$$\varepsilon_{sm} = \varepsilon_{s2} - \beta_t \cdot (\varepsilon_{sr2} - \varepsilon_{sr1}) \quad (4)$$

3. Determination of curvature:

$$\kappa = \frac{|\varepsilon_{sm}| + |\varepsilon_c|}{d} \quad (5)$$

The diaphragm wall, as well as the railway tunnel structure, is divided into segments with an almost constant degree of reinforcement. For each of these segments the stiffness is determined assuming complete cracking. The very low stiffness at the lower outer corner of the tunnel cross-section (Fig. 6) is most noticeable. This is attributed to a very low degree of reinforcement that is thought to be due to an initial design as a hinged corner. While the ends of the tapered walls transfer loads into the ground, the base slabs between them only have to counteract water pressure.

Using the model described above, the following stages of loading are simulated:

1. Initial conditions (Loading stage 0)
2. Construction of diaphragm walls (Loading stage 1)

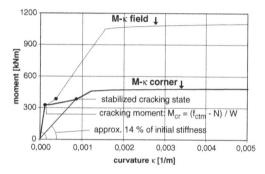

Figure 6. Moment vs. curvature for base slab of suburban railway tunnel.

To account for the loosening of the soil, the stiffness of the ground elements is reduced proportionally to the ratio of active earth pressure vs. earth pressure at rest.

3. Construction of jet-grouted base (Loading stage 2) und successive excavation of pit including drawdown of the water table and installation of several layers of bracing. Each stage is modelled by deactivating and activating the respective elements (Loading stages 3–6)

Several of these loading stages are shown in Fig. 7. For the calculations the commercial program system SOFISTIK, Version 21, was used. Since the non-linear behaviour of the reinforced concrete elements is reduced to a linear reduced stiffness, and since the soil is not loaded up to its capacity, an incremental iterative computation procedure without line-search algorithm is sufficient. The computation time for one complete calculation took only a few minutes.

2.4 *Results of simulation*

In this context, the presentation of the results of the numerical simulation focuses more on the southern

109

Loading stage 1-2:
Construction of diaphragm walls and jet-grouted base

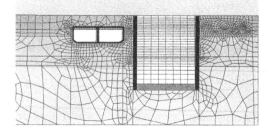

Loading stage 3:
Initial excavation

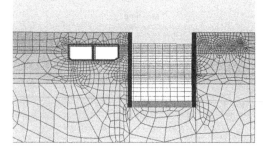

Loading stage 4-6:
Construction of cover slab and segment-wise excavation

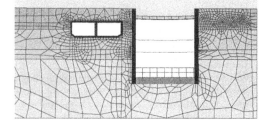

Figure 7. Different stages of loading.

diaphragm wall S. For an exhaustive discussion of the results for diaphragm wall W also see Speier, Görtz and Rath (2005).

The deformation of the diaphragm wall only provokes a slight loosening of its backfill, which in turn leads to a decrease of the earth pressure beneath the railway tunnel base and settlements. The latter cause additional bending loading in the old tunnel structure.

Fig. 8 presents the calculation results. The diagram above shows the computed vertical deformations of the railway tunnel at loading stage 6 "complete excavation of pit". Deformations are computed assuming completed cracking according to the model described

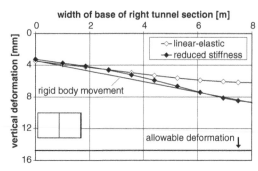

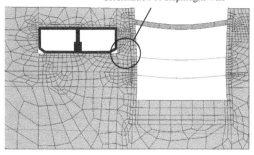

Figure 8. Comparison of vertical deformations of base of the right suburban railway tunnel section after complete excavation of pit assuming full and reduced stiffness, respectively (above). Deformed system assuming reduced stiffness (below).

above, as well as uncracked concrete elements, for the sake of comparison. Both are shown in relation to the initial conditions. It can be concluded that taking into account the reduction of stiffness due to cracking yields to an increase in deformation of approximately 30%. However, deformations do not reach the maximum allowable value of 15 mm.

The additional deformations mostly are rigid body movements and therefore induce just a slight increase in bending loading. These additional bending moments can be carried safely by the tunnel structure (Fig. 8). Although the loosening of the soil provokes a decrease in contact pressure, it is still sufficiently high to prevent a detachment of the tar board sealing from the wall.

2.5 Comparison with in situ measured deformations

Throughout all of the construction a detailed monitoring system is installed to measure the deformations of the pit walls, just as of the surroundings. All measurements are performed by the Gesellschaft für Informationsmanagement (gim) mbH. The essential parts of the measurement program are:

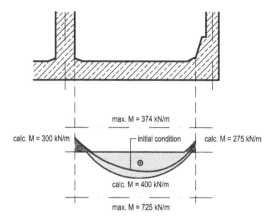

Figure 9.   Distribution of moment in tunnel base.

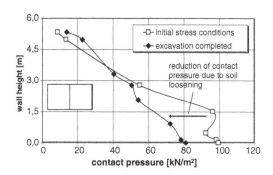

Figure 10.   Contact pressure on exterior of northern tunnel wall.

– to measure the deformations of the diaphragm walls using vertical inclinometers installed in the walls in combination with a tachymetric control of wall head deformations;
– to geodetically monitor deformations of adjacent structures, as well as of the ground surface;
– to automatically register changes in vertical position and wall inclination of the suburban railway tunnel using water levels and inclinometers.

Taking into account seasonal variation in the field conditions, measurement inside the railway tunnel began 140 days before the beginning of the construction according to Schran (2003). Fig. 11 shows a comparison of the measured and the simulated vertical deformation of the tunnel.

The diagram in Fig. 11 reflects the actual state of construction and measurement in October 2005. It can be seen that during the construction of the diaphragm walls settlements occurred in the order of about 1 mm which lay below the accuracy of the numeric model. Variations in measuring devices also have to be taken

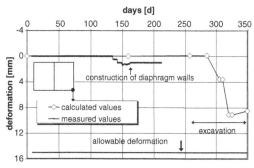

Figure 11.   Comparison of calculated and measured vertical deformations, respectively, of suburban railroad tunnel.

into account, e.g. in Marte and Semprich (1998). Anyhow, measured settlements are very low. They also correspond with the findings presented in Mayer and Kudella (1999), which show that the loosening effect decreases with an increasing distance to the wall.

Deformations will be measured continuously. The actual comparison between measured and predicted deformations will be presented in the framework of the conference. In case of their increase, actual and computed deformations will correspond even more closely.

3   SUMMARY

In Berlin the metro line station "Brandenburger Tor" is under construction. Starting in 2007, it will serve as a temporary terminus for the metro shuttle connecting the Brandenburg gate to the new central station Lehrter Bahnhof and the Reichstag.

The construction site is located in the inner city area near the river Spree. Below a layer of fill and debris approximately 3.3 meters thick the ground relevant for construction consists of glacially deposited soils. The uppermost strata mostly consist of fine to medium sands with some gravel. Below these layers there lies the ground moraine of the Saale glaciation in form of either a heavily overconsolidated glacial till ("boulder marl") or a layer of boulders at the former base of the moraine. This base is located at depths between 16 m and 26 m below street level. Underlying the marl, there are overconsolidated glaciofluviatile sands and gravels. The groundwater table is high at about 3.2 m below ground surface.

As a first step of construction two deep excavation pits are being built using the cut-and-cover method. Between these the ground will be stabilized by freezing and the platform building of the station will be driven by conventional mined excavation. The station is situated close to a number of existing buildings, noticeably the directly adjacent suburban railway station "Unter den Linden", constructed in the nineteen thirties. Prior to construction of the new station, the effects on the existing sensitive structures had to be analysed closely.

111

During the ongoing construction, the prognosticated behaviour is monitored by in situ measurements.

The numerical model utilised for the prognoses has to serve two purposes: On the one side they have to be capable to realistically describe the complex soil-structure-interaction with regard to load bearing and deformations. On the other side it has to be kept simple to yield practicable results within the tight time frame of an ongoing construction project.

Whereas the soil is represented by an elastoplastic material law with the Mohr-Coulomb failure criterion, the highly non-linear behaviour of reinforced concrete elements is modelled by a reduced stiffness, which is derived from the moment-curvature-relationship.

Using this model it can be shown that the suburban railway tunnel can safely accommodate the slight increase in deformations due to the construction of the excavation pit. These deformations will have no detrimental effect on the serviceability or the load bearing capacity of the structure. The verification of the numerical model by comparison with measured deformation will continue throughout the construction process.

REFERENCES

Assmann P.: Der geologische Aufbau der Gegend von Berlin. Herausgegeben vom *Senator für Bau- und Wohnungswesen*, Berlin, 1957

BBI Baugrund Berlin Ingenieurgesellschaft: Baugrundachten für die U-Bahnlinie U5-Abschnitt Pariser Platz-Bahnhof Berliner Rathaus (Stadtbezirk Mitte von Berlin), 24.01.1996

Cepek, A.G. (Leitung und Redaktion) et al.: Geologische Karte der Deutschen Demokratischen Republik – Karte der quartären Bildungen; 1:500.000; Berlin, 1973

Deutscher Ausschuss für Stahlbeton: Erläuterungen zu DIN 1045-1. (2003), *Heft 525 des Deutschen Ausschusses für Stahlbeton*, Berlin

Drucker, D.C.; Prager, W.: (1952) Soil Mechanics and Plastic Analysis or Limit Design, Q. *Appl. Math, Vol. 10, No. 2*, pp. 157–165

Marte, R.; Semprich, S.: Untersuchung zur Meßgenauigkeit von Inklinometermessungen. *Bautechnik 75*, Heft 3, March 1998

Mayer, P.-M.; Kudella, P.: Verformungen und Spannungsänderungen im Boden durch die Schlitzwandherstellung. *VDI-Berichte 1436: Tiefe Baugruben – Neue Erkenntnisse bei ungewöhnlichen Baumaßnahmen*, Berlin 1999

Ottoson, N.: A Failure Criterion for Concrete. *ASCE Vol. 103 EMA4 (1977)*, pp. 527–535

Schran, U.: Untersuchungen zu Verschiebungen von Schlitzwänden beim Unterwasseraushub in Berliner Sanden. *Veröffentlichungen des Grundbauinstitutes der Technischen Universität Berlin*, Heft 33, 2003

Speier, L.; Görtz, S.; Rath, T.: Einsatz von GFK-Bewehrung bei der Ausführung des U-Bahnhofs Brandenburger Tor. *Bautechnik 82*, Heft 9, September 2005

3. Soil improvement

Numerical Modelling of Construction Processes in Geotechnical Engineering for Urban Environment – Triantafyllidis (ed)
© 2006 Taylor & Francis Group, London, ISBN 0 415 39748 0

Vibro stone column installation and its effect on ground improvement

F. Kirsch

GuD Geotechnik und Dynamik Consult GmbH, Berlin, Germany

ABSTRACT: Results of installation accompanying measurements of the in situ stresses, the ground stiffness and the ground surface vibrations in various test fields are presented. Following the installation, the stone columns are subjected to group loading whilst the instrumentation remains active including load pressure cells and settlement transducers. The results show significant alteration of the stress state and the soil stiffness during penetration of the depth vibrator and the subsequent stone column installation. Two major effects can be distinguished: the displacement of the ground due to the creation of the stone column body and the ground vibration with associated changes within the soil due to the movements of the depth vibrator. The results of the measurements are used to model the installation effects in numerical analyses using the finite element method. Comparison between the numerical results and the findings of the measurements as well as the load test results are used to calibrate this numerical model. Finally parametric studies are undertaken to estimate the additional improvement effect of the installation process on the total ground improvement factor for vibro stone column measures.

1   INTRODUCTION

Developed in the 1950ies as an extension of the at that time well established vibratory ground improvement method, the first vibro stone column project is believed to be the foundation measure for a locomotive shed in Braunschweig, Germany (Kirsch 1993). Since that time depth vibrators are used to install stone columns in cohesive soils by displacing and if possible compacting the in situ soil.

Usually the fields of application of vibro stone columns and deep vibratory ground improvement are distinguished depending on the grain size distribution of the soil. It is apparent that there is no distinct partition but a range in which densification of the soil is still possible, but additional material is usually added to ensure an effective and economical improvement.

Generally, it is assumed that the surrounding soil maintains its original strength and stiffness parameters whilst the improvement is dominated by the highly compacted stone column material. Any improvement of the in situ soil acts as a hidden safety in the system.

The design of vibro stone column measures does not take into account the improvement of the surrounding soil apart from a certain increase in the stress state typically using earth pressure at rest assumptions. Furthermore, the design parameters such as stiffness and shear strength are derived from investigations prior to the column installation and are usually not crosschecked with additional testing.

On the other hand it is known from experience that one can measure a significant loss of cone penetration resistance immediately after the stone column installation and a subsequent rise often above original values after some time. Therefore it is deemed necessary to put focus on the investigation of alterations in the soil properties during vibratory stone column installation in order to draw a realistic picture of their load carrying behaviour.

2   MEASUREMENTS DURING STONE COLUMN INSTALLATION

2.1   *Ground movement*

It is apparent that the installation of a grid of stone columns in mainly uncompactible soils results in a significant surface heave, which was monitored e.g. by Gruber (1994). He reports that the volume of the surface heave was between 90% and 100% of the installed stone column volume. Figure 1 shows the surface

Figure 1.   Surface heave after the installation of five 0.8 m diameter stone columns.

heave after the installation of five 0.8 m diameter stone columns in a 1.6 m diagonal grid. One should keep in mind that the influence on the stress state becomes essential when this movement is restrained e.g. by a stiff layer on top.

2.2 *In situ stresses*

Measurements of the stress field surrounding vertical displacement elements such as piles or stone columns are reported by many authors. Whilst Gruber (1994) reports no significant increase of the horizontal stresses after the installation of stone columns, Watts et al. (2000) show an increase of up to 60 kPa of the horizontal stresses during column installation. Randolph et al. (1979) gave an explanation of the processes during pile driving within the framework of critical state soil mechanics during cavity expansion, which was used by Cunze (1985) to predict the increase in pore water pressure due to pile driving.

The development of additional stresses due to stone column installation was addressed by Jacobi (1998) using simple but effective measurement devices to find, that the installation of three vibro stone columns in front of the gauge leads to additional stresses in a silty soil of up to 100 kPa as shown in figure 2.

The results of pore pressure measurements during the installation of stone columns in two test fields, each consisting of 25 columns in a square pattern, were presented by Kirsch (2004). Figure 3 shows the outline of the test fields.

The earth and pore water pressure cells (E1 and E2) and the Ménard pressuremeters (P1 and P2) were set up prior to the column installation. The installation sequence was chosen moving towards the measurement devices.

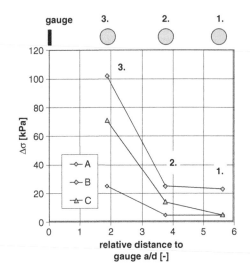

Figure 2. Increase of total stresses due to the subsequent installation of three stone columns with a diameter d = 0.8 m in falling distance a to the gauge for three different series A, B, C.

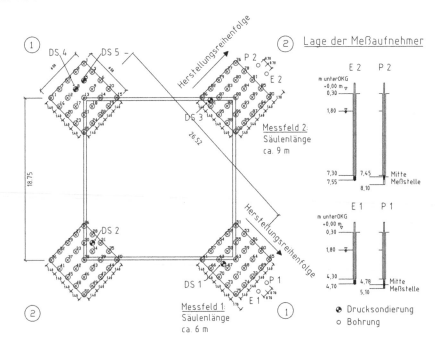

Figure 3. Outline of test fields 1 and 2.

The comparison between the predicted (according to Cunze 1985) and the measured pore water pressure increase due to the stone column installation is shown in figure 4. Apart from a reasonable match, it becomes apparent, that there is a considerable scatter in the results.

Looking at the stress development during the installation of each single column within the above test fields one can distinguish different phases: 1-Lowering of the vibrator, 2-alternating lift and sag cycles compacting the stone fill supported by air pressure, 3-stop to refill the material lock inside the vibrator (see figure 5).

The example shown in figure 5 depicts the total horizontal stresses measured in a distance of 7.33 m in a depth of 4.7 m and the effective horizontal stresses calculated by subtraction of the measured pore water pressures during the installation of a stone column having 0.8 m diameter and a length of approx. 5.5 m. Thus the gauge was installed in weak to stiff sandy silt of medium plasticity (MI). Obviously the maximum stress peak is reached, when the depth of the vibrating tip of the vibrator reaches the level of the gauge. The lift and sag cycles invoking the lateral displacement of the stone column material can be easily identified in the stress development. After completion of the stone column the stress state both in terms of pore water pressure and effective stresses remains higher than before column installation.

Horizontal stress increases at a given point measured during the installation of the whole group of 25 columns in each of the two test fields are shown in figure 6. Displayed are the effective horizontal stresses in relation to the initial stresses and normalized by

the vertical overburden pressure leading to a specific k-value. Also shown is the installation sequence with locations of the column installation getting closer to the place of the gauges.

It becomes evident that the stress state rises to a value of up to 1.6 times the initial stresses due to

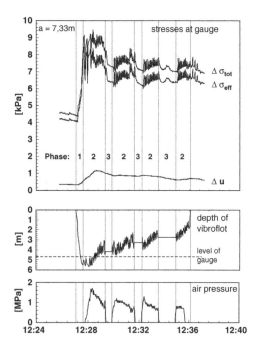

Figure 5. Additional horizontal stresses and pore water pressures at a depth of 4.7 m during the installation of a stone column (d = 0.8 m) in a distance of a = 7.33 m.

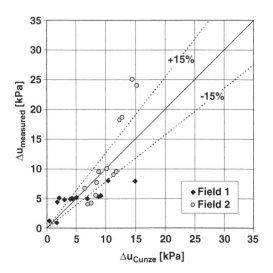

Figure 4. Measured and predicted pore water pressure increase.

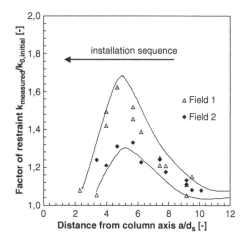

Figure 6. Factor of restraint measured during the installation of stone columns ($d_S = 0.8$ m).

117

the ground displacement when the location of column installation gets closer to the measurement location. Once a critical distance of about four to five times the column diameter d_S is reached the displacing virtue is superimposed by stress reducing effects which can be addressed as remoulding and dynamic excitation leading to a considerable loss of strength.

2.3 Stiffness development

Looking at the development of the in situ stiffness the above test fields were also instrumented with pressuremeter cells in each of the two fields. Figure 7 shows the measured Ménard-moduli in relation to the distance from the latest installed column with an installation sequence approaching the pressuremeter cell.

We find that the stiffness rises to a maximum of 2.5 times the initial stiffness in a distance of about 4 to 5 times the column diameter d_S. Installation of stone columns closer than $4 \cdot d_S$ lead to a loss of stiffness even below initial stiffness showing a potential liquefaction of the soil due to the dynamic excitation.

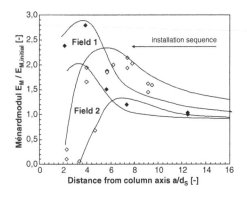

Figure 7. Development of ground stiffness during the installation of stone columns ($d_S = 0.8$ m).

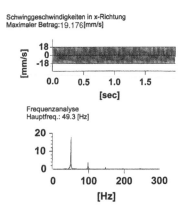

Figure 8. Typical signal of ground vibration induced by stone column installation (z: vertical).

2.4 Ground vibration

We found that there is a critical distance from the location of the column installation of approximately five times the column diameter. This distance corresponds to the distance where the dynamic excitation of the soil rises rapidly following an exponential law. Figure 8 shows the typical response of the ground surface to the excitation of a depth vibrator during the installation of a stone column. In figure 9 the maximum particle velocity is drawn as a function of the distance to the column axis normalized by the column diameter.

2.5 Summary of installation effects

Summing up, we can distinguish different effects:

- Stress and stiffness increases in the surrounding soil due to vibro stone column installation can be verified by in situ measurements.
- These increases can be expected to be permanent in soils which do not tend to creep, i.e. having a significant fraction of sandy soil.
- The soil adjacent to the column is displaced and remoulded during column installation.
- The soil displacement leads to an increase in the stress state and the stiffness in a distance between $4 \cdot d_c$ and $8 \cdot d_c$ around the columns and the column group, respectively.
- Dynamic excitation close to the column reduces stress and stiffness increases.

These findings have important impact on stone column design and practical application.

- Installation effects should be included in design and analysis in order to model the behaviour under loading correctly.

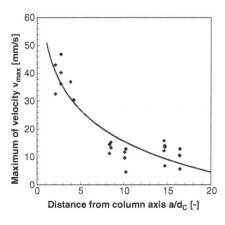

Figure 9. Maximum particle velocity v_{max} at ground level during the installation of vibro stone columns ($d_c = 0.8$ m) at a distance a [m].

118

- Due to the stress increase vibro stone column installation influences existing structures within a distance of approx. 8 to 10 times the column diameter.
- The rise of the stress level is reduced within a distance of less than 4 times the column diameter due to remoulding and liquefaction effects.
- Column installation in the direct vicinity to sensitive structures should always be by moving towards the structure.

3 SIMULATION OF INSTALLATION EFFECTS

3.1 Methodology

There are a couple of attempts published in literature to simulate the effects which are induced by a depth vibrator in the soil (e.g. Fellin 2000). Due to the complexity of the associated phenomena, the models used are rather simple in terms of geometrical conditions. Schweiger (1989) applies a strain field in order to model the volume input of column installation. Most authors model the displacing effect of column installation by a cylindrical cavity expansion starting from an arbitrary diameter to the final column (e.g. Debats et al. 2003).

In the following chapters, the individual effect of simulating a cavity expansion in the numerical model is shown. Additionally, the above finding of an enhancement zone in the far field as a global installation effect is introduced in the analysis to come to a numerical procedure which accounts for installation effects of a group of stone columns. Figure 10 illustrates the two main installation effects of vibro stone columns.

3.2 Cavity expansion

To show the effects incorporated in the modelling of a cylindrical cavity expansion, the installation of a single stone column in a layered soil is modelled in a 3-D elasto-plastic continuum. Table 1 shows the material data, whilst the finite element model is shown in figure 11. In order to avoid singularities the cavity expansion close to the column toe starts with a zero expansion rising to the chosen value within the first three elements above the column toe.

The expansion starts from a initial radius of the column of 0.3 m. Figure 12 shows the plots of radial stresses σ_x, radial displacements u_x and plastic strains ε_{pl} due to a cavity expansion of 0.1 m radius

Table 1. Soil parameters for cavity expansion example.

Layer	Depth [m]	E [kPa]	φ' [°]	c' [kPa]	ν [−]	ψ [°]	γ [kN/m³]
1	0–3	500	10	15	0.3	$0.3 \cdot \varphi$	9.0
2	3–7	750	10	15	0.3	$0.3 \cdot \varphi$	9.0
3	7–10	1000	10	15	0.3	$0.3 \cdot \varphi$	9.0
4	10–20	1500	10	15	0.3	$0.3 \cdot \varphi$	9.0

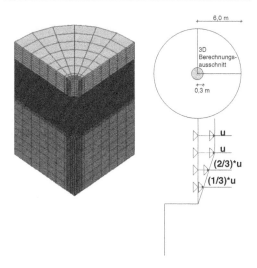

Figure 11. 3-D FE model for cavity expansion.

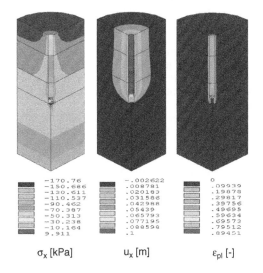

σ_x [kPa] u_x [m] ε_{pl} [-]

Figure 12. Radial stresses σ_x, radial displacements u_x and plastic strain ε_{pl}.

Figure 10. Global and individual installation effect of vibro stone column ground improvement.

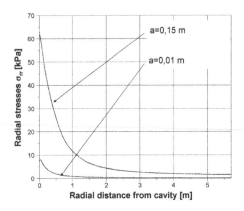

Figure 13. Radial stresses σ_x in relation to the distance from the cavity surface for an expansion of 0.01 m and 0.15 m respectively.

amplification. In Figure 12 the radial stresses in a radial slice in a depth of 3 m in soil layer 2 are plotted for two exemplary radius magnifications of a = 0.01 m and a = 0.15 m. According to Baguelin et al. (1978) the stress increase in an elasto-plastic medium due to cavity expansion can be calculated as:

$$\sigma_{rr} = \left(\sigma_F + c\cdot\cot(\varphi)\right)\cdot\left(\frac{G}{\left(\sigma_{h0} + c\cdot\cot(\varphi)\right)\cdot\sin(\varphi)}\cdot\frac{\left(r_0+a\right)^2 - r_0^2}{\left(r_0+a\right)^2}\right)^{\frac{1-K_a}{2}}$$
$$- c\cdot\cot(\varphi) \qquad (1)$$

with

$$\sigma_F = \sigma_{h0}\cdot\left(1+\sin(\varphi)\right) + c\cdot\cos(\varphi)$$
$$K_a = \frac{1-\sin(\varphi)}{1+\sin(\varphi)}$$

With the above parameters from table 1 we calculate:

• a=0,01 m: $\sigma_{rr}=\sigma_x=28$ kPa,
• a=0,15 m: $\sigma_{rr}=\sigma_x=71$ kPa.

These values correspond reasonably well with the calculated values in figure 13. The deviation results from the fact that in the numerical analysis a non-associated flow rule is incorporated using the angle of dilatancy ψ.

There are other possibilities to model the stress increase in the soil e.g. the application of a temperature gradient to the column material in addition to the definition of a temperature expansion coefficient or the application of a strain or a stress field to the finite element net. Kirsch (2003) showed that the cavity expansion described here is superior to the other methods due to numerical stability.

3.3 Global and individual installation effects and the influence on the ground improvement

In addition to the above examined cavity expansion and in order to model the soil displacing effects of

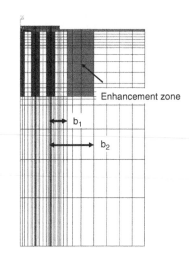

Figure 14. Enhancement zone in the numerical model.

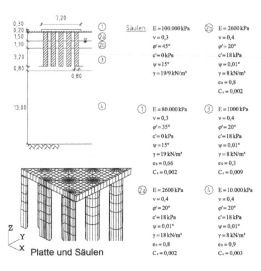

Figure 15. Geometry and parameters of the test field.

vibro stone column installation with subsequent stress increase, we learnt from the in situ measurement that there exists an enhancement zone around a column or a column group, respectively. In the numerical model this zone is defined by the two distances b_1 and b_2 from the outmost column centre as shown in figure 14.

To illustrate the effect of a stiffness increase in the enhancement zone, it is necessary to take a look at the load settlement behaviour of the column group and vary the stiffness in the enhancement zone. Therefore the situation shown in figure 15 shall be simulated, which corresponds to the test field as presented in the previous chapter. A group of 25 columns in a layered soil is loaded by a square shallow foundation

120

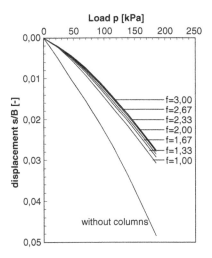

Figure 16. Load settlement response of the square footing (B = 7.2 m) considering global installation effects for different stiffness increase factors.

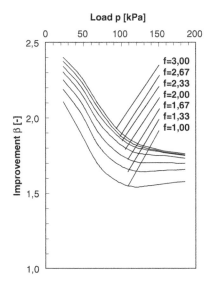

Figure 17. Improvement factor β as defined in eq. 2 for different stiffness increase factors f in the enhancement zone.

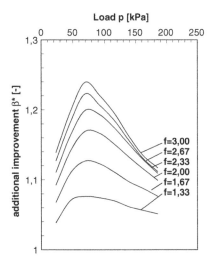

Figure 18. Improvement factor β^* as defined in eq. 2 for different stiffness increase factors f in the enhancement zone.

of B = 7.2 m up to 180 kPa. The improvement due to vibro stone columns is defined by the improvement factor β being a settlement reduction factor. The additional improvement due to stiffness increase $f = E/E_{init}$ is defined by β^*, whilst the additional improvement due to the cavity expansion a is denominated as β^{**}.

$$\beta = \frac{s_{without\ improvement}}{s_{with\ improvement}} \quad ; \quad \beta^* = \frac{s_{with\ improvement}\,(a=0, f=1)}{s_{with\ improvement}\,(a=0, f>1)};$$

$$\beta^{**} = \frac{s_{with\ improvement}\,(a=0, f=1)}{s_{with\ improvement}\,(a>0, f=1)} \quad (2)$$

The results of the numerical simulation in terms of the load settlement response are shown in figure 16. Eight different situations are analysed: The original situation without ground improvement by vibro stone columns; Vibro stone column improved soil beneath the footing without the simulation of installation effects neither individual (cavity expansion a = 0) nor global (enhancement zone, f = 1) and six computations with different stiffness increase factors f = 1.33 ... 3.0 in the enhancement zone with $b_1 = 2 \cdot d_C$, $b_2 = 5 \cdot d_C$ to model the global installation effect. The resulting improvement factors β from eq. (2) are shown in figure 17, whilst the improvement factors β^* from eq. (2) are shown in figure 18.

Obviously, the improvement factor due to the global installation effect with a stiffness increase in the enhancement zone is dependant on the loading state, which can be explained by the fact that at higher loadings the plastic deformation plays a predominant role and stiffness increase has a minor influence. Depending on the stiffness increase factor, the

additional improvement β^* reaches values of up to 1.25, meaning that the consideration of the global installation effects leads to a rise in the calculated improvement of 25% of the calculated improvement factor without taking into account any installation effects.

So far we only incorporated the global installation effect. Now we take a look at the influence of the individual installation effect being the cavity expansion as presented in chapter 3.2. We analyse the same situation from the test field.

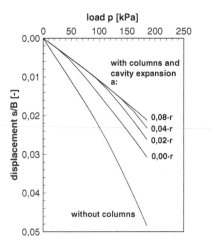

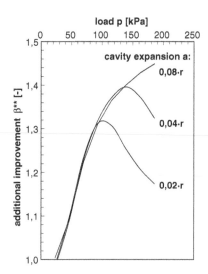

Figure 19. Load settlement response of the square footing (B = 7.2 m) considering individual installation effects for different cavity expansion values a.

Figure 21. Improvement factor β^{**} as defined in eq. 2 for different cavity expansion values a for each column.

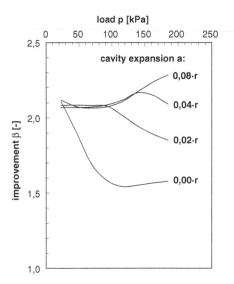

Figure 20. Improvement factor β as defined in eq. 2 for different cavity expansion values a for each column.

Figure 19 shows the load settlement response of the square footing for situations without ground improvement stone columns, with columns however only considering the individual installation effect by applying a cavity expansion a of 0%, 2%, 4% and 8% of the final column diameter $d_C = 0.8$ m. Figures 20 and 21 depict the corresponding improvement factors β and β^{**} from eq. (2).

It turns out that the consideration of the individual installation effect by applying a cavity expansion

between 0% and 8% of the final column diameter leads to additional improvement factors of up to 45% of the improvement without any installation effects. The additional improvement is highly dependant on the loading stage rising with higher loading and falling again after reaching a maximum at loading values which are again dependant on the cavity expansion value. This behaviour can be explained by taking into account the nature of the load carrying mechanisms of stone columns. At early loading stages, the stone column material reaches its plastic limit and the column starts to bulge causing lateral support in the ground. The higher the initial horizontal stress state, which is influenced by the value of the cavity expansion, the better this lateral support prevents the column from bulging, leading to reduced settlements. Once the limit state in the surrounding soil is reached this additional support plays a minor role leading to lower additional improvement factors at higher loading stages.

This explains that both individual and global installation effects play a positive role by reducing the settlements under load, but do not influence the ultimate load of the structure. This effect can be compared with prestressing in reinforced concrete. To illustrate this effect, figure 22 shows the load settlement response of a single column $d_C = 1.0$ m loaded by a circular plate d = 2.0 m.

3.4 *Comparison with analytical design*

The measurements in the above described test fields gave reason to set the cavity expansion to model individual installation effects to a value of a = 4% of the final column diameter of $d_C = 0.8$ m in order to match

122

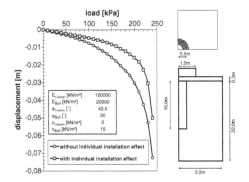

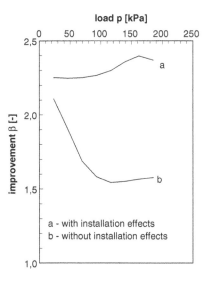

Figure 22. Calculated load settlement response of a single column with and without the consideration of installation effects.

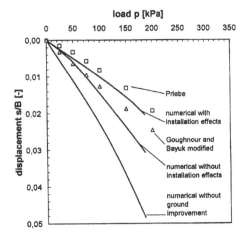

Figure 23. Analytical and numerical results of the load settlement behaviour of a group of 25 stone columns.

Figure 24. Improvement factor β for the with and without the consideration of installation effects in the numerical analysis.

as $k = 1$ gives a reasonable picture of the actual situation. Therefore both the method of Priebe (2003) and the method of Goughnour and Bayuk (1979) with the modification by Kirsch (2004) are able to predict the settlements with reasonable accuracy.

The consideration of installation effects in numerical analysis leads to a steady distribution of the improvement factor β along the different loading stages as shown in figure 24.

4 SUMMARY AND CONCLUSIONS

The extensive instrumentation of two test fields each consisting of 25 vibro stone columns allowed the assessment of the influence of the stone column installation on the soil properties.

Two effects can be distinguished. An individual installation effect causing the rise of the stress state and a global installation effect in an enhancement zone around the columns and the column group respectively.

Both effects can be modelled numerically making use of analogical procedures such as the cavity expansion and the stiffness increase.

The isolated and combined analysis of the additional improvement due to the installation effects showed that within a group of 25 columns the consideration of the global installation effect raises the improvement by approx. 5% to 25% whilst the individual installation effects raise the improvement to values about 40% higher than without installation effects. The combination of both the global and the individual

the stress increase surrounding the column. The stiffness increase factor accounting for global installation effects was measured as $f = 2.0$ in a zone limited by $b_1 = 2 \cdot d_C = 1.6$ m and $b_2 = 5 \cdot d_C = 4.0$ m around the column group.

In order to judge the results of the numerical analysis, a comparison is made with the results of analytical design methods. Priebe (2003) allows the computation of settlements of a stone column group in a floating situation meaning that the columns are not necessarily set on a hard layer. The routine of Gouhnour and Bayuk (1979) was extended by Kirsch (2004) to calculate the settlements of a group of columns. Both methods are compared with the calculated load settlement response in figure 23.

It can be seen that the assumptions made in analytical procedures about the horizontal stress ratio chosen

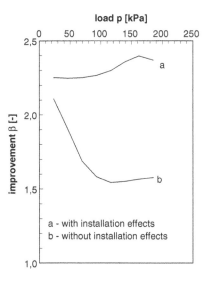

123

installation effect leads to a more steady distribution of the improvement along the different loading stages.

REFERENCES

Antoine, P.C. et al. 2003. Study of deep compaction in order to its application to deep foundations. *Proc. 13th Europ. Conf. Soil Mech. Geotech. Engg.* Vol. 2: 443–448.

Baguelin, F. et al. 1978. *The Pressuremeter and Foundation Engineering.* TransTech Publications.

Cunze, G. 1985. *Ein Beitrag zur Abschätzung des Porenwasserüberdruckes beim Rammen von Verdrängungspfählen in bindigen Böden.* Dissertation. Uni Hannover.

Fellin, W. 2000. *Rütteldruckverdichtung als plastodynamisches Problem. Advances in geotechnical engineering and tunneling 2.* Rotterdam: Balkema.

Goughnour, R.R and Bayuk, A.A. 1979. Analysis of stone column – soil matrix interaction under vertical load. *Coll. Int. Renforcements des Sols.* Paris. 279–285.

Gruber, F.J. 1994. *Verhalten einer Rüttelstopfverdichtung unter einem Straßendamm.* Dissertation. TU Graz.

Jacobi, C. 1998. *Kurzbericht Hildesheim Wohnbebauung Pieperstraße.* unpubl.

Kirsch, F. 2003. *Die numerische Analyse von Baugrundverbesserungsmaßnahmen*, 2. Zwischenbericht, IGB·TUBS, unpubl.

Kirsch, F. 2004. *Experimentelle und numerische Untersuchungen zum Tragverhalten von Rüttelstopfsäulengruppen.* Dissertation. TU Braunschweig.

Kirsch, F. et al. 2004. Berechnung von Baugrundverbesserungen nach dem Rüttelstopfverfahren. In DGGT (ed.), *Vorträge der Baugrundtagung 2004 in Leipzig.* 149–156. Essen. VGE.

Kirsch, K. 1993. Die Baugrundverbesserung mit Tiefenrüttlern. In K. Englert & M. Stocker (eds.), *40 Jahre Spezialtiefbau*: 219–255. Düsseldorf: Werner.

Priebe, H. 2003. Zur Bemessung von Stopfverdichtungen – Anwendung des Verfahrens bei extrem weichen Böden, bei schwimmenden Gründungen und beim Nachweis der Sicherheit gegen Gelände- oder Böschungsbruch. *Bautechnik 80.* 380–384.

Randolph, M.F. et al. 1979. Driven piles in clay – the effects of installation and subsequent consolidation. *Géotechnique 29.* No. 4: 361–393.

Schweiger, H.F. 1989. *Finite Element Analysis of Stone Column Reinforced Foundations.* Dissertation. University of Swansea.

Watts, K.S. et al. 2000. An instrumented trial of vibro ground treatment supporting strip foundations in a variable fill. *Géotechnique 50.* No. 6: 699–708.

Numerical Modelling of Construction Processes in Geotechnical Engineering for
Urban Environment – Triantafyllidis (ed)
© 2006 Taylor & Francis Group, London, ISBN 0 415 39748 0

On modelling vibro-compaction of dry sands

G. Heibrock, S. Keßler & Th. Triantafyllidis

Institute of Soil Mechanics and Foundation Engineering, Ruhr-University Bochum, Germany

ABSTRACT: Due to the complex interaction between probe and soil, modelling of vibro-compaction is still a challenging task. The paper discusses the demands on suitable constitutive models and proposes the application of an explicit calculation scheme overcoming the shortcomings resulting from accumulation of calculation errors when using implicit approaches. The calculated stress-strain behaviour is discussed with respect to real behaviour. Results indicate that the proposed model is capable to mirror densification effects known from experience.

1   INTRODUCTION

A variety of soil improvement techniques has been developed to increase soil strength and reduce settlements (Mitchell and Gallagher 1998). Among these, Vibro-Compaction has become a premier technique for densifying sand deposits. Although these methods have been successfully used for decades, design still refers to empirical knowledge describing the post construction density of the granular soil as a function of spacing and pattern of compaction points.

2   THE BASIC PROCESS

Vibro-compaction methods are characterized by the insertion of long probes into the ground followed by compaction of the soil caused by vibration during withdraw of the probe. Techniques differ in induced polarization of vibrations (horizontal or vertical) and whether backfill is used or not. The following sections deal with vibrocompaction or vibroflotation as referred to in the recent years. Figure 1 shows a schematic drawing of the vibroflotation process. The crane mounted vibroflot (vibrator incorporated in the lower end of a steel probe with extension tubes) is usually jetted to the bottom of the soil layer to be compacted (1). The unbalance rotates around the vertical axis of the vibrator, generating lateral and torsional vibrations. Lower end vibroflot diameters range from 300–400 mm. Vibrators typically operate at frequencies of 30–50 Hz, generating displacement amplitudes of 6–30 mm (vibrator in air) and dynamic forces of 150–450 kN (Degen and Hussin 2001). Penetration speeds range from 1–2 m/min. At final depth, jetting is reduced. The compaction process starts when the probe is hold steady (2). During compaction a funnel is forming at the surface which is usually

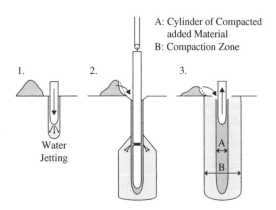

Figure 1.   Schematic drawing of vibrocompaction process.

compensated by adding granular material. A typical volume of added material is about 1.0–1.5 m³ per meter of depth. The material is flowing down the probe and is compacted at the base of the vibrator. When the required compaction level is achieved the probe is raised in 0.3–1.0 m stages maintaining the probe steady for 30–60 seconds at each level (3). A pilgrim-step scheme may be used to support material downward flow. This procedure is repeated until the compaction point is built up to ground level. Depending on soil conditions and vibro equipment compaction reaches a radius about 1.5–4.0 m from the probe.

3   DENSIFICATION MECHANISM

It is common sense that the compaction effect in dry soils is achieved by reducing the friction between soil particles and then allowing them to redeposit under gravity and vibration into a more dense condition (Slocombe et al. 2000; Greenwood and Kirsch 1983). Some authors attach some importance to a cavity expansion effect (Fellin 2000; Cudmani et al. 2003). A material flow downward the vibratory probe arises during the compaction phase due to the volume deficit arising from densification in the vibrator surrounding zone. They presume that the horizontally vibrating probe pushes this material sideways into the surrounding soil. In more or less saturated soils additional effects from increasing pore pressures occur, which are not accounted for in the present paper. There are only very few studies exceeding this phenomenological description of the densification mechanism. The basic approach is to identify different compaction zones around the probe by defining some kind of energy measure needed to destabilize the soil structure. Rodger (1979) uses acceleration levels, Green (2001) and Massarsch (2003) use predicted particle velocities v as a function of distance from the probe (vibro-probe compaction method). For this purpose first order methods and empirical or measured damping and attenuation functions are applied. A one dimensional approach is used assuming that the compaction effect is governed by radially propagating vertically polarized shear waves. The measure related to the compaction effect is the shear strain γ resulting from the shear waves (Dowding 1996):

$$\gamma = \frac{v}{v_s} \qquad (1)$$

where: γ = Shear strain from S_v-wave,
        v = particle velocity of S_v-wave,
        v_s = Shear wave velocity of soil.

The particle velocity v is predicted based on empirical attenuation functions taking into account the effect of "geometrical" or attenuation and material damping. An iterative procedure is used to determine the resulting shear strains γ. Based on the estimated particle velocity v and shear modulus degradation curves (Ishibashi and Zhang 1993), γ is computed from (Green 2001):

$$\gamma = \frac{v}{v_{s,max} \cdot \sqrt{\dfrac{G}{G_{max}}}} \qquad (2)$$

where: $v_{s,max}$ = small strain shear wave velocity of the soil ($\gamma = 10^{-6}$),
        G = Shear modulus,
        G_{max} = small strain shear modulus.

Using threshold concepts, specific shear strains are correlated to specific effects, e.g. liquefaction ranges or deformation modes. Massarsch (2003) uses these shear strains to identify the three compaction zones

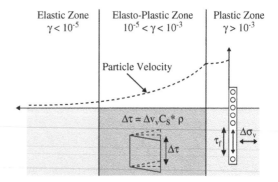

Figure 2.   Shear-Strains as a function of distance from the vibro-probe (adopted from Massarsch 2003, modified).

for the vibro probe method shown in Figure 2. None of the mentioned models takes into account the probe-soil interaction and therefore assumptions are needed to estimate velocities in at least one point in the vicinity of the probe. In addition, there is no real check of the assumption that a one dimensional approach is sufficient to cover the governing effects. Measurements performed at vibro compaction sites show that horizontal accelerations approach the same order of magnitude than the vertical accelerations (Massarsch 2003).

Due to the complex interaction of the probe and the soil resulting in multidimensional loading of the soil only few attempts have been made to develop models for vibrocompaction. Fellin (2000) focuses on quality control of vibrocompaction and uses spring dashpot models to describe the probe soil interaction. Density dependent model constants (shear modulus) allow backanalysis of acceleration measurements from the tip of the probe. However, highly nonlinear soil behaviour limits the potential to predict compaction effects from the acceleration data.

Cudmani et al. (2003) use a hypoplastic constitutional equation (von Wolffersdorff 1996) with intergranular strains (Niemunis 2003) to describe the soil behaviour. Using assumptions concerning the probe-soil interaction, they performed 2-D calculations describing vibrocompaction as a symmetric, dynamic stress controlled expansion of a cylinder. They state that the results showed deficits of the model resulting from plane strain conditions. As a consequence, they use a stress controlled numeric element test to calculate compaction as a function of estimated vibrator induced stresses and time.

A different approach to describe the compaction process in granular systems can be found in Richard et al. (2005). They summarize the actual knowledge regarding compaction processes from a discrete element point of view. Since these studies are based on tapping experiments with isotropic granular media

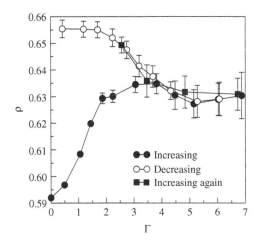

Figure 3.   Chicago-Tapping-Experiment.

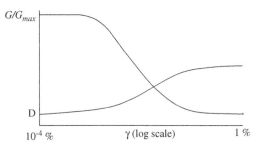

Figure 4.   Typical shear-modulus degradation curves.

4   DEMANDS ON A SUITABLE CONSTITUTIVE MODEL

From the discussion above three main motivations arise to use a numerical model describing vibrocompaction:

- To overcome the difficulty to derive case specific estimates for the probe-soil interaction and resulting particle velocities in the vicinity of the probe,
- To describe the full 3-D stress-strain path resulting from the probe-soil interaction and therefore to check whether dominating components can be identified justifying the use of lower dimension models,
- To calculate the densification effect as a function of the induced loading cycles.

From this, the constitutive formulation obviously should be able to predict a realistic wave propagation and therefore should result in shear modulus and damping curves as shown in Figure 4.

In addition, the formulation should be able to describe the accumulation of deformations due to the cycling loading. Elastoplastic models with a single yield surface are not suitable since they predict no irreversible strain in the elastic regime. For example, a multi-surface (Nakano and Nakai 2003) or hypoplastic constitutive equations overcome this shortcoming. The constitutive equation should also be able to describe dilatancy effects.

Additional difficulties arise from the large number of cycles (about 2000–4000 per compaction stage) that have to be taken into account. Besides excessive calculation times, numerical errors from an implicit calculation strategy will accumulate to orders of magnitude that question the results in general.

An explicit model suggested by Wichtmann (2005) avoids the shortcomings from the implicit solution strategy. It describes the accumulation of irreversible strains for cyclic loading paths as a function of the number of loading cycles. Figure 5 shows the basic idea of the explicit calculation scheme.

The accumulation of strain due to an increment of ΔN loading cycles with the amplitude ε^{ampl} results

placed in closed boxes with free surfaces transfer of results from these investigations to geotechnical applications is somewhat limited. Nevertheless, some interesting parallels to observations made in geotechnical contexts can be found. Figure 3 shows results from a tapping experiment in a 1.9 cm diameter and 1m high tube filled with 2 mm diameter glass beads subject to vertical shaking. The initial packing was a low density configuration. At each value of the ratio $\Gamma = a/g$ of the tap peak acceleration a and gravity acceleration g the system was tapped 10^5 times followed by an additional series of tapping with increased Γ. The density increases up to tapping intensities Γ smaller than three. Larger intensities show a small decrease in density. This is interpreted as void creation due to a too large agitation. Reducing Γ results in again increasing densities. The lower left branch of the compaction curve is interpreted as "shaking out the voids", i.e. particles fill large voids. This rearrangement affects only a few particles next to the void. The upper left branch is described as "collective reorganization" based on the idea that further increase in density requires rearrangement of larger numbers of particles connected to each other. The two mentioned processes exhibit different kinetics, where the first is commonly described as diffusion process (Richard et al. 2005).

Even having in mind that the described behaviour is closely related to the experimental setup, the observation that when choosing too large agitation the sample cannot be compacted to maximum density seems to be connatural to the observation in geotechnics that when choosing too large strain amplitudes during compaction the initial contractant behaviour changes to dilatant behaviour and therefore a slight decrease in density.

in an irreversible strain $\mathbf{D}^{acc}\Delta N$. This implies that the accumulation rate $\mathbf{D}^{acc}$ can be considered as constant during the cycles under consideration (sequential update of amplitudes). The accumulation rate itself is calculated from

$$\mathbf{D}^{acc} = \mathbf{m} \cdot f_{ampl}\, f_N\, f_p\, f_Y\, f_e\, f_\pi \qquad (3)$$

The scalar functions $f_{ampl}, f_N, f_p, f_Y, f_e, f_\pi$ describe the influence of the strain amplitude ε^{ampl}, the Number of cycles N, the average mean pressure p_{av}, the void ratio e and the polarization π of the strain loop. The unit tensor $\mathbf{m}$ points in the direction of accumulation in the strain space. The constitutive equation used to relate stresses and strains is a viscoplastic type equation:

$$\mathring{\mathbf{T}} = \mathbf{E} : (\mathbf{D} - \mathbf{D}^{acc} - \mathbf{D}^{pl}) \qquad (4)$$

wherein $\mathring{\mathbf{T}}$ is the Zaremba-Jaumann rate of the Cauchy stress, $\mathbf{D}$ denotes the total strain rate, $\mathbf{E}$ is a pressure dependent elastic stiffness and $\mathbf{D}^{pl}$ the plastic strain rate.

The general scheme of the explicit calculation strategy takes two major steps:

1. Calculate implicitly a sufficient number (typically 2–20) of load cycles using a suitable constitutive

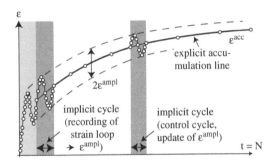

Figure 5. Explicit calculation scheme for total strains and general trend (adopted from Wichtmann 2005).

equation (see discussion above, in our case we used a hypoplastic approach) and record the strain path during the last loading cycle,

2. Determine the accumulation rate $\mathbf{D}^{acc}$ using equation (3). In the subsequent cycles only the general trend of the accumulation is calculated. This trend is depicted with the thick line in Figure 5.

Details concerning the explicit strategy can be taken from Wichtmann (2005). The following section describes the results from a first 3-D FE-calculation serving the first half of the above mentioned step no 1.

We assume that the material flow is needed to keep a good contact between probe and soil ensuring optimum energy transfer to the surrounding soil but has a minor effect on the compaction process itself.

5 FE-MODEL

First trials with plain-strain models (vertical and horizontal orientated) showed that results were dominated by consequences resulting from restrictions of plane-strain conditions. Therefore a full 3-D approach was performed. To keep calculation time at reasonable dimensions, we are looking at a disk shaped section located at 14.25 m depth. The 23.60 m diameter model disk is composed of one row of elements of 0.50 m height (see Figure 6). Vertical and horizontal fixities are applied to the outer circumference nodes, geostatic stresses σ_u and σ_l to the upper and lower plate surfaces. The plate size is sufficient to ensure that no reflections from the boundaries will influence the region of interest i.e. the three meter diameter range around the probe during the calculated initial 5 cycles.

The 0.30 m diameter and quasi rigid probe (linear-elastic stiffness of $E = 2.1 \cdot 10^7$ kPa) is placed at the center of the disk. Mohr-Coulomb type interface-elements allow relative displacements of probe- and soil-nodes. Due to the high vibration forces even in dry soils a narrow zone of almost fluidized soil surrounding the probe will form during most of the compaction

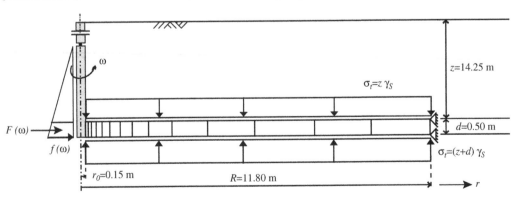

Figure 6. FE-Model of disk located at 14.25 m depth.

time. Due to this a no-friction condition is assumed for the interface, only normal stresses are transferred to the soil. Different from real vibrators no fins are taken into account. Consequences resulting from this assumption still remain to be analysed.

The driving dynamic force was calculated for a probe generating dynamic forces of 400 kN assuming a linear distribution of this force along the vibrator (see left hand side of Figure 6). A vibrating frequency of 54 Hz was used.

The hypoplastic constitutive model (Kolymbas 1991; von Wolffersdorff 1996), extended by intergranular strains (Niemunis and Herle 1997; Gudehus 1996) is used. This model meets the requirements listed in section 4 (Niemunis 2003). The Material parameters representing a medium sand listed in Table 1 have been used in the analysis. The compaction starts from loose conditions at $e_0 = 0.90$.

6 RESULTS

Figure 7 shows the trace of a point located in the center of the vibroflot in the r-θ-plane (for orientations see Figure 11).

Table 1. Material parameters of the used Hypoplastic Material Law with intergranular strain.

φ_c	ν	h_S	e_{d0}	e_{c0}
31.2°	0.38	591000 kPa	0.577	0.874
e_{i0}	α	β	m_2	m_5
1.005	0.12	1.0	1.45	2.9
R	β_r	χ	–	–
10^{-4}	0.2	6.0	–	–

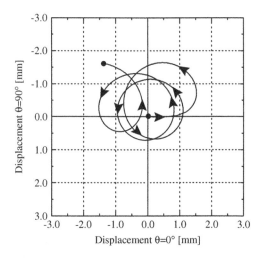

Figure 7. Trace of vibrator center in the r-θ-plane.

The calculated displacement amplitude equals 3 mm at a frequency of 54 Hz which corresponds to one fifth to one tenth of the measured amplitudes of high dynamic force vibrators suspended in air and therefore the result shows a reasonable size. The vibrator rotates on its axis which differs from reality, where fins prevent this behaviour.

As discussed in section 3, vibrator induced acceleration and particle velocities can be considered as important indicators for the compaction range. Figures 8 and 9 show the maximum values of the absolute accelerations/velocities from the five simulated cycles. The typical decrease with increasing distance from the probe can be observed. A direct comparison to measured velocities is of limited usefulness since the calculated values depend on the specific boundary conditions differing from those valid for the measured values. However, measured accelerations during placement of stone columns (similar vibrators) showed values at about 3 g at 1 m distance from the probe (Baez 1995).

This again is at the same order of magnitude than the calculated accelerations. Horizontal accelerations

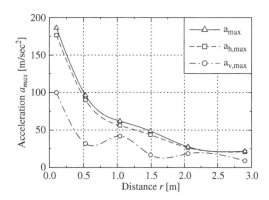

Figure 8. Calculated max. Accelerations.

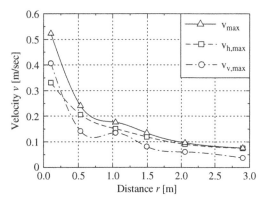

Figure 9. Calculated max. absolute Velocities.

129

dominate, which is reasonable from the rotating vibrator action. In contrary, induced horizontal and vertical "particle-velocities" show values of the same order of magnitude.

Figure 10 shows the densification effect, as a function of distance and the number of cycles. Despite of the fact that the maximum densification is calculated next to the vibrator interpretations from CPT-tests at vibro-compaction sites show maximum densities at distances of about 0.5–1.0 m from the vibrator, the general trend corresponds to real behaviour. The evaluation of the order of magnitudes as well as the analysis of differences regarding the maximum densification effect is somewhat restricted by the small number of calculated cycles. The trend of decreasing compaction rates with increasing number of cycles can clearly be depicted.

Remembering that induced shear strains govern the densification effect the order of magnitude of calculated shear-strains is of special interest (Figure 12, acting orientations can be taken from Figure 11). The dominating shear strains can be observed in $\varepsilon_{r\theta}$-plane. Maximum $\varepsilon_{r\theta}$ shear-strain decreases with increasing distance from $1 \cdot 10^{-3}$ at 0.1 m distance

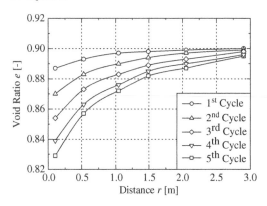

Figure 10. Calculated void ratio e vs distance for 1st to 5th cycle.

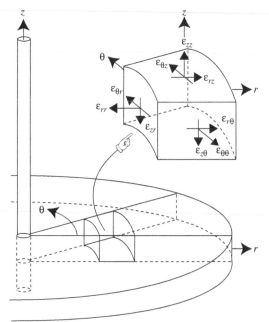

Figure 11. Activity orientation of Shear-Strains.

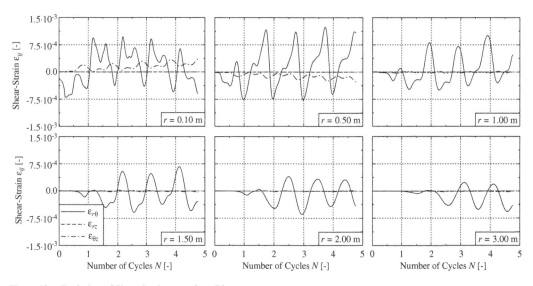

Figure 12. Evolution of Shear-Strains at various Distances r.

to $2.5 \cdot 10^{-4}$ at 3 m distance. Furthermore, the motion of the vibroflot induces deviatoric shear strains which can be calculated from

$$\varepsilon_{dev} = \varepsilon_{rr} - \varepsilon_{\theta\theta} \qquad (5)$$

or

$$\varepsilon_{dev} = \varepsilon_{rr} - \varepsilon_{zz} \qquad (6)$$

Figure 13 documents the deviatoric strains calculated from equation (5) which show values almost twice as large as the shear strains. This clearly means that shear-strains are not sufficient to describe the loading cycle. Increasing permanent deviatoric strains in $\varepsilon_{\theta\theta}$ and ε_{zz} plane result from formation of funnel like settlements next to the vibrator.

Assuming that the model reflects real 3-D behaviour – which is not obvious since the disk is free to distort frictionless relative to its upper and lower boundary – the results indicate that a simplifying one dimensional approach (like those mentioned in section 3) is not sufficient to describe the compaction effect. This means that horizontally polarized shear waves, propagating from the vibrator as well as horizontally and vertically polarized deviatoric strains resulting from p-waves, all three result in significant compaction effects during vibrocompaction of dry sands. However a 3-D model describing a larger section in depth is needed to verify this result.

Figures 14 and 15 show the $\varepsilon_{r\theta}$-shear-strain and $\sigma_{r\theta}$-shear-stress hysteresis loops and the resulting shear modulus calculated at 0.1 m and 0.5 m distance from

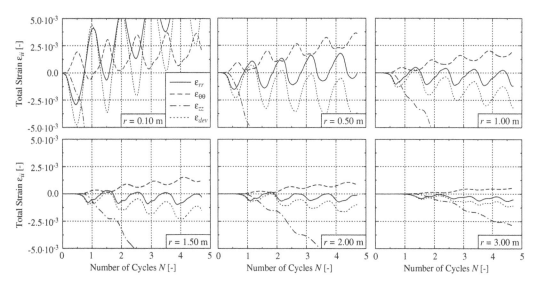

Figure 13. Evolution of Total Strains at various Distances r.

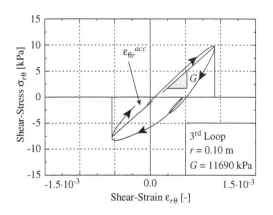

Figure 14. Calculated $r\theta$-shear-strain shear-stress hysteresis loop at 0.1 m distance from the probe.

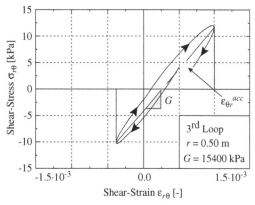

Figure 15. Calculated $r\theta$-shear-strain shear-stress hysteresis loop at 0.5 m distance from the probe.

the probe. At 14.25 m depth, the model shows a decrease in shear modulus as required. At larger distances, the calculated shear modulus remains almost constant which means that the model shows elastic behaviour (with respect to shear stiffness) at distances equal or larger than 1.5 m (Triantafyllidis 2000) from the vibrator. Taking into account that shear strain amplitudes show values of about $5 \cdot 10^{-4}$ at these distances, a further increase with distance would be expected. This indicates that model parameters describing the behaviour at relatively small strains may have to be readjusted.

7 CONCLUSIONS

Improvement of constitutive models during the last decade allows for modelling of compaction of dry sands. While elastoplastic models are not suitable, multi-surface elastoplastic or hypoplastic constitutive equations offer the capability to mirror the key influences on the densification mechanism:

- realistic wave propagation (shear modulus degradation curves and probe-soil energy transfer),
- accumulation of strains due to relatively small deformations,
- dilatancy effects.

Results from 3-D FE-modelling vibrocompaction of a disk shaped section at 14 m depth confirm this finding. They show that compaction results from horizontally polarized propagating shear waves as well as from horizontally and vertically polarized deviatoric strains resulting from p-waves. Modelling larger sections and a larger number of cycles (need of infinite boundaries) will allow for direct comparison of predicted and measured quantities. Especially the model behaviour next to the vibrator needs a more careful investigation. Parameter studies will investigate the models' capabilities to mirror different site specific soil conditions.

An explicit calculation scheme can be used to overcome problems resulting from accumulation of numerical errors when using implicit strategies to calculate several thousands of loading cycles. The influence of multi-polarization of loading cycles on the compaction process remains to be analysed.

REFERENCES

Baez, J. (1995). *A Design Model for the Reduction of Soil Liquefaction by Vibro-Stone-Columns*. Ph. D. thesis, The University of Southern California.

Cudmani, R., T. Meier, and W. Wehr (2003). Entwicklung und Verifikation eines Verfahrens zur Bemessung von Rütteldruckverdichtungsmaßnahmen. In J. Grabe (Ed.), *Bodenverdichtung - Experimente, Modellierung,*

Geräteentsicklung, Baustellenberichte, F + E-Bedarf, pp. 61–82.

Degen, W. and J. Hussin (2001). Soil-densification: Short course notes. In *Geo-Odyssey Conference*, pp. 89. ASCE.

Dowding, C. H. (1996). *Construction vibrations*. Prentice Hall.

Fellin, W. (2000). *Rütteldruckverdichtung als plastodynamisches Problem*. Ph. D. thesis, University of Innsbruck, Institute of Geotechnics and Tunneling.

Green, R. A. (2001). *Energy-Based Evaluation and Remediation of Liquefiable Soils*. Ph. D. thesis, Virginia Polytech Institute and State University.

Greenwood, D. and K. Kirsch (1983). Specialist ground treatment by vibratory and dynamic methods. In *Proceedings of the international conference on advances in piling and ground treatment for foundations*.

Gudehus, G. (1996). A comprehensive constitutive equation for granular materials. *Soils and Foundations 36*, 1–12.

Ishibashi, I. and X. Zhang (1993). Unified Dynamic Shear Moduli and Damping Ratios of Sand and Clay. *Soils and Foundations 33*(1), 182–191.

Kolymbas, D. (1991). An outline of hypoplasticity. *Archis of Applied Mechanics 61*, 143–151.

Massarsch, K. R. (2003). Effects of Vibratory Compaction. In *TransVib-Conference, Brussels*.

Mitchell, J. and P. Gallagher (1998). Guidelines for Ground Improvement of Civil Works and Military Structures and Facilities. *Publication No. ETL 1110-1-185US Army Corps of Engineers*.

Nakano, K. and A. Nakai (2003). A Description of Mechanical Behaviour of Clay and Sand based on Evolutions of Soil Structure and Overconsolidation. In *Proceedings of the First Japan - U.S. Workshop on Testing, Modeling, and Simulation*, Geotechnical Special Publication No. 143, pp. 136–153.

Niemunis, A. (2003). *Extended Hypoplastic Models for Soils*. Habil. thesis, Ruhr-University Bochum, Institute of Soil Mechanics and Foundation Engineering.

Niemunis, A. and I. Herle (1997). Hypoplastic model for cohesionless soils with elastic strain range. *Mechanics of Cohesive-Frictional Materials 2*(279–299).

Richard, P., M. Nicodemi, R. Delannay, P. Ribière, and D. Bideau (2005). Slow relaxation and compaction of granular systems. *Nature 4*, 121–128.

Rodger, A. A. (1979). Vibrocompaction of cohesionless soils. *Cementation Research Limited, Int. Rep., R. 7/79*. Summarized in Greenwood (1983).

Slocombe, B., A. Bell, and J. Baez (2000). The densification of granular soils using vibro methods. *Géotechnique 50*(6), 715–725.

Triantafyllidis, T. (2000). Abschätzung von Setzungen beim Einbau von Rüttelinjektionspfählen. In T. Triantafyllidis (Ed.), *Böden unter fast zyklischer Belastung: Erfahrungen und Forschungsberichte*, Volume 32 of *Schriftenreihe des Institutes für Grundbau und Bodenmechanik der Ruhr-Universität Bochum*, pp. 39–57.

von Wolffersdorff, P.-A. (1996). A hypoplastic relation for granular materials with a predefinied limit state surface. *Mechanics of Cohesive-Frictional Materials 1*(251–271).

Wichtmann, T. (2005). *Explicit accumulation model for noncohesive soils under cyclic loading*. Ph. D. thesis, Ruhr-University Bochum, Institute of Soil Mechanics and Foundation Engineering.

Numerical Modelling of Construction Processes in Geotechnical Engineering for Urban Environment – Triantafyllidis (ed)
© 2006 Taylor & Francis Group, London, ISBN 0 415 39748 0

FE simulations of the installation of granular columns in soft soils

T. Meier[1] & R. Cudmani[2]

[1]*Institute for Soil and Rock Mechanics, University of Karlsruhe*
[2]*Ed. Züblin AG, Stuttgart*

ABSTRACT: A parametric study is conducted with the FEM and a visco-hypoplastic constitutive equation for the description of the mechanical behaviour of soft soils, in order to identify the relevant quantities leading to critical deflections (buckling) of an auger during the installation of sand-cement columns. It is concluded that the buckling load of an auger (or a pile-like structure) depends strongly on the magnitude and shape of the initial geometrical imperfections and the stiffness of the soil. The use of the undrained shear strength to judge the stability of slender pile-like structures in soft soils, as e.g. recommended by Eurocode 7, appears to be physically questionable.

1 INTRODUCTION

The combined soil stabilization with vertical columns (CSV) is a technique where a dry sand-cement mixture is forced into soft ground by means of a continuous flight auger. Due to capillarity and excess pore water pressure caused by the cavity expansion in the soil under undrained conditions, pore water flows into the columns, allowing the mixture to set and the adjacent soil to densify.

In very soft soil difficulties during the installation of the columns might be encountered. At the time when the auger is pushed into a stiffer underlying stratum and the soil does not provide enough lateral support, significant deflections of the auger from the reference position may occur. In extreme cases this may lead to a blending of the sand-cement mixture and the soil.

The objective of this work is to examine under which circumstances buckling of the auger during the installation of the columns can occur.

2 CONSTITUTIVE MODEL

In all calculations presented in this paper, the mechanical behaviour of the soft soil is modelled using the visco-hypoplastic constitutive equation after (Niemunis 2003). It is an extension of hypoplasticity (Wolffersdorff 1996) allowing for creep, relaxation and rate-dependence of the material response. This visco-hypoplastic model has been validated in many cases, including e.g. predictions of deformations due to open pit mining (Karcher 2003) or the construction process of diaphragm walls (Schäfer and

Triantafyllidis 2004). The mechanical behaviour is described by the following equations:

$$\overset{\circ}{\mathbf{T}} = f_b(\lambda, \varphi_c) \, \hat{\mathcal{L}}(\mathbf{T}, e) : (\mathbf{D} - \mathbf{D}^v) \tag{1}$$

$$\mathbf{D}^v = -D_r \, \vec{\mathbf{B}} \, OCR^{-1/I_v} \tag{2}$$

$$\vec{\mathbf{B}} = \frac{\hat{\mathcal{L}}^{-1} : \hat{\mathbf{N}}}{||\hat{\mathcal{L}}^{-1} : \hat{\mathbf{N}}||} \tag{3}$$

$$OCR = p_e/p_e^+ \tag{4}$$

$$p_e = p_{e0} \left(\frac{1+e}{1+e_{e0}} \right)^{-1/\lambda} \tag{5}$$

$$p_e^+ = p \left[1 + \left(\frac{q}{M(\mathbf{T}) \, p} \right)^2 \right] \tag{6}$$

In the basic rate-type equation (1) the objective stress rate $\overset{\circ}{\mathbf{T}}$ is a function of an incrementally linear stiffness tensor $\hat{\mathcal{L}}$, derived from the original hypoplastic formulation, the elastic strain rate $\mathbf{D} - \mathbf{D}^v$ and a barotropy factor f_b. The viscous strain rate $\mathbf{D}^v$ (2) represents the non-linear part of the model with its intensity determined by a reference creep rate D_r, the overconsolidation ratio OCR and the viscosity index I_v after (Leinenkugel 1976). $\vec{\mathbf{B}}$ is the direction of the viscous flow derived from hypoplasticity (3). p_e is the isotropic equivalent pressure, which depends on the void ratio e according to Butterfield's compression law (5) (Butterfield 1979). In order to allow for non-isotropic stress conditions, the yield surface function of the Modified Cam Clay model (MCC) (Roscoe and Burland 1968) is consulted. For an arbitrary stress

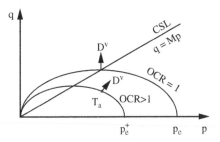

Figure 1. Yield surface of the MMC.

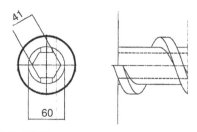

Figure 2. Mechanical drawing of the auger.

Table 1. Most important model parameters.

λ	Compression index after Butterfield
κ	Swelling index after Butterfield
φ_c	Critical friction angle
I_v	Leinenkugel's viscosity index
D_r	Reference creep rate
(p_{e0}, e_{e0})	Point on the reference isotach in the e–$\ln p$-diagram

condition $\mathbf{T}_a$ the corresponding isotropic pressure p_e^+ can be calculated with (6) (cf. Fig. 1). Contrary to MCC, in the framework of hypoplasticity the critical stress ratio M is not a material parameter but depends on $\mathbf{T}$ according to (Matsuoka and Nakai 1977).

3 FINITE ELEMENT MODELS

The real auger consists of segments of three metres length each. In the 2D simulations the auger is considered as a beam. Neglecting the helix, the flexural stiffness of the auger is $I_a \approx 4.7 \cdot 10^{-7}\,\mathrm{m}^4$, that of the connector $I_c \approx 1.7 \cdot 10^{-7}\,\mathrm{m}^4$ (Fig. 2). With these values the buckling load without lateral support of the soil is $F_B = 11.1\,\mathrm{kN}$ in case of an auger consisting of three segments and a connector length $l_c = 10\,\mathrm{cm}$. For $l_c = 30\,\mathrm{cm}$ the buckling load reduces to $F_B = 9.7\,\mathrm{kN}$. The static system as well as the first three eigenmodes of buckling are depicted in Figure 3.

In order to study the influence of the soil surrounding the auger on the buckling load, three different FE models were developed:

1. 2D plane strain model including the installation process (cavity expansion) and the granular column around the auger
2. Simplified 2D plane strain model without cavity expansion and granular column
3. 3D model, again without cavity expansion and granular column

In all models undrained conditions are assumed. There is no interface between the auger and the soil, i.e.

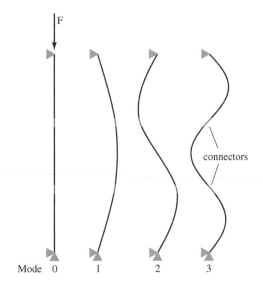

Figure 3. Static system and eigenmodes of the buckle analysis.

relative displacements are not possible. The models are described briefly in the following subsections.

3.1 2D plane strain model #1

The finite element mesh of this 2D model is shown in Figure 4. The lateral boundaries of the calculation domain are 20 m away from the auger and constrained in the horizontal direction, whereas the lower boundary is vertically constrained. The model consists of approximately 5400 4-node plane strain continuum elements. The auger is modeled with 120 2-node beam elements. In order to determine the buckling load and to enable deformations beyond, it is necessary to apply initial geometrical imperfections to the auger, i.e. small lateral deflections Δx relative to the reference position in 1-direction. The simulation is carried out in several steps:

1. Expansion of the cavity from an initial width of two centimetres to 23 cm. The resulting state of the

134

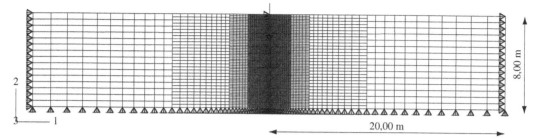

Figure 4. 2D the plane strain model.

soil (effective stress, excess pore water pressure)
are transferred to another (undeformed) mesh and
serve as the initial state for the next calculation step.
2. In the new mesh the auger, enclosed by a granular column modelled with the aid of a hypoplastic
constitutive equation, is activated inside the cavity.
3. Eventually the displacement-controlled axial loading is applied to the top of the auger.

3.2 2D plane strain model #2

The geometry and boundary conditions are the same as
before. However, there is no granular material around
the auger and the excess porewater pressures and
changes of state due to the expansion of the cavity
are not considered. The axial loading of the auger is
carried out in the same manner as in model #1.

3.3 3D model

In order to be able to examine how far the plane strain
considerations described above are justifiable, a 3D
model (Fig. 5) was developed. The height of the model
is the same as in the preceding models ($h = 8$ m), the
outer radius is $r_o = 10$ m. The lower boundary is fixed
in the vertical direction, the outer one in radial direction. The auger is modelled as a steel tube (linear
elastic). In order to avoid a fixed support at the toe of
the auger, only one single node is fixed in the vertical
direction (cf. Fig. 6). The resulting small eccentricity
of the support serves as initial imperfection. At the
surface the whole section of the auger is forced down
uniformly.

3.4 Results of the comparative calculations

1. The simulations performed with the plane strain
model #1 take relatively long and are somewhat
pedestrian, since the cavity expansion and the axial
loading of the auger are two independent calculations. However, the comparative calculations with
the plane strain model #2 show that there is practically no influence of the change of state due to the

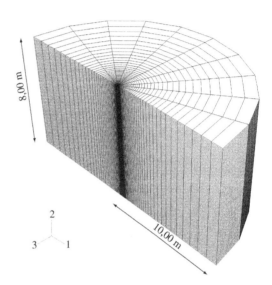

Figure 5. 3D model.

Figure 6. Vertical support of the auger in the 3D model.

cavity expansion and of the granular body around
the auger on the buckling load.
2. The calculations with the 3D model yield that, when
reaching the buckling load in the soil, the shape of
the auger does not correspond to eigenmode 1 but
3 (Fig. 3). The evolution of the deformation of the
auger during loading is depicted in Figure 7. Recent

model and in situ tests confirm this result (Vogt and Vogt 2005).

3. If the shape of the deformed auger as obtained from the 3D calculations (corresponding to eigenmode 3, cf. Fig. 3) is applied as initial geometrical imperfection in the 2D simulations, a comparable buckling load results.

Thus we act on the assumption, that with the aid of a parametric study using plane strain model #2, it is possible to identify the relevant parameters with respect to the buckling load.

4 PARAMETRIC STUDY

The influence of the following quantities on the buckling load F_B of the auger was examined.

1. Surrounding Soil:
   - Compressibility: $\lambda = 0.15/\lambda = 0.45$
   - Critical Friction Angle: $\varphi_c = 30°/\varphi_c = 40°$
   - Void ratio: $e = 8/\bar{e} \approx 18$

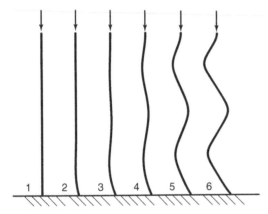

Figure 7. 3D calculation – evolution of the deformation of the auger.

2. Auger:
   - Flexural Stiffness of the connectors: $I_c = 1.7 \cdot 10^{-7} \, \text{m}^4 / I_c = 0.5 \cdot 10^{-7} \, \text{m}^4$
   - Length of the connectors: $l_c = 0.1 \, \text{m}/l_c = 0.3 \, \text{m}$
   - Static system: Toe (= tip) horizontally restrained (fixed)/unrestrained (free)
   - Shape of initial imperfections: shape of buckling mode 1/3
   - Maximum amplitude of initial imperfections: $|\max(\Delta x)| = (1, 3, 5) \, \text{cm}$

The calculation results are summarized in Table 2. In the scope of this study it was not possible to perform calculations for all combinations of the quantities mentioned above. Based on the numerical results the following relevant quantities with respect to the buckle load can be identified:

Compressibility of the soil: The comparison between #16 and #18 shows the significant influence of the stiffness of the soil, expressed as λ, on F_B.

Geometrical imperfections of the auger: For the same $|\max(\Delta x)|$ the choice of the mode of the initial deflections has a clear influence on F_B. The same holds true for a change in $|\max(\Delta x)|$ for a given mode.

Flexural stiffness of the auger and the connectors: The significant influence of I_a and I_c on F_B is self-explanatory.

The shear strength of the soil, expressed as the critical friction angle φ_c, has little influence on F_B (cf. #15 and #16). The constraint of the toe in horizontal direction played a minor role as well (cf. #13 and #14).

5 CONCLUSIONS

Three different finite element models were developed, which made possible the identification of the factors influencing the buckling load of a flight auger in the subsoil. In agreement with experiments the results of

Table 2. Calculation results.

#	$e_0 = 8$	$\bar{e}_0 = 18$	Imp. mode	$|\max(\Delta x)|$ in m	I_c in $10^{-7} \, \text{m}^4$	l_c in m	φ_c	Toe	λ	F_B^* in kN
1	x	–	1	0,01	1,7	0,1	40°	Free	0,45	164
2	–	x	1	0,01	1,7	0,1	40°	Free	0,45	165
3	–	x	–	–	1,7	0,1	40°	Free	0,45	>400
4	–	x	3	0,01	0,5	0,3	40°	Free	0,45	102
5	–	x	1	0,01	0,5	0,3	40°	Free	0,45	142
13	–	x	3	0,03	0,5	0,3	30°	Free	0,45	71
14	–	x	3	0,03	0,5	0,3	30°	Fixed	0,45	72
15	–	x	3	0,05	0,5	0,3	30°	Fixed	0,45	59
16	–	x	3	0,05	0,5	0,3	40°	Fixed	0,45	60
17	–	x	3	0,05	1,7	0,3	40°	Fixed	0,45	82
18	–	x	3	0,05	0,5	0,3	40°	Fixed	0,15	110

the 3D model show that the auger deforms according to the third (buckling) eigenmode. In the worst case the buckling load of the embedded auger was five to six times larger than in the air. Along with the flexural stiffness of the auger, the main factors influencing the buckling load are the flexural stiffness of the connectors, the initial geometrical imperfections of the auger and the compressibility of the surrounding soil. The critical friction angle, and thus the shear strength of the soil, did not show any strong influence. Thus, according to the results of the present study, the stiffness and not the undrained shear strength (c_u) of the soil is the decisive quantity. For this reason, the assessment of the susceptibility to buckling of slender embedded structures in terms of c_u appears to be questionable (after e.g. EN 1997-1:2003 §7-8(5) no proof of safety against buckling for piles embedded in a soft soil with a characteristic undrained shear strength $c_u > 10$ kPa is required). Since the relationship between the stiffness and the undrained shear strength can be quantitatively very different for different soft soils, the use of c_u instead of a stiffness quantity (e.g. initial stiffness of the soil) is unjustified.

REFERENCES

Butterfield, R. (1979). A natural compression law for soils. *Géotechnique* 29(4), 469–480.

Karcher, C. (2003). *Tagebaubedingte Deformationen im Lockergestein.* Ph.D. thesis, Veröffentlichungen des Instituts für Bodenmechanik und Felsmechanik/Universität Karlsruhe. Heft 160.

Leinenkugel, H. J. (1976). *Deformations- und Festigkeitsverhalten bindiger Erdstoffe. Experimentelle Ergebnisse und ihre physikalische Deutung.* Ph.D. thesis, Veröffentlichungen des Instituts für Bodenmechanik und Felsmechanik/Universität Karlsruhe. Heft 66.

Matsuoka, H. and T. Nakai (1977). Stress-strain relationship of soil based on the SMP. In S. Murayama and A. Schofield (Eds.), *Constitutive equations for soils. Proc. of Speciality Session 9, IX ICSMFE*, Tokyo, pp. 153–162. Japanese Society of Soil Mechanics and Foundation Engineering.

Niemunis, A. (2003). *Extended hypoplastic models for soils.* Politechnika Gdanska, Monografia nr 34.

Roscoe, K. and J. Burland (1968). On the generalized stress-strain behaviour of wet clay. In Heyman and Leckie (Eds.), *Engineering plasticity.* Cambridge University Press.

Schäfer, R. and T. Triantafyllidis (2004). Influence of the construction process on the deformation of diaphragm walls in soft clayey ground. *Tunnelling and Underground Space Technology 19*(4–5), 475ff.

Vogt, N. and S. Vogt (2005). Knicken von Pfählen in weichen und breiigen Böden. In J. Stahlmann (Ed.), *Pfahl- Symposium*, Braunschweig, pp. 197–218.

Wolffersdorff, P. v. (1996). A hypoplastic relation for granular materials with a predefined limit state surface. *Mech. Cohes.-Fric. Mater. 1*, 251–271.

4. Tunnelling

Numerical Modelling of Construction Processes in Geotechnical Engineering for
Urban Environment – Triantafyllidis (ed)
© 2006 Taylor & Francis Group, London, ISBN 0 415 39748 0

Large scale three-dimensional finite element analysis of underground construction

F.H. Lee & K.K. Phoon
National University of Singapore, Singapore

K.C. Lim
Tritech Consultants Pte. Ltd., Singapore

ABSTRACT: This paper discusses the possibility of using three-dimensional finite element analyses to analyse and design underground construction works. Two common problems associated with three-dimensional analyses mean very large memory requirement and long computing times. To overcome these problems, the Authors outline some recent developments in iterative solution algorithm which can lead to large savings in memory and time. It is shown that these iterative methods offer viable and faster alternatives to Gaussian elimination approach for solving very large finite element equations in geotechnical engineering, such as those arising from three-dimensional analyses, with significantly reduced memory requirement. The convergence characteristics of ill-conditioned coupled-flow problems can be improved by using a generalized Jacobi preconditioner. Drained problems can be efficiently solved using the preconditioned conjugate gradient method with the standard Jacobi preconditioner. On the other hand, undrained problems are better solved as "nearly-impermeable" consolidation problems, using quasi-minimal residual method with the generalized Jacobi preconditioner. The advantages of iterative methods increase with the size of the problem. For very large problems, the speed-up can be very significant. The applicability of iterative methods is illustrated by the study of a tunneling project involving earth pressure balance machine. The details of face pressure application, tunnel convergence around shield and lining installation were modeled. The results show that large-scale three-dimensional analyses with iterative solution cannot only be implemented on relatively modest computation platforms with reasonable turnaround times, but they are also able to illuminate ground mechanisms which cannot be reflected by two-dimensional analyses.

1 INTRODUCTION

Two-dimensional finite element analyses are now routinely conducted, and a number of commercial softwares are available. However, real problems, such as those encountered in underground construction works, are often intrinsically three-dimensional (3D) in nature. The need to conduct increasingly realistic analysis of geotechnical problems in the construction industry has led to the development of 3D finite element (FE) codes with special elements and facilities to model soil-structure interaction features and construction aspects in three dimensions, instead of merely idealizing the problem into a plane section. However, many realistic 3D analyses require huge memory and excessive time to run on personal computers (PCs). This paper reviews some recent technological advances for solving large sets of FE equations. The requirements of realistic 3D analyses are first illustrated through a few problems. As will be seen, these requirements underscore the need for a new class

of memory- and time-efficient solution algorithms. Next, the formulation, implementation and performance of two Krylov subspace iterative solvers are introduced. The performance of the iterative solvers is then compared against that of a frontal solver, the latter being taken as being representative of direct solution algorithms. Finally, one of the proposed algorithms is applied to an earth pressure balance tunnelling problem as an illustration.

2 REQUIREMENTS OF 3D ANALYSIS

It is now increasingly recognised that many underground construction problems are intrinsically 3D in nature and would benefit from 3D analysis. For instance, Lee et al. (1998) suggested that many building basement excavations are significantly influenced by the effects from their corners. This is especially so for those excavations where the hard stratum into which the retaining wall is socketed lies at a large depth, compared to the span of the wall. In an increasingly congested urban environment, wall spans of deep basement excavations are likely to become more restricted while the need to provide more carpark space is likely to drive basement excavations deeper. In such circumstances, corner effects in basement excavations can only become more significant with time.

Apart from basements, many underground constructions have long trench-like or tunnel-like configurations; it is often tempting to presume that they can be considered as reasonably plane strain problems. Or can they? In the discussion below, two real constructions will be considered to illustrate that long span constructions are also not necessarily plane strain in nature, when one starts to consider the working stress conditions, with construction sequence factored in.

The first case refers to a deep trench excavation with multiple-level strutting as shown in Fig. 1. An examination of this case shows that, in spite of its long span, 3D effects cannot be avoided, even if local effects are to be neglected. The trench excavation lies on a gentle curvilinear alignment. The excavation was retained by diaphragm wall panels supported by multiple struts without any global walers, apart from a

Figure 1. Curved diaphragm wall supporting deep trench excavation.

ground beam along the top of the wall. Each individual wall panel was supported by its own strut(s). When the line of diaphragm wall panels on the inside of the curve deflects into the excavation, circumferential strain is induced along the wall line. As the wall on the inner side of the curve deflects into the trench, the wall line extends, thereby inducing in-plane tension in the diaphragm wall panels. Since there are no steel bars in between diaphragm wall panels, inter-panel cracking may occur. This will not occur in the wall on the outer side of the curve, which would experience compression when it deflects into the excavation. This is a problem which can only be demonstrated with a 3D analysis which models the joints between diaphragm wall panels. The total stress analysis of this site, which was conducted using the software ABAQUS, requires more than 2.5 GB of RAM (minimum) and requires about 5 days to complete.

The second example refers to a NATM tunnel with forepoling to form an umbrella pipe arch, see Figs. 2a–b. The aim is to investigate the interaction between the pipe arch and the ground with a view to ascertaining the maximum step length behind the tunnel face which can be open at any one time. This is again a 3D problem. The effective stress analysis of this problem, considering coupled groundwater flow, was conducted using ABAQUS. The minimum RAM required was 3 GB and each run takes about 2 weeks to complete.

These two cases underscore the fact that many underground construction problems are intrinsically 3D in nature. However, what is also clear is that 3D modeling requires a huge amount of computing memory and time. The version of ABAQUS used for the two problems mentioned above was able to address a maximum of 3 GB RAM using Windows XP on a PC. In any case, Windows XP can utilize a maximum of only about 3.6 GB RAM since it is still a 32-bit operating system. This is likely to limit the size and complexity of the problem that can be solved in 3D using a PC. Since the typical computer platform that a consultant

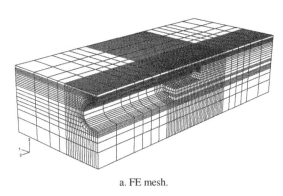

a. FE mesh.

b. Forepoling.

Figure 2. NATM tunnel problem.

can afford is a PC, this puts a lot of the 3D analysis out of reach of many consultants. In order to bring 3D analysis within reach of many consultants, more memory- and time-efficient ways of solving large complex 3D problems need to be found.

3 SOLUTION OF FINITE ELEMENT EQUATIONS

3.1 *Direct solution*

FE analysis often leads to a set of linear algebraic equation of the form

$$Ax = b \tag{1}$$

where $A \in \Re^{N_d \times N_d}$, $x \in \Re^{N_d}$ and $b \in \Re^{N_d}$. The matrix A has dimensions N_d-by-N_d, where N_d denotes the number of degrees-of-freedom. $\Re^{N_d \times N_d}$ denotes vector space of real N_d-by-N_d matrices and $\Re^{N_d}$ denotes the vector space of real N_d-vectors. The global matrix A resulting from drained and undrained geotechnical problems can be shown to be positive definite (Chan 2002). In contrast, consolidation problems based on Biot's (1941) formulation generally lead to indefinite A matrices.

The most widely used algorithms for solving Eq. 1 are refined variants of the Gaussian elimination approach, such as the bandwidth solver (Zienkiewicz 1989), Frontal solver (e.g. Irons 1970, Britto & Gunn 1987) and the multi-frontal solver (Hibbitt et al. 1997). Such solvers are very efficient for small and medium-sized problems where the matrix is still fairly dense and the time spent on factoring is roughly equivalent to the time spent on solving the system iteratively. Gaussian elimination solution methods usually have clear and consistent convergence characteristics which are not problem dependent (e.g. Irons 1970, Britto & Gunn 1987). If the problem involved has a condition number which is less than the limit set by the precision of the code and computer, the time needed to obtain the solution is almost a constant. What is compromised as the condition number increases is the precision to which the solution can be obtained. In other words, as the condition number increases, so do the round-off errors. If the condition number exceeds the allowable limit, then precision is compromised to the extent that the problem is perceived to be a singular problem by the code and solution will normally not proceed.

For 3D problems involving large matrices, even just storing the front or semi-bandwidth can require a large amount of memory. Even with out-of-core access, memory capacity may not even be able to cope with minimum storage required to run the analysis. In addition, in multi-clocked hierarchical computer systems such as PCs, the large amount of indirect addressing used by such algorithms tends to lead to a rather low cache data re-use rate, thereby causing a drop in CPU efficiency.

3.2 *Krylov subspace iterative methods*

In contrast to Gaussian elimination methods, iterative methods may not require assembly of the global stiffness matrix, thereby significantly reducing computer memory usage. Barrett et al. (1994) showed that, properly preconditioned, Krylov subspace iterations could provide efficient and simple "general purpose" viable alternatives to direct solvers. In contrast to direct methods, iterative methods converge iteratively towards the correct answer, the iterations being terminated when the required accuracy is reached. Thus, whereas direct methods control the solution time and allow accuracy to degrade as the condition number increases, iterative methods often control the accuracy of the solution and allow the solution time to vary with the condition number. Many iterative methods have convergence characteristics which vary substantially with condition number. In general, the larger the condition number, the more likely is the failure to converge. However, this is only a general guideline and the type of problem also has a major effect.

For positive definite A-matrices, such as those resulting from drained and undrained geotechnical problems, and many structural problems, the conjugate gradient (CG) method, first proposed by Hestenes & Stiefel (1952), has been shown to be highly memory- and time-efficient (e.g. Papadrakakis 1993, Mitchell & Reddy 1998, Wang 1996, Lim et al. 1998). Fox & Stanton (1968) and Fried (1969) pointed out that the assembly of the global stiffness matrix is not essential in CG methods and that the matrix-vector operations can be performed at element level. Such element-by-element (EBE) implementation of CG methods can drastically reduce memory requirements in large 3-D finite element analyses.

To accelerate convergence, iterative methods usually incorporate a preconditioning process with a preconditioner matrix. A commonly used, simple and inexpensive preconditioner is the Jacobi Preconditioner (e.g. Smith 2000, Saad 1996), which is the principal diagonal of the stiffness matrix. Hereafter, it is designated as EPCG. The Jacobi preconditioner can be applied to the individual element matrices before CG iterations. This allows the preconditioning process to be removed from PCG iterations, with additional savings in operation counts per iteration.

More complex EBE-based preconditioners have also been developed from approximate Cholesky factorisation (Winget & Huges 1985) and LU splitting techniques (Nour-Omid & Parlett 1985). Other preconditioners of similar form include the Crout and Gauss-Seidel EBE preconditioner (Papadrakakis 1993). However, Smith & Wong (1889) and Dayde et al. (1997) showed that the reduction in the number of iterations through use of a product or polynomial-form preconditioner does not always translate into greatly reduced total CPU time since the time per iteration is substantially increased. Mitchell & Reddy (1998) developed a recursively-defined preconditioner based on successive p-refinement, which can be implemented with EBE strategies. However, substantial working storage is needed for the hierarchical vector spaces and the lowest-p order global matrix, even for 3-D problems with highly regular domains. For realistic 3-D geotechnical analyses, the storage demand is likely to be increased even further by the presence of complex geometries, multiple material zones and irregular soil stratifications. For these reasons, the Jacobi preconditioner remains widely used in EBE strategies (e.g. Wang 1996, Smith & Griffiths 1997, Lim et al. 1998).

For consolidation problems, the A-matrix is indefinite and convergence cannot be guaranteed with CG methods (e.g. Paige & Saunders 1975, Golub & Van Loan 1989, Barrett et al. 1994, Smith 2000). Krylov subspace methods such as Minimal Residual (MINRES) or Symmetric LQ (SYMMLQ) (e.g. Paige & Saunders 1975) can achieve iterative solutions of indefinite systems. However, MINRES tends to be very vulnerable to rounding off errors (Saad & van der Vorst 2000). In addition, the construction of an effective preconditioner for MINRES remains an open problem. SYMMLQ methods (e.g. Paige & Saunder 1975) employed a LQ decomposition of the tridiagonal matrix T and minimised the norm of the error instead of minimising the norm of the residual. The advantage of SYMMLQ method over MINRES lies in regular short recurrences, minimal overhead and economic storage. However, SYMMLQ often converge much slower than MINRES for ill-conditioned systems (van der Vorst 2002).

Freund & Nachtigal (1994a, 1994b) proposed a Quasi-Minimal Residual (QMR) method for solving symmetric indefinite systems. QMR has the advantage over MINRES or its variants because it can be combined with an indefinite preconditioner, which is readily available within the system matrix and is thus more practical. QMR requires slightly more storage than MINRES but the additional storage vectors is marginal. On the flip side, QMR method converges in considerably less iterations than MINRES (Freund & Nachtigal 1994b). Thus QMR appears to be a more robust method for solving symmetric indefinite system.

4 PERFORMANCE OF JACOBI PRECONDITIONING

The performance of Jacobi preconditioning in PCG and QMR applied to geotechnical analysis have been studied by Lee et al. (2002), using a relatively small problem consisting of about 1000 degrees-of-freedom. In their study, the number of iterations needed to achieve "convergence", N_c, is taken to be the number of iterations needed for the relative residual norm $R_r^{(n)}$, defined by

$$R_r^{(n)} = \frac{\left\| b - Ax^{(n)} \right\|}{\left\| b - Ax^{(0)} \right\|} \tag{2}$$

to fall below 1×10^{-6}. In Eq. 2, the subscripts $^{(0)}$ and $^{(n)}$ denote the initial value and value after n iterations, of the variable. Fig. 3a shows the variation of N_c with condition number for drained and undrained cases studied. In this figure, N_c has been normalized by the number of degrees-of-freedom N_d. As can be seen, although N_c/N_d, increases with the condition number, the rates of increase for drained, undrained and consolidation analyses are distinctly different.

For the drained cases analysed, N_c/N_d increases gradually from about 10% to about 30%. Both the EPCG and EQMR algorithms require roughly the same number of iterations for the same condition number. However, each EQMR iteration involves slightly more operations than each EPCG iteration; hence, EPCG is faster for this class of problems. For the undrained problems, the points are banded differently from those of the drained cases, with N_c/N_d increasing much more rapidly with condition number than the drained problems. For consolidation problems, the points fall largely into a third band between that of the drained and undrained analyses, Fig. 3b.

Lee et al. (2002) further showed that the convergence characteristic of PCG and QMR algorithms depends not only on the condition number, but also on the eigenvalue distribution and the stiffness ratios between different eigenmodes. The convergence behaviour of drained problems affected by "material ill-conditioning" arising from large stiffness ratios between the different material zones are readily improved by Jacobi preconditioning. This is explainable by the fact that the stiff and soft material zones occupy different parts of the FE domain and are thereby linked to different degrees-of-freedom. By normalising the stiffness coefficients for the degrees-of-freedom by their respective diagonal entries, Jacobi preconditioning, in effect, homogenises the various material zones' stiffness.

For undrained cases modelled using a nearly incompressible pore fluid, the Jacobi preconditioner appears to be much less effective. The number of iterations to converge is far higher than the drained problems, even though the condition number, after preconditioning, may be similar. This is because the material ill-conditioning of an undrained problem

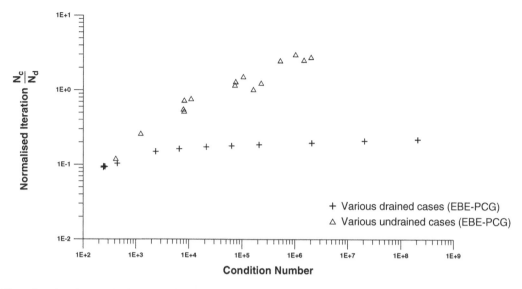

Figure 3a.   Iteration number for EPCG algorithm on drained and undrained problems.

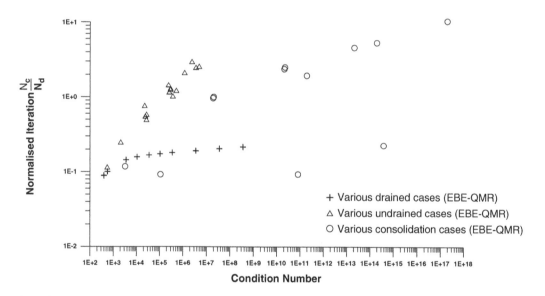

Figure 3b.   Iteration number for EQMR on drained, undrained and consolidation problems.

arises from the large stiffness ratios between compression/dilatation and shear flow eigenmodes. In such situations, Jacobi preconditioning changes the eigenvalues for these modes by approximately the same ratio, thus yielding little or no compression to the eigenvalue distribution.

For consolidation problems, some eigenvalues are displacement-dominated, whereas others are excess pore pressure dominated. The Jacobi preconditioner compresses the displacement-dominated eigenvalues in a similar manner to the drained cases. However, the pore pressure eigenvalues appear to be over-scaled.

Of the three classes of problems considered, the convergence characteristics of drained problems are most readily improved by the standard Jacobi preconditioner. Undrained problems are commonly modeled as a single-phase, nearly incompressible problem or a dual-phase problem involving an almost incompressible pore water phase. Lee et al. (2002) showed that this "nearly-incompressible" approach leads to a form of global stiffness matrix A which is highly resistant to standard Jacobi preconditioning.

Consolidation problems are more resistant to Jacobi preconditioning than drained problems but are less like undrained problems, owing to the large differences between the magnitudes of the displacement-related and excess-pore-pressure-related eigenvalues. Lee et al. (2002) also showed that, by under-scaling the preconditioning coefficients for the excess-pore-pressure degrees-of-freedom or over-scaling the preconditioning coefficients for the displacement degrees-of-freedom, the convergence characteristics of consolidation problems under Jacobi preconditioning can be markedly improved to approach that of drained problems. This approach was subsequently analyzed rigorously by Phoon et al. (2002) in their development of an efficient generalized Jacobi (GJ) preconditioner for consolidation problems.

The development of efficient Jacobi preconditioners for consolidation problems also led to an efficient way for solving undrained problems. The latter can be formulated as "nearly-impermeable" consolidation problems instead of "nearly-incompressible" problems. The effectiveness of this approach was demonstrated by Phoon et al. (2003), who showed the "nearly-impermeable" formulation gives a much faster convergence than the "nearly-incompressible" formulation. Since the "nearly-impermeable" formulation has additional pore pressure degrees-of-freedom, it results in a larger global A matrix but this is of no major consequences since the global A matrix is not assembled in an EBE approach anyway.

5 COMPARISON OF FRONTAL, PCG AND QMR ALGORITHMS

In addition to small idealised problems, Lim (2004) also benchmarked the performance of the EBE-PCG (EPCG) and EBE-QMR (EQMR) algorithms against the frontal solver on larger finite element domains, the latter being a representative version of the Gaussian elimination approach.

Figs. 4 and 5 show the finite element mesh of a single lined-tunnel and a twin lined tunnel mesh respectively. The single-tunnel problem consists of 3120 20-noded brick elements and about 40,000 degrees-of-freedom. The twin-tunnel problem consists of 9920 20-noded brick element and about 115,000 degrees-of-freedom. Drained, undrained and consolidation analyses were conducted. For consolidation analyses, the lateral boundaries and ground surfaces are assumed to be free-draining with hydrostatic conditions, except for the plane of symmetry in the single tunnel problem, which is impermeable. The physical properties and the

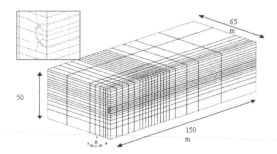

Figure 4. Single tunnel mesh.

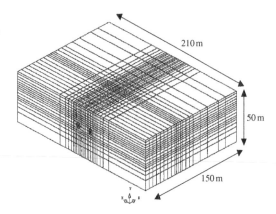

Figure 5. Twin-tunnel mesh.

test configurations of the two meshes are tabulated in Table 1.

The tunnelling process was simulated by removing soil elements, followed by installation of concrete lining. This is not entirely realistic but it does not detract from the objective of the study, which is to compare computing time of various algorithms.

The soil was modeled as a Mohr Coulomb model with the soil parameters shown in Table 2 while the lining properties are reflected in Table 3. The water table is assumed to be located 10 m below ground surface. The comparative study was conducted on a personal computer with a 1.4 GHz AMD CPU and 1 GB RAM.

The test results are summarised and shown in Fig. 6. As can be seen, the frontal solver is faster for the undrained and low permeability consolidation problems of the single tunnel mesh. In all other cases, EPCG or EQMR is faster. Owing to limitation of memory, the Frontal method is able to solve in-core only for the single tunnel mesh which has 3120 elements. The larger problem could not be solved by the frontal algorithm owing to insufficient RAM. As such, the frontal solution timings for the twin tunnel mesh of 9960 elements were estimated based on the method recommended by Britto & Gunn (1990). They suggested the

Table 1. Features of finite element meshes.

	No. of elements	NDF (Drained/undrained element)	NDF (Consolidation element)	MAX. Front-width (Drained/undrained/consolidation)
Single tunnel	3120	40005/40005	43533	1650/1650/1916
Twin tunnel	9920	115131	115131	7461/7461/8445

Table 2. Soil parameters used for Mohr Coulomb model.

Types of domain	E' (kPa)	ν	c' (kPa)	ϕ	γ_{unit} (kN/m^3)	k (m/s)	Time /increment (sec)
Undrained	27	0.3	5	30	20/10	NA	NA
Drained	27	0.3	5	30	20/10	NA	NA
Consolidation case I	27	0.3	5	30	20/10	1.E-3	1.73E4
Consolidation case II	27	0.3	5	30	20/10	1.E-9	0.2

Table 3. Typical concrete parameters (isotropic elastic model).

Concrete properties	E' (kPa)	ν	G (kPa)	γ_{conc} (kPa)	k (m/s)
Concrete in drained/undrained Case	28E3	0.25	11.2E3	24	NA
Concrete in consolidation case I	28E3	0.25	11.2E3	24	1.0E-3
Concrete in consolidation case II	28E3	0.25	11.2E3	24	1.0E-9

timing for the frontal solver could be reasonably computed by multiplying a timing factor with the product of the square of the maximum frontwidth and the total degrees of freedom.

As Fig. 6 shows, the relative efficiency of EPCG and EQMR is dependent on mesh size and the type of domain analysis (i.e. drained, undrained or consolidation). For larger 3-D problems such as the twin tunnel mesh, EPCG or EQMR is more efficient compared to the frontal solver. For each problem size, the drained cases will always outperform the consolidation cases, followed by the undrained cases. In the analysis, undrained cases are the slowest to converge and to reach the ideal solutions. These results suggest that, for undrained problems and ill-conditioned consolidation problems (i.e. those involving very low permeability materials), the breakeven size of problem for the EPCG and EQMR algorithms is about 40,000 degrees-of-freedom. Below the breakeven size, frontal solution is faster. Above the breakeven size, EPCG and EQMR are faster. For drained problems and well-conditioned consolidation problems (i.e. those involving high permeability materials), the breakeven point appears to be significantly lower, perhaps in the region of 10,000 degrees-of-freedom or even less. The advantages of EPCG and EQMR increase with the size of the problems. For very large problems involving hundreds of thousands or even millions of degrees-of-freedom, the speed-up offered by EPCG and EQMR algorithms can easily reach 10 to 100 times compared to Gaussian solvers.

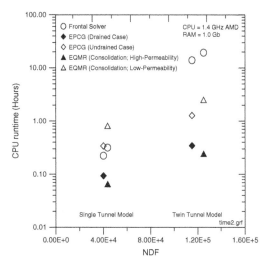

Figure 6. CPU runtimes for various solvers.

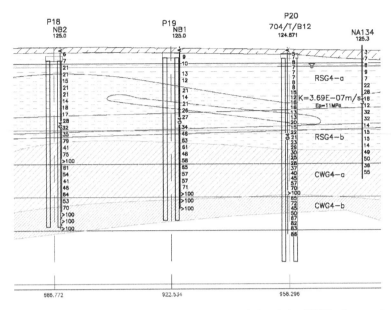

Figure 7. Soil profile along C704. Borehole identifiers are shown above the boreholes, SPT N-values are shown next to each borehole. The parameters k and E_p denote permeability and pressuremeter modulus, respectively.

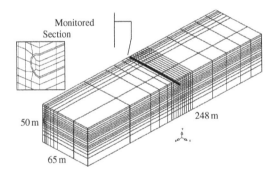

Figure 8. FE mesh for EPB tunneling.

6 CASE STUDY – EARTH PRESSURE BALANCE (EPB) TUNNELLING

6.1 Modelling aspects

The tunnelling project studied herein is the North East line (NEL) tunnel, contract C704, where twin rail tunnels were driven using two LOVAT RME257SE Earth Pressure Balanced (EPB) machines. Fig. 7 shows the sub-surface soil profile along the tunnel alignment. The springline of the tunnel varies from a depth of 18 m to 21 m below the ground surface. In the finite element modelling, for ease of geometrical modelling, the springline depth was modelled 21 m from ground surface.

The tunnel extrados diameter was 6 m and the EPB shield was 9 m in length. As Fig. 8 shows, taking

advantage of symmetry, only half of the tunnel and ground domain was modelled. The finite element mesh extended laterally for a distance of approximately 10 times the diameter, D, from the centre of the tunnel. This is to ensure that the lateral boundaries have no significant effects on the results (Oteo & Segaseta 1982). The longitudinal boundary is set at 20 D i.e. approximately 120 m ahead and behind the tunnel face at the monitored section. The finite element mesh used in this analysis consisted of 3120 20-noded brick elements.

The soil domain was modelled using 20-noded linear strain brick (LSB) elements with pore pressure degrees of freedom at the vertices. The vertical sides of the mesh were laterally restrained against transverse movement whilst the base is completely fixed. Following pre-construction standpipe readings, the in-situ groundwater table was taken to lie at a depth of 5 m below the ground surface. The vertical plane of symmetry, base as well as the initiating and terminating faces of the mesh were assumed to be impermeable, so as to allow any tunnel drawdown effects to be manifested.

The vertical side of the mesh, which runs parallel to the tunnel axis and opposite of the vertical plane of symmetry is assumed to be a hydrostatic drainage (i.e. recharge) boundary throughout the analysis.

The ground domain was sub-divided into three soil layers, the parameters and in-situ conditions of which are shown in Table 4. Conventional laboratory tests (Dames & Moore 1983) showed that the granitic saprolites found in Singapore generally behave in a manner that is akin to an over-consolidated soil with

148

Table 4. Typical G4 soil parameters found in C704.

Sub-layer	RS-G4a	RS-G4b	CW-G4a
Depth (m)	$0 \sim 7.5$	$7.5 \sim 40$	$40 \sim 50$
SPT N Values	11	31	50
C_u (kPa)	73 ± 26	80 ± 29	150 ± 26
c' (kPa)	19.8	19.6	20
ϕ' (deg)	19.2	24.2	30.5
C_c	0.26	0.24	0.16
C_r	0.041	0.054	0.0164
OCR	2.1	1.25	1.0
e_0	0.99	0.82	0.49
k (m/s)	2.16×10^{-7}	1.46×10^{-6}	1×10^{-8} $\sim 1 \times 10^{-7}$
K_0	0.86	0.65	0.51

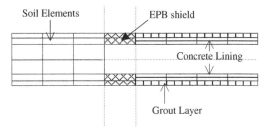

Figure 9a. FE construction sequences for EPB modeling (Stage A).

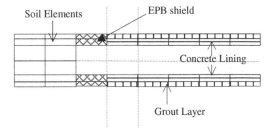

Figure 9b. FE construction sequences for EPB modeling (Stage B).

over-consolidation ratio (OCR) of about 3. This includes its tendency to dilate in consolidated undrained tests conducted under in-situ the effective stress levels. The moderately high OCR of 3 indicate that while compression and shear yielding are possible, it is also rather unlikely apart from isolated regions of stress concentration around the tunnel heading. For this reason, it is quite possible that the ground behaviour is dictated by elastic rather large-scale plastic behaviour.

The concrete tunnel lining was assumed to be impervious, as the concrete lining has permeability that is typically at least two orders of magnitude lower than that of the surrounding soil (Fitzpatrick 1980). The concrete lining is modeled as an elastic material with $E = 28$ GPa, Poisson's ratio $= 0.25$ and coefficient of permeability of 1×10^{-12} m/s. In this study, the weight of the lining and EPB machine was modelled using 3D brick elements having full shield weight or lining weight. The lining is assumed to be continuous in the analysis. In practice, the lining used consisted of precast segments. This may have a lower stiffness than a continuous lining. To assess the significance of lining stiffness on ground response, sensitivity studies were conducted to determine the effects of the stiffness of lining. However, the results of the lining sensitivity study are not reported in here.

The tunnel excavation and lining installation consists of three repeated steps. The first step involves excavation of the tunnel face which takes place concurrently with the extension of the thrust rams (piston jacks) against the tunnel lining. The piston jacks were then retracted, followed by the construction of the segmental lining by a rotary erector. This cycle of activities is then repeated, with the piston jacks being extended against the new lining and the shield advancing again. In an ideal scenario, once the lining is grouted, no further ground movement should occur. However, ground movement may occur over the span of the TBM machine, which typically has a diameter slightly smaller than that of the cut cavity.

To facilitate finite element modelling, a simplified sequence was simulated as shown in Fig. 9. As shown in this figure, at any stage of tunnelling, the space which is to be occupied by the tunnel can be sub-divided into three regions. The first comprised the original soil domain which is ahead of the tunnel face and yet to be excavated. The second consisted of the space occupied by the TBM machine. To enable ground movement into the cut cavity, the excavated soil is replaced by "shield machine elements" which have properties shown in Table 5 which have reduced Young's modulus compared to the soil. The external compressible shield elements allow the surrounding soil to move inwards and partially take up the gap between the machine and the cut cavity. They are necessitated by the lack of gap elements in the program. Finally, the shield machine elements are followed by a span of tunnel lining elements which consisted of an elastic layer of brick elements surrounded by another layer of more compressible "grout" elements. The purpose of this grout layer is to allow stress relieve into the shield and tail voids. It is evident that the properties of the shield, lining and grout elements are not readily measured or defined since they are likely to be highly dependent upon operating parameters and workmanship.

The advancement of the tunnel and lining are simulated in the analysis using the following stages:

1. The soil elements ahead of the EPB shield were removed and face pressure was applied. In C704,

Table 5. Material properties of "Sheild" elements.

Type	Young's modulus, E_s (GPa)	Shear modulus, (GPa)	Poisson's ratio, ν	k_x	k_y
Inner Shield (internal rim)	28	11.2	0.25	1×10^{-12}	1×10^{-12}
Outer shield to simulate the gap due to face overcut	2.8	1.2	0.25	1×10^{-12}	1×10^{-12}

the applied face pressure due to EPB advancing was about 100 kPa to 200 kPa (Shirlaw 2002). In this study, a face pressure of 100 kPa was adopted in the analysis. Parametric study on the sensitivity of the face pressure will be shown later. At the excavated boundary, pore pressure was fixed at atmospheric pressure.

2. The excavated soil elements were replaced by the EPB shield elements (see Fig. 9a). A rate of advance of 4 m/day was adopted, this being typical of the rate of advance in this segment of tunnel.

3. During tunnel lining installation, the "shield" elements were replaced by concrete lining and grout layer elements.

7 RESULTS

7.1 Effects of pore-pressure fixity

In the numerical simulation, the excavated tunnel face and side over the unlined length of the tunnel was assumed to be a flow boundary where the pore pressure was maintained at atmospheric pressure. This pore pressure fixity condition was released as the lining elements were inserted to simulate the sealing of the lined portion. This assumption is reasonable as EPB machines usually inflict some degree of over-cutting, which is defined as the difference between the cross-sectional area of the tunnel cut by the machine and that of the tailskin shield. In C704, the over-cutting is about 0.5 % of the face area (Shirlaw et al. 2001). This is approximately equivalent to an all-round 75-mm gap between the excavated tunnel and the tail-skin shield. Grouting was only conducted at the tailend of the shield and was mainly used to fill up the overcut and tailvoid gaps. Furthermore, the stiff to very stiff residual soil is stable. Thus, the excavated tunnel is unlikely to have collapsed inwards significantly. Thus, pore pressure along the tunnel wall of the unlined segment is likely to be atmospheric.

To determine the effects of pore pressure fixity on the ground response, a case with no pore pressure fixity was simulated and compared with the baseline case. In both cases, the pore pressure fixity at the face remains the same and the only difference lies in the presence or otherwise of pore pressure fixity at the unlined tunnel periphery. As shown in Fig. 10,

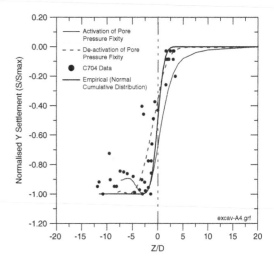

Figure 10. Effect of pore pressure fixity on longitudinal settlement trough.

removing the pore pressure fixity leads to much reduced settlement and a virtual absence of rebound at the lower end of the trough length.

Moreover, as Fig. 11 shows, the drawdown effect arising from the pore pressure fixity is also evident from the sharp drop in pore pressure at a point located 2 m above the crown of the tunnel. On the other hand, removing the pore pressure fixity leads to an initially small drop in pore pressure, due to lateral relieve of earth pressure ahead of the tunnel, followed by a slight increase due to the influence of the face pressure.

7.2 Effects of TBM weight

Figure 12 shows the effect of including the TBM weight into the computation. As can be seen, for $-0.5 < S/S_{max} < 0.0$, the normalised results for both cases analysed does not show much deviations. Some discrepancies are present further behind the tunnel face; the results computed without considering TBM weight deviating away from the cumulative normal distribution curve. However, the differences remain fairly small indicating that the weight of the TBM machine does not have an influence over the shape of the settlement profile in the longitudinal direction.

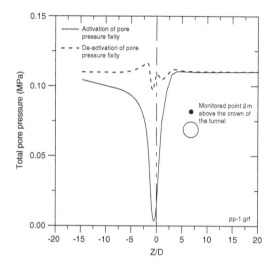

Figure 11. Effect of pore pressure fixity on pore pressure fluctuations just above tunnel crown.

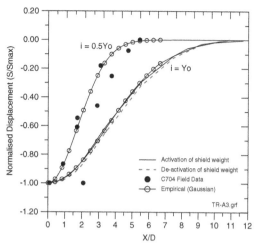

Figure 13. Effect of including TBM weight on settlement trough (cross-section).

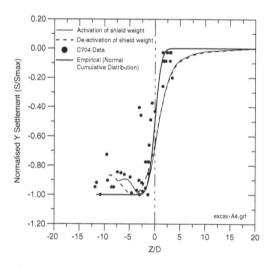

Figure 12. Effect of including TBM weight on settlement trough (longitudinal direction).

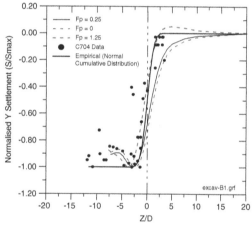

Figure 14. Effect of face pressure on settlement trough (longitudinal direction).

In the transverse direction, the differences are even smaller, with both cases showing essentially similar settlement profiles, Fig. 13.

7.3 Effects of face pressure

Results were computed for 3 different face pressures, namely 10 kPa, 100 kPa and 500 kPa. The total over-burden pressure at the springline of the tunnel is about 400 kPa. In the following discussion, the normalised face pressure F_p will be used, this being defined as the ratio of the face pressure to the total overburden pressure.

As Fig. 14 shows, the effect of face pressure is more significant. Using a higher face pressure leads to larger heave ahead of the tunnel and larger settlement behind the TBM. This is attributable to the fact that using a higher face pressure inhibits the movement of soil ahead of the tunnel face towards the rear of the TBM. As shown in this figure, using a higher face pressure not only increases the compressive stresses prior to arrival of the tunnel heading but also increases the deviator stress and reduces the compressive stress after the tunnel heading has passed. Thus, while face pressure is useful for enhancing stability of the tunnel face, using a higher face pressure may actually aggravate the settlement of the ground surface, in particular the differential settlement, as seen from the maximum gradient of the longitudinal settlement profile. The influence of the face pressure on the cross-sectional settlement profile is much less. This is to be expected since the face pressure is only applied in the longitudinal direction.

8 CONCLUSION

The conclusions from the foregoing discussion can be summarized as follows:

a. Realistic three-dimensional finite element analysis has still not reached a point of development where it can be used routinely in engineering design. A major problem is the computer memory and time needed to solve such problems. There is a need to develop more efficient solution algorithms.
b. Krylov-iterative methods potentially offer viable and faster alternatives to Gaussian elimination approach for solving very large finite element equations in geotechnical engineering, such as those arising from three-dimensional analyses, with significantly reduced memory requirement.
c. The convergence characteristics of ill-conditioned consolidation problems can be significantly improved by using simple and appropriate preconditioners.
d. The advantages of such iterative methods increase with the size of the problem. For very large problems, the speed-up and memory savings can be very significant.
e. With efficient solution algorithms, realistic three-dimensional geometry, material zones and construction sequences can potentially be simulated on relatively modest PC platforms.

ACKNOWLEDGEMENT

This study was conducted as part of the PhD study of the third Author, who would like to acknowledge the financial support from the National University of Singapore Research Scholarship during his PhD study.

REFERENCES

Barrett, R, Berry, M, Chan, T, Demmel, J, Donato, J, Dongarra, J, Eijkhout, V, Pozo, R, Romine, C & van der Vorst, H. 1994. *Templates for the Solution of Linear Systems: Building Blocks for Iterative Methods*. Philadelphia: SIAM Press.

Britto, AM & Gunn, MJ 1987. *Critical State Soil Mechanics via Finite Elements*. Ellis Horwood Ltd.: Chichester, West Sussex.

Britto, AM & Gunn, MJ 1990. *Crisp90 User's and Programmer's Guide*. Cambridge University Engineering Department, Soil Mechanics Group.

Chan, SH, Phoon, KK & Lee, FH 2001. A modified Jacobi preconditioner for solving ill-conditioned Biot's consolidation equations using symmetric quasi-minimal residual method. *International Journal for Numerical and Analytical Methods in Geomechanics* 25: 1001–1025.

Chan, SH 2002. Iterative solution for large indefinite linear systems from Biot's finite element formulation. Ph.D. Thesis, National University of Singapore.

Dayde, MJ, L'Excellent, JY & Gould, NIM 1997. Element-by-element preconditioners for large partially separable optimisation problems. *SIAM Journal of Scientific Computing* 18 (6): 1767–1787.

Fox, RL & Stanton, EL 1968. Developments in structural analysis by direct energy minimization. *American Institute for Aeronautics and Astronautics Journal* 6: 1036–1042.

Freund, RW & Natchigal, NM 1994a. An implementation of the QMR method based on coupled two term recurrences. *SIAM, Journal of Scientific computing* 15: 313–337.

Freund, RW & Nachtigal, NM 1994b. A new Krylov-subspace method for symmetric indefinite linear system. In *Proceedings of the 14th IMACS World Congress on Computational and Applied Mathematics*, Atlanta, USA, Ames WF (ed), 11–15 July 1994, 1253–1256.

Fried, I 1969. More on gradient iterative methods in finite element analysis, *AIAA J.* 7: 565–567.

Golub, G & Van Loan, C 1989. *Matrix Computations*, 2nd Edition, The John Hopkins University Press, Baltimore.

Hestenes, MR & Stiefel, E 1952. Methods of conjugate gradient for solving linear systems. *Journal of Research of the National Bureau of Standards*: 409–436.

Hibbitt, Karlsson & Sorensen, INC 1997. *Abaqus/Standard User's Manual*, Volume I, Chapter 8.

Irons, BM 1970. A frontal solution program for finite element analysis. *International Journal for Numerical Methods in Engineering* 12: 5–32.

Lee, FH, Phoon, KK, Lim, KC & Chan, SH 2002. Performance of Jacobi preconditioner in Krylov subspace solution of finite element equations. *International Journal for Numerical and Analytical Methods in Geomechanics* 26(4): 341–372.

Lim, KC, Lee, FH & Phoon, KK 1998. Three-dimensional analysis of twin tunnels. In *Proceedings of the 8th KKNN Seminar on Civil Engineering*, 30 November–1 December 1998, NUS, Singapore, 452–457.

Lim, KC 2004. Three-dimensional finite element analysis of earth pressure balance tunneling. PhD thesis. National University of Singapore.

Mitchell, JA & Reddy, JN 1998. A multilevel hierarchical preconditioner for thin elastic solids. *International Journal for Numerical Methods in Engineering* 43: 1383–1400.

Nour-Omid, B & Parlett, BN 1985. Element preconditioning using splitting techniques. *SIAM Journal of Scientific and Statistical Computing* 6(3): 761–770.

Oteo, CS & Sagaseta, C 1982. Prediction of settlements due to underground openings. In *Int. Symp. on Numerical models in Geomechanics*, Zurich, 653–699.

Paige, CC & Saunders, MA 1975. Solution of sparse indefinite systems of linear equations, *SIAM Journal of Numerical Analysis* 12: 617–629.

Papadrakakis, M 1993. Solving large-scale linear problems in solid and structural mechanics. *Solving Large-scale Problems in Mechanics – The Development and Application of*

Computational Solution Methods, Papadrakakis M (ed). John Wiley: Chichester, 1–32.

Phoon, KK, Toh, KC, Chan, SH & Lee, FH 2002. An efficient diagonal preconditioner for finite element solution of Biot's consolidation equations. *International Journal for Numerical Methods in Engineering*; 55(4): 377–400.

Phoon, KK, Chan, SH, Toh, KC & Lee FH 2003. Fast iterative solution of large undrained soil-structure interaction problems. *International Journal for Numerical and Analytical Methods in Geomechanics* 27(3): 159–181.

Saad, Y 1996. *Iterative method for sparse linear systems*. PWS Publishing Company: Boston.

Saad, Y & van der Vorst, HA 2000. Iterative solution of linear systems in the 20-th century. *Journal of Computational and Applied Mathematic* 123:1–33.

Shewchuk, JR 1994. *An introduction to the conjugate gradient method without the agonizing pain*. Release version 1.25, August 4, 1994. School of Computer Science, Carnegie Mellon University.
FTP: warp.cs.cmu.edu. (IP Adddress: 128.2.209.103), filename <quake-papers/painless-conjugate-gradients.ps>.

Shirlaw, JN, Ong, JCW, Rosser, RB, Osborne, NH, Tan, CG & Heslop PJE 2001. Immediate Settlements due to tunnelling for the North East Line. In *Proceedings of Underground Singapore 2001*, 29th–30th November, 2001.

Sleijpen, GLG & van der Vorst, HA 2000. Differences in the effects of rounding errors in Krylov solvers for symmetric indefinite linear systems. *SIAM, Journal on Matrix Analysis and Applications* 22(3): 726–751.

Smith, IM 2000. A general-purpose system for finite element analyses in parallel. *Engineering Computations* 17(1): 75–91.

Smith, IM & Griffiths, DV 1997. *Programming the Finite Element Method*, 3rd edn. John Wiley: Chichester.

Smith, IM & Wong, SW 1989. PCG methods in transient FE analysis. Part I: First order problems. *International J. for Numerical methods in engineering* 28: 1557–1566.

Van der Vorst, H 2002. *Lecture notes: Iterative methods for large linear systems*. June 24, 2002, 195pp. http://www.math.uu.nl/people/vorst/lecture.html,

Wang, A 1996. *Three Dimensional finite element analysis of pile groups and pile-rafts*. Ph.D. Thesis, University of Manchester.

Winget, JM & Hughes, TJR 1985. Solution algorithms for non-linear transient heat conduction analysis employing element-by-element iterative strategies. *Computer Methods in Applied Mechanics and Engineering* 52: 711–815.

Zienkiewicz, OC & Taylor, RL 1999. *The Finite Element Method, Vol. 1: Basic Formulation and Linear Problems*, 4th edn. McGraw-Hill: London.

Numerical Modelling of Construction Processes in Geotechnical Engineering for
Urban Environment – Triantafyllidis (ed)
© 2006 Taylor & Francis Group, London, ISBN 0 415 39748 0

Iterative solution of intersecting tunnels using the generalised Jacobi preconditioner

K.K. Phoon & F.H. Lee
National University of Singapore, Singapore

S.H. Chan
C T Toh Consultant, Malaysia

ABSTRACT: This study presents settlement analyses of intersecting tunnels using the 3-D finite element method. This study demonstrates that the reinforcing or strengthening effects due to the presence of existing tunnel(s) are significant on the problem being investigated. Any existing tunnel prior to the new tunnelling has to be incorporated into analyses if more realistic assessments on the tunnelling impacts are required. This study also shows that when multiple closely spaced tunnels are constructed, the assumption that each tunnel behaves independently and the movements that would have occurred for each tunnel acting independently can be superimposed, is unlikely to provide good estimations. Note that the linear superposition technique has been commonly used in practice to reduce the complexity of very large problems that involve construction of multiple tunnels. Consequently, the entire problem has to be modelled and large-scale computing is unavoidable. The symmetric quasi-minimal residual method (SQMR) incorporating the generalised Jacobi (GJ) preconditioner is shown to be very economical in solving such large-scale finite element problems on personal computers.

1 INTRODUCTION

The extension of underground transportation systems in urban areas often involves construction of new tunnels below existing tunnels at various angles of intersection (Cooper & Chapman 1999). To ensure that the existing tunnel continues to comply with serviceability requirements, it is important to simulate the influence of construction on the behaviour of the existing tunnel. Intersecting tunnels are irreducible 3-D problems – their soil-structure interaction mechanisms and stress-deformation fields can only be modelled realistically using a 3-D finite element mesh. If the surrounding soil exhibits time-dependent behaviour, consolidation effects have to be incorporated into the analyses as well.

To date, the cost of 3-D finite element analyses incorporating consolidation is generally prohibitive to average practitioners in terms of run time and memory requirements. The main objective of this study is to demonstrate that element-by-element iterative solution strategies with proper preconditioning can be very economical for such practical large-scale problems. The symmetric quasi-minimal residual (SQMR) method, Freund & Nachtigal (1994), is used because the coefficient matrix for the finite element formulation of Biot's consolidation problem is indefinite. In many soil-structure interaction problems

involving material zones of widely differing stiffness and low permeability, the coefficient matrix is very ill-conditioned (Lee et al. 2002), but a recently developed generalised Jacobi (GJ) preconditioner has been shown to be exceedingly effective in mitigating this undesirable aspect (Phoon et al. 2002). It will be shown that the settlement of perpendicular tunnels can be analysed on a modest personal computer platform at a relatively reasonable cost.

2 GENERALISED JACOBI PRECONDITIONER

A 3-D finite element analysis, that requires the solution of a very large system of algebraic equations, is typically needed to maintain reasonable realism and accuracy. When large scale computing is involved, iterative solution methods such as Krylov subspace techniques can be very effective in the presence of proper preconditioning.

A diagonal (Jacobi) preconditioner is simple to construct, trivial to invert and most important of all, readily applicable to large-scale "element-by-element" or EBE implementation. Unfortunately, the standard Jacobi preconditioner is only mediocre at best for consolidation problems (Lee et al. 2002). Chan et al. (2001) demonstrated that it is possible to achieve significant improvement by merely changing some diagonal entries while maintaining the simple diagonal form. However, the performance of the modified Jacobi preconditioner outside the scope of study could not be assured because the preconditioner was essentially constructed from a heuristic basis.

This section develops a generalised Jacobi preconditioner systematically from a more rigorous mathematical basis.

2.1 Finite element discretisation

Finite element discretisation of Biot's consolidation equations leads to the following symmetric indefinite linear systems:

$$A\mathbf{x} = \mathbf{b} \tag{1}$$

where

$$A = \begin{bmatrix} K & L \\ L^T & -C \end{bmatrix} \in \Re^{n \times n} \tag{2}$$

$K = K^T \in \Re^{m \times m}, L \in \Re^{m \times (n-m)}, C = C^T \in \Re^{(n-m) \times (n-m)}$
K = effective stress stiffness matrix, L = link matrix, C = flow matrix, $\mathbf{x}$ = nodal displacement and excess pore pressure vector, and $\mathbf{b}$ = nodal load vector. The coefficient matrix A in Equation (1) is symmetric but indefinite. Hence, the popular preconditioned conjugate gradient (PCG) method is not guaranteed to converge in this instance.

2.2 Murphy's block diagonal preconditioner

For nonsingular indefinite matrices of saddle-point form ($C = 0$), Murphy et al. (2000) noted that preconditioners incorporating an exact Schur complement lead to preconditioned matrices with exactly three distinct eigenvalues. This is an important observation since it implies that Krylov subspace methods will converge in three iterations if such a preconditioner is applied.

The above theoretical framework can be generalised for Biot's system of equations, where the (2,2) block or flow matrix, -C, is small but not identically zero. The proposed generalisation is:

$$P_\alpha = \begin{bmatrix} K & 0 \\ 0 & \alpha(L^T K^{-1} L + C) \end{bmatrix} \tag{3}$$

where α is a non-zero scalar that may possibly be negative. It must be acknowledged that α has been shown to have almost no impact on the number of iterations when the (1,1) block in Equation (3) is exact (Fischer et al. 1998). It is tempting to conclude that α is also not effective in practical implementations where inexpensive approximations of the (1,1) block have to be used. However, the theoretical behaviour of the exact form would serve to mislead, rather than illuminate the behaviour of the approximate form in this instance.

2.3 Diagonal approximation

The preconditioner P_α is of block diagonal form and not a simple diagonal form. Although the exact form has a highly attractive eigenvalue clustering property, it is too expensive for practical use since it is not readily invertible and not convenient for EBE implementations. As noted above, a diagonal form is highly desirable for large-scale computing and it is not unreasonable to assume that a diagonal approximation to P_α would still produce some eigenvalue clustering, albeit in a more diffused way. The following simple diagonal approximation to P_α is proposed in this:

$$\hat{P}_\alpha = \begin{bmatrix} \text{diag}(K) & 0 \\ 0 & \alpha \, \text{diag}\{L^T[\text{diag}(K)]^{-1}L + C\} \end{bmatrix} \tag{4}$$

where diag(.) = diagonal matrix consisting of leading diagonal entries in argument. Note that Equation (4) is identical to the standard Jacobi preconditioner insofar as the displacement DOFs are concerned, i.e. for $i = 1$ to m:

$$\hat{P}_{ii} = k_{ii} \tag{5a}$$

For $i = m + 1$ to n

$$\hat{P}_{ii} = \alpha \left[\left(\sum_{j=1}^{m} \frac{l_{ji}^2}{k_{jj}} \right) + c_{ii} \right] \tag{5b}$$

where l_{ij}, k_{jj} and c_{ii} = entries in L, K and C referenced by the global indexing system of A. To store $\hat{P}_\alpha$ efficiently using only one n-dimensional vector in an EBE implementation, Equation (5b) can be further approximated as:

$$\hat{P}_{ii} = \alpha \left[\left(\sum_{j=1}^{m} \frac{\left(\sum_e l_{ji}^e \right)^2}{k_{jj}} \right) + c_{ii} \right] \approx \alpha \left[\left(\sum_{j=1}^{m} \frac{\sum_e \left(l_{ji}^e \right)^2}{k_{jj}} \right) + c_{ii} \right] \tag{6}$$

where l_{ij}^e = entry in the L-block of the eth finite element referenced globally. Details of this generalised Jacobi (GJ) preconditioner are given elsewhere (Phoon et al. 2002).

3 INTERSECTING TUNNELS

The extension of underground transportation or/and sewerage systems in urban areas often entails situations where new tunnels are constructed close to existing tunnels. In the design and construction of new tunnels, it is important to ensure that any existing underground tunnels in close proximity to the proposed tunnels can continue to operate safely, both during and after construction. For sensitive infrastructures, such as mass rapid transit tunnels and tracks which have very low tolerance for ground movements, stringent criteria and regulations are often laid down by local authorities for safety assurance. For example, the Land Transport Authority (LTA) of Singapore specified that the total movement in the structure or track of MRT tunnels arising from nearby construction activities should not exceed 15 mm in any direction (Lee et al. 1998). In such circumstances, a more realistic assessment of tunnelling effects and a better understanding of interaction behaviours between closely spaced tunnels is of significant practical value in design and construction aspects.

Most of the existing numerical investigations on perpendicular (or crossing at a skew angle) tunnels are based upon 2-D finite element analyses through some justifiable simplification and clever manipulation (e.g. Ghaboussi et al. 1983, Saitoh et al. 1994, Samuel et al. 1999). The approximation techniques are assumed to be necessary due to the excessive storage requirements and long computational time

in performing full 3-D finite element analyses. In reality, perpendicular tunnels are 3-D in geometry; their interaction mechanisms and stress-deformation fields can only be modelled realistically using 3-D procedures. If the surrounding soil exhibits time-dependent behaviour, consolidation effects have to be incorporated into tunnelling analyses as well. Nonetheless, the cost of modeling perpendicular tunnels using 3-D finite element analyses with consolidation is generally prohibitive to average practitioners and sometimes even to researchers in research centres. Coupled with the symmetric QMR method and EBE strategies, the GJ preconditioner allows modeling of perpendicular tunnels even on relatively moderate computing platforms such as personal computers.

3.1 Problem configuration

Figure 1 shows the finite element mesh used for simulating construction of perpendicular tunnels using the earth pressure balance (EPB) tunnelling method. The mesh consists of 9,040 20-noded brick elements

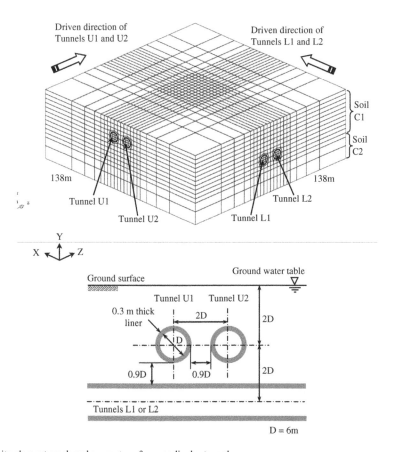

Figure 1.   Finite element mesh and geometry of perpendicular tunnels.

157

Table 1. Material properties for perpendicular tunnels.

Modified Cam Clay model

Material	Depth (m)	λ	κ	M	Effective Poisson's ratio, v'	e_{cs}	Over-consolidation ratio, OCR	Bulk unit weight, γ_{bulk} (MN/m³)	Coefficient of permeability, k (m/s)	K_0
Soil C1	0-30	0.1	0.02	1.2	0.3	1.8	3	18.5×10^{-3}	1×10^{-8}	0.8

Mohr-Coulomb model

Material	Depth (m)	Effective cohesion, c' (MN/m²)	Effective angle of friction, ϕ' (°)	Effective Young's modulus, E' (MN/m²)	Effective Poisson's ratio, v'	Bulk unit weight, γ_{bulk} (MN/m³)	Coefficient of permeability, k (m/s)	K_0
Soil C2	30–48	15×10^{-3}	30	100	0.25	18.5×10^{-3}	5×10^{-7}	0.8

Liner elastic model

Material	Effective Young's modulus, E' (MN/m²)	Effective Poisson's ratio, v'	Bulk unit weight, γ_{bulk} (MN/m³)	Coefficient of permeability, k (m/s)
Liner	20×10^3	0.2	24.0×10^{-3}	1×10^{-12}
Grout	1×10^3	0.2	24.0×10^{-3}	1×10^{-12}

coupled with excess pore pressure degrees-of-freedom (DOFs), which give rise to 111,735 DOFs. All the tunnels are identical, with an inner diameter (D) of 6 m and a concrete liner thickness of 0.3 m. The springline for tunnel U is located 2D below ground surface. The horizontal and vertical spacings between tunnel axes are 2D. The excavated diameter is 7 m with a clearance of 0.2 m between the cutting edge and liner; this clearance would accommodate ground deformations resulting from stress relief (caused by soil excavation), and will be filled with grout afterwards. The modified Cam Clay and Mohr-Coulomb models are used for soils C1 and C2, respectively. Liner and grout are modelled using the linear elastic model. The material properties are summarised in Table 1.

Numerical simulation of construction of a single tunnel is performed in 21 increment blocks, each of which consists of three increments. At the ith increment block, soil is excavated for ith segment, liner and grout are installed for $(i − 1)$th segment, pressure is applied against the excavated surfaces of ith segment, and the previously imposed pressure on $(i − 1)$th segment is released as shown in Figure 2. The tunnel advancement rate is dependent on the installation of liner and grout, i.e. 6 m/day. The time increment is equally divided among the increments within an increment block. The face pressure applied is equal to 50% of the geostatic horizontal stress at the centreline of the tunnel.

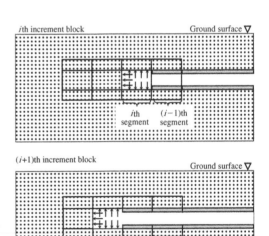

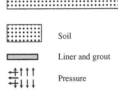

Figure 2. Construction procedure of a tunnel in numerical simulation.

Finite element analyses are carried out using NUSCRISP, which is a fast version of CRISP program (Britto & Gunn 1987), which has been fitted with the SQMR-GJ iterative solver, in addition to its original frontal direct solver. Owing to the element-by-element (EBE) implementations, the SQMR-GJ solver requires 205 MB of memory, in comparison with 5,151 MB of the frontal solver, for solution at double precision.

4 INTERACTION BETWEEN TUNNELS

The interaction of perpendicular tunnels is often analysed using the "green field" subsoil movement that is induced by tunnelling in the absence of existing tunnel, e.g. Kimmance et al. (1996). To investigate the influence of existing tunnel(s), three cases are considered. Case 1 simulates the "green field" condition that involves no existing tunnel, Case 2 involves an existing tunnel, and Case 3 involves two existing tunnels.

4.1 Tunnel settlement

Figure 3 presents the evolution of the subsurface/ tunnel settlement, caused by the construction of Tunnel L1, for Cases 1 to 3. The monitoring point is as indicated in the schematic picture attached. The computed settlements are qualitatively similar ("S" curves) to the monitoring results obtained in the construction of new Heathrow Express tunnels, which cross beneath the Piccadilly Line at a skew angle of 70°, as shown in Figure 4.

It is clear that the presence of the existing tunnel(s) invariably reduces the subsurface/tunnel settlement. Case 1 yields the largest settlement whilst Case 3 predicts the smallest settlement. The calculation of the existing tunnel settlement using the "green field" assumption is very conservative. The tunnel settlements of Cases 2 and 3 are respectively 50% and 44% of that obtained from the "green field" site. The reinforcement effect arising from the presence of existing tunnel(s) is quite significant.

It can be observed from Figure 3 that the rate of subsurface/tunnel settlement induced before the excavation face passes beneath Tunnel U2 is greater than that after the face passes beneath the tunnel. The plot shows that only about 20 to 30% of the total settlement took place after the excavation face has passed beneath the tunnel axis. Similar results were reported by Cooper & Chapman (1999) for NATM shotcrete "half face" tunnelling method. In their case, about 23% of the total settlement took place before the passage of the "upper half face", and about 47% of the total settlement took place between the passage of the "upper half face" and "lower half face". The remaining 30% took place after the passage of the "lower half face". Note that the latter stage is approximately equivalent

to the passage of the excavation face (full face) of the reference example in this study. The computed settlement results seem able to capture the field measurements qualitatively.

4.2 Superposition

When two or more tunnels are constructed, it is commonly assumed that each tunnel acts independently and the superposition technique can be used in predicting the ground movements as well as movements of existing tunnels if they are present (e.g. Ghaboussi & Ranken 1977, Mair & Taylor 1997). However, for closely spaced tunnels, the superposition technique may not be adequate.

Three cases of tunnelling, i.e. Cases 3 to 5 are used for study. Note that in these three cases, Tunnels U1 and U2 would act as existing tunnels during construction of Tunnels L1 or/and L2. The difference between Cases 3 to 5 is: Case 3 involves the construction of Tunnel L1 only, Case 4 involves the construction of Tunnel L2 only, whereas Case 5 involves the construction of Tunnels L1 and L2. The final movement obtained in Case 3 will be superimposed to that of Case 4, these combined movements will then be compared with the total final movements obtained in Case 5 (Figure 5).

It can be observed that additional movements due to interaction can be quite significant. For those tunnels with shorter pillar width, more significant interaction effects would be expected. Some degrees of asymmetry are also observed. Mair & Taylor (1997) gave a good explanation for the occurrence of the interaction effects as well as the asymmetry: "...when tunnels are very closely spaced, the ground in the region where the second tunnel is to be constructed will already have been subjected to appreciable shear strains associated with construction of the first tunnel, resulting in reduced stiffness, and hence a higher volume loss is likely for the second tunnel...". As a result, larger movements would be induced on the existing tunnel during construction of the second tunnel and asymmetry can often be observed.

Superposition methods cannot reflect these interaction effects, thus may not give adequate results when dealing with closely spaced tunnels. For construction of multiple tunnels, all tunnels have to be included in the modeling if more realistic assessments are required.

4.3 Runtime

Figure 6 shows the evolution of cumulative iteration count and computer runtime for Case 3. It is clear that both cumulative iteration count and computer runtime curves increase consistently with the increment number. The computer runtime is about 0.9 hr per increment for a Pentium II, 450 MHz desktop

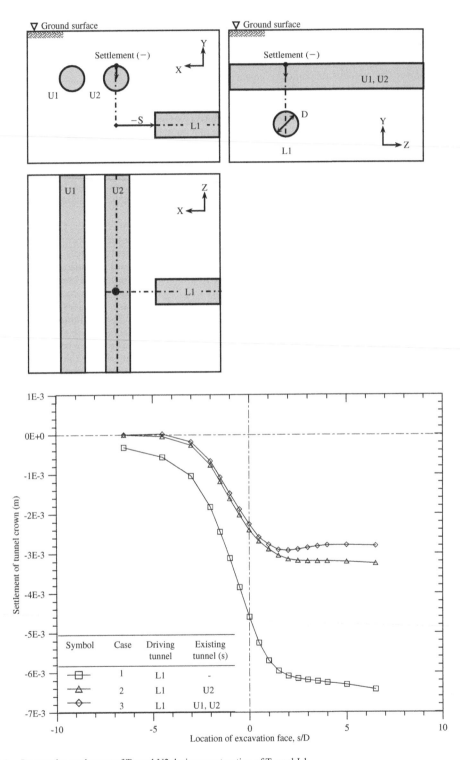

Figure 3.   Progressive settlement of Tunnel U2 during construction of Tunnel L1.

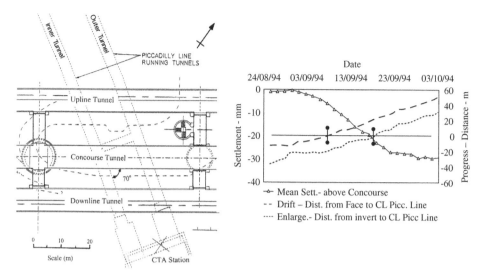

Figure 4. Progressive settlement of Piccadilly Line above Concourse (Source: Cooper & Chapman 1999).

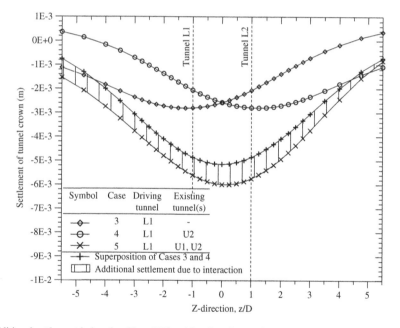

Figure 5. Additional settlement induced on Tunnel U2, arising from interaction effects of Tunnels L1 and L2.

PC with 384 MB memory. The plot also shows the "real" construction time, assuming an average tunnel advancement rate of 6 m/day. Note that the curve is not linear, because the distance traversed by each increment (equal to one finite element layer) is not uniform. It seems possible to conduct numerical analyses incorporating the latest on-site information such as construction time, applied face pressure, etc. along the tunnel axis for monitoring purposes, as well as to predict upstream problems using a simple PC.

161

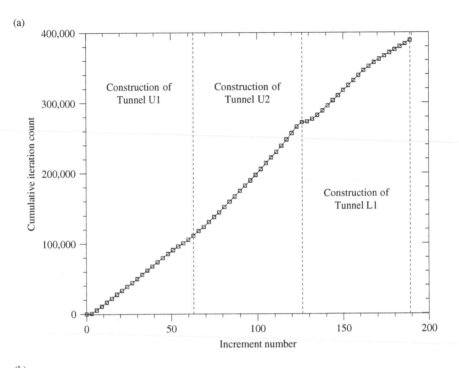

(a)

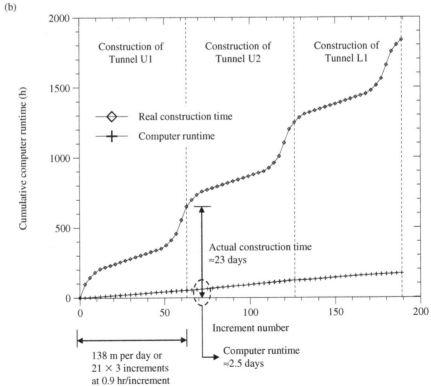

(b)

Figure 6. Evolution of (a) cumulative iteration count; and (b) cumulative computer runtime, for Case 3 (Pentium II, 450 MHz PC).

162

5 CONCLUSIONS

Numerical results show that the reinforcing or strengthening effects due to the presence of existing tunnel(s) are quite significant in predicting tunnel settlement. This study also shows that when multiple, closely spaced tunnels are constructed, the assumption that each tunnel behaves independently and the movements that would have occurred for each tunnel acting independently can be superimposed, is unlikely to provide good estimations. note that the linear superposition technique has been commonly used in practice to reduce the complexity of very large problems that involve construction of multiple tunnels. Consequently, the entire problem has to be modelled and large-scale computing is unavoidable. For such large-scale finite element analyses, the SQMR incorporating the GJ preconditioner is capable of providing economical solutions using relatively modest computational resources such as personal computers.

ACKNOWLEDGEMENT

This study was conducted as part of the PhD study of the third Author, who would like to acknowledge the financial support from the National University of Singapore Research Scholarship during his PhD study.

REFERENCES

Britto AM & Gunn MJ 1987. *Critical state soil mechanics via finite elements*. Chichester: Ellis Horwood.
Chan SH, Phoon KK & Lee FH 2001. A modified Jacobi preconditioner for solving ill-conditioned Biot's consolidation equations using symmetric quasi-minimal residual method. *International Journal for Numerical and Analytical Methods in Geomechanics*; 25:1001–1025.
Cooper ML & Chapman DN 1999. Settlement, rotation and distortion of Piccadilly Line tunnels at Heathrow. In *Proceedings of the International Symposium on Geotechnical Aspects of Underground Construction in Soft Ground*, 19–21 July 1999, Tokyo, Japan.
Fischer B, Ramage A, Silvester DJ & Wathen AJ 1998. Minimum residual methods for augmented systems. *BIT*; 38(3):527–543.
Freund RW & Nachtigal NM 1994. A new Krylov-subspace method for symmetric indefinite linear system. In *Proceedings of the 14th IMACS World Congress on Computational and Applied Mathematics*, 11–15 July 1994, Atlanta, USA.
Ghaboussi J & Ranken RE 1977. Interaction between Two Parallel Tunnels, *International Journal for Numerical and Analytical Methods in Geomechanics*; 1: 75–103.
Ghaboussi J, Hansmire WH, Parker HW & Kim KJ 1983. Finite Element Simulation of Tunneling over Subways. *Journal of Geotechnical Engineering*; 109(3): 318–334.
Kimmance JP, Lawrence S, Hassan O, Purchase NJ & Tollinger G 1996. Observations of deformations created in existing tunnels by adjacent and cross cutting excavations. In *Proceedings of the International Symposium on Geotechnical Aspects of Underground Construction in Soft Ground*, 15–17 April 1996, London, UK.
Lee FH, Yong KY, Quan CN & Chee KT 1998. Effect of Corners in Strutted Excavations: Field Monitoring and Case Histories, *Journal of Geotechnical and Geoenvironmental Engineering*; 124(4), 339–349.
Lee FH, Phoon KK, Lim KC & Chan SH 2002. Performance of Jacobi preconditioner in Krylov subspace solution of finite element equations. *International Journal for Numerical and Analytical Methods in Geomechanics*; 26(4): 341–372.
Mair RJ & Taylor RN 1997. Theme Lecture: Bored Tunnelling in the Urban Environment. In *Proceedings of the Fourteenth International Conference on Soil Mechanics and Foundation Engineering*, 6–12 September 1997, Hamburg, Germany, 2353–2385.
Murphy MF, Golub GH & Wathen AJ 2000. A note on preconditioning for indefinite linear systems. *SIAM Journal on Scientific Computing*; 21(6):1969–1972.
Phoon KK, Toh KC, Chan SH & Lee FH 2002. An efficient diagonal preconditioner for finite element solution of Biot's consolidation equations. *International Journal for Numerical Methods in Engineering*; 55(4): 377–400.
Saitoh AK, Gomi & Shiraishi T 1994. Influence Forecast and Field Measurement of a Tunnel Excavation Crossing Right above Existing Tunnels. In *Proceedings of the International Congress on Tunnelling and Ground Conditions*, 3–7 April 1994, Cairo, Egypt, 83–90.
Samuel HR, Mair RJ, Lu YV, Chudleigh I, Addenbrooke TI & Readings P 1999. The Effects of Boring a New Tunnel under an Existing Masonry Tunnel. In *Proceedings of the International Symposium on Geotechnical Aspects of Underground Construction in Soft Ground*, 19–21 July 1999, Tokyo, Japan, 263–268.

Numerical Modelling of Construction Processes in Geotechnical Engineering for
Urban Environment – Triantafyllidis (ed)
© 2006 Taylor & Francis Group, London, ISBN 0 415 39748 0

Numerical modelling of a reinforcement process by umbrella arch

S. Eclaircy-Caudron, D. Dias & R. Kastner
URGC Géotechnique, INSA de Lyon, Villeurbanne, France

L. Chantron
Centre d'Etude des Tunnels (CETU), Bron, France

ABSTRACT: This article presents the numerical modeling of tunnel reinforcement by an umbrella arch. The models are based on a real tunnel project, near Besançon (FRANCE) that includes such process and whose construction began in July 2005. The simulation of the boring process is done taking into account the real phases of the underground work. Several approaches are compared: 2D and 3D approaches. The comparisons permit a better understanding of the behavior of this reinforcement technique and show difficulties to represent such devices in 2D. Numerical models are achieved with the finite element software CESAR-LCPC.

1 INTRODUCTION

The construction of a tunnel induces a modification of the initial stress field in the ground. This unbalance results in movements of soil like convergence of the cavity, pre-convergence ahead of the face and extrusion of the face and settlements.

During the excavation, if the initial stress field in the ground, the strength and deformability of the soil allow it, an arch effect is created and the stresses in the soil mass are canalized on the shape of the cavity. If the strength of the soil is not sufficient to stabilize the cavity, lining supports are set up in order to limit soil movements and thus, to avoid the failure of the structure.

The lining support can include shotcrete, ribs and/or radial bolting. In certain cases, when the ground has low mechanical characteristics, additional methods like ground reinforcement or ground improvement at some stage of the excavation is necessary. Several reinforcement systems have been developed. Pelizza et al. (1998) presented the different methods of soil and rock improvement used to permit safe tunnelling in difficult geological conditions. Lunardi (2000) classified the support methods into three groups: pre-confinement, confinement and presupport. Each group exerts a different kind of effect on the cavity.

In difficult geological conditions, the control of deformation may be a serious problem. Without support or adapted treatment, the ground tends to move into the opening (tunnel face failure, tunnel face extrusion). In order to cope with such problems, a preconfinement may be required. A preconfinement action can increase the formation of an arch effect in the ground ahead of the tunnel face. The pre-confinement can be achieved by reinforcement or protective intervention ahead the tunnel face. The umbrella arch method and the face bolting are included in protective interventions or pre-support methods. These elements of support and forepoling introduce different types of three dimensional complex soil–structure interactions. Consequently, it is difficult to model analytically these complicated phenomena. Moreover, although the umbrella arch method is widely used, there are no simple approximations to simulate this method by numerical analyses. The design of an umbrella arch is still based today on empirical considerations or on simplified schemes.

Peila et al. (1997) developed design charts which permit to design the steel pipes of an umbrella arch. Several numerical studies carried out in 2D and in 3D, were focused on the way of taking into account umbrella arch. Uhtsu et al. (1995) have studied by 2D and 3D models the prediction of ground behaviour due to umbrella arch tunnelling in urban areas under relatively shallow overburden. Tan et al. (2003) modelled the steel pipes individually to study the behaviour of an excavation in soft ground using pipe roof method as a pre-support. Several numerical modellings were carried out in 2D and in 3D also on the way of taking into account the face bolting (Chungsik Yoo, 2002; Al Hallak, 1999; Dias, D. 1999). More complex the projects (geometry of the underground work, behaviour of the soil mass, digging phases...) and more 2D and 3D sophisticated numerical models are necessary to understand the phenomena which will happen.

An umbrella arch consists on setting up, prior to the excavation of the tunnel, steel pipes into pre-drilled

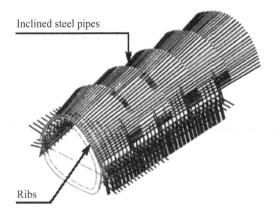

Figure 1. An umbrella arch (CETU, 1998).

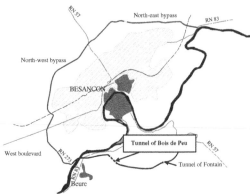

Figure 2. The situation of the project (CETU, 2003).

holes around the section which will be excavated, as shown in Figure 1. The pipes are generally slightly tilted compared to tunnel walls. They take supports on ribs which are set up progressively during the digging. The excavation geometry is not constant due to the pipes' tilt. The support behind the face is completed by shotcrete implemented between the ribs.

If necessary, the pipes can be filled with concrete, connected or jointed. In this case, a transversal and longitudinal effect must be considered for the reinforcement by umbrella arch.

Support and forepoling devices are various. In this article, we present the numerical modelling of a digging of a tunnel reinforced by an umbrella arch and face bolting. The aim of this study is to find a 2D representative model by comparison with 3D numerical analysis and to show the influence of the umbrella arch method on the soil movements. Numerical models have been achieved in 2D and in 3D with the finite element software CESAR-LCPC. Due to the geometrical and mechanical complexity of the studied problem, there are few theoretical and numerical analyses regarding the effect of the umbrella arch in reducing ground movements. Most of them have shown consequent reductions on displacements. G. J. Bae et al. (2005) developed a mathematical framework based on a homogenization technique similar to the Lee et al. (2000) to simulate in three dimensions the elastic and elastoplastic behaviour of tunnels reinforced by a grouted pipe-roofing.

The presented study is based on a real project, a tunnel near Besançon, in France, that includes such forepoling and whose construction began in July 2005.

In a first part, this communication introduces the tunnel project on which it is based. Then, in a second part, the numerical models are presented. And, in a third part, the results obtained in terms of movements and stresses in pipes in 3D and 2D are analysed and compared in order to show the influence of the

umbrella arch on induced displacements and to assess the quality of the carried out 2D model.

2 PRESENTATION OF THE REAL PROJECT

2.1 General presentation

The real project, on which this study is based, is a tunnel situated near Besançon (France). This tunnel is part of a project of the southeastern town bypass. Figure 2 locates the project. This bypass includes various civil engineering structures like bridges and two tunnels. Tunnel excavation began in July 2005. It is carried out full face by drill and blast or road header in clay. This tunnel is composed of two tubes. The length of the tunnel is about 520 metres and the height of cover varies between 8 metres and 140 metres.

2.2 Geotechnical properties

An exploration gallery was dug in 1995 in order to assess the mechanical properties of the ground. It has a width of 3 m and a height of 3,5 m. Various laboratory and in situ tests were carried out. The results lead to conclude that the tunnel is situated in a disturbed area. Eighteen geological units are identified. A geological cross section is showed in Figure 3. Among these eighteen units, four sorts of materials can be distinguished: clays, marls, limestones and interbeddings of marls and limestones. The exact position of these different units is difficult to know before the digging. Finally, following the observation made in this exploration gallery, many uncertainties are remaining. Therefore, it was decided to apply the "interactive design" method during the digging in order to adapt the lining support to the real ground conditions. In the framework of this method, an important number of experimental measurements is foreseen.

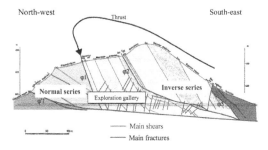

North-west Thrust South-east

Normal series | Inverse series
Exploration gallery

Main shears
Main fractures

Figure 3. The geological cross section.

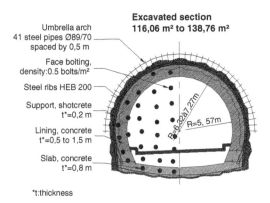

Excavated section
116,06 m² to 138,76 m²

Umbrella arch
41 steel pipes Ø89/70
spaced by 0,5 m

Face bolting,
density:0.5 bolts/m²

Steel ribs HEB 200

Support, shotcrete
t*=0,2 m

Lining, concrete
t*=0,5 to 1,5 m

Slab, concrete
t*=0,8 m

R=5, 57m

*t:thickness

Figure 4. Cross section of the studied lining support (CETU, 2003).

2.3 Monitoring

Monitoring includes traditional measurements like convergences measurements and specific measurements like the monitoring of the face bolting in order to assess the strain of the soil mass ahead of the face.

Convergences measurements are realized in all supports. The space between two monitoring sections depends of the supports. For a complex support, the monitoring sections must be close-together. The number of convergence measurement sections is equal to 30 for the two tubes.

Three cross sections will be instrumented by radial extensometers of twelve meters length and with six measurement points per extensometer. These extensometers will permit to know the soil mass strain around the cavity. Moreover, in two cross sections, pressure and strain sensor will be set up in order to asses the pressure exerted by the soil on the support and the current stress in the rib.

2.4 Geometry

Four kinds of supports are foreseen in the project phase. Three of them could be adapted to the real

Table 1. Geometrical parameter of pipes.

	Value	Units
Pipe diameter (out/in)	0.089/0.070	m
Number of pipes per cross section	41	–

Table 2. Geometrical parameter of ribs.

	Value	Units
Type	HEB200	–
Cross section	0.0078	m
Space	1.5	m

conditions met during the entire construction period according to the measured and analysed behaviour of the tunnel and surrounding ground. The last one is systematically foreseen in the argillaceous material. In this communication, only this latter is studied. It is the more complex lining support. It is composed of a forepoling by umbrella arch and face bolting. It is illustrated in Figure 4. The argillaceous material is situated near one of tunnel ends. For this support, the maximum height of cover is equal to 25 m.

This mixed support is composed of shotcrete with a thickness of 0,2 m and steel ribs. It is later completed by a concrete lining. Moreover, a forepoling by umbrella arch and bolting is set up. This umbrella arch is constituted of forty-one steel pipes drilled into the soil mass every nine meters with a length of 18 m, a covering of 9 m and a tilt of six degrees which induces variable excavated section and variable lining thickness. The umbrella arch studied is composed of disjointed pipes: thus, the transversal effect will be reduced.

The geometrical parameters of the pipes and the ribs are presented in Tables 1 & 2.

The face bolting is also set up every nine metres with a length of 18 m and a covering of 9 m without tilt. The bolts are installed in the ground ahead of the face and are destroyed as the excavation advances. These bolts are made of fiber glass. This kind of bolt has a high tensile strength but can be destroyed easily during excavation thanks to its brittle behaviour. The tunnel support includes also a concrete invert arch set up at the tunnel advance.

2.5 Stages of construction

In clay, the tunnel is dug with a constant step of 1.5 m. Every 9 metres the steel pipes of the umbrella arch and the bolts are introduced in the tunnel face. A rib is set up at each phase of digging and shotcrete is set up between ribs. The invert is realised after every 3 metres of excavation.

Table 3. Properties of the soil mass.

	Value	Units
Density ρ	2300	Kg/m^3
Young's modulus E	80	MPa
Poisson ratio υ	0.3	–
Cohesion C	0.15	MPa
Friction angle ϕ	17	degrees
Dilatancy angle ψ	0	degrees
Earth pressure ratio k_0	0.7	–

Table 4. Mechanical parameters of structural elements.

	Density ρ (Kg/m^3)	Young's modulus E (MPa)	Poisson ratio υ
Pipes	7850	210,000	0.2
Bolts	–	20,000	0.2
Shotcrete	2450	7,000	0.2
Concrete	2450	32,000	0.2

3 THE NUMERICAL MODELS

A linear elastic and perfectly plastic behavior with a Mohr Coulomb failure criterion was adopted to represent the soil behaviour. The height of cover is taken constant and equal to twenty five metres. It corresponds to the maximum height of cover met by the tunnel in the argillaceous unit considered. The time effects are not studied in this analysis. Therefore, the definitive lining is not modelled. Two cases were compared: a first model, simple and which converges where the umbrella arch is not activated (taken as a reference case) and a second where it is activated, which allows seeing the influence of the arch on soil movements induced by the digging. Due to the fact that with the real geotechnical properties of the project and without the umbrella, the underground work cannot reach an equilibrium, we modify the cohesion (multiplied by 6) to see the influence of the umbrella arch. The geotechnical parameters introduced in computations are resumed in Table 3.

In the numerical models, the ribs are not individually simulated. They are considered in a homogenised section made up of ribs and shotcrete with a linear elastic behaviour. The homogenised section has a thickness of 0,247 m and an elastic modulus equal to 10107 MPa. The mechanical properties of the various structural elements are resumed in Table 4. For the umbrella arch pipes, we consider a linear elastic behaviour.

In this paper, focusing on the influence of the umbrella arch, the influence of the face bolting on the soil displacement will be not presented.

Figure 5. Elementary grid.

3.1 Three dimensional models

In the three dimensional models, we are able to simulate the real excavation geometry, the digging steps and take into account the umbrella arch of each pipe. The invert is represented.

3.1.1 The grid

Taking into account of the structural elements inducing a complex grid. The section is not constant. But, the covering between two umbrella arches being constant and equal to 9 m, the geometry is periodic with a period equal to 9 m. This geometry is modeled by assembly of elementary sections of 9 m length. The elementary section was also set up in several stages. It is an assembly of five grids. These five grids are illustrated in Figure 5.

The final grid, represented in Figure 6, is composed of two parts. The first part contains four elementary sections. The other represents the ground ahead of the tunnel face. The numerical model comprises 40,469 nodes and 12,184 elements.

The umbrella arch pipes are simulated by linear structural elements (1494 structural elements are used). Beam elements, which can sustain bending moments and normal stresses, are used to model the arch umbrella pipes. An illustration of these elements is given in Figure 7.

3.1.2 The initial and boundary conditions

A geostatic initial stresses field is applied. The earth pressure ratio is equal to 0,7. On the lateral far field boundaries and at the lower far field boundary, the displacements are fixed perpendicularly to the boundaries. On the upper far field boundary, displacements are free.

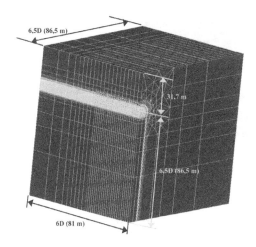

Figure 6. Final grid.

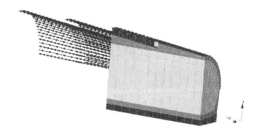

Figure 7. Beam elements used to simulate the umbrella arch.

3.1.3 Simulation of the sequential excavation

The simulated phases are the following:

1 – Definition of the natural ground pressure by applying the initial stress field.

2 – If the arch is simulated, set up of the first umbrella arch on 18 m length. First phase of digging of 1,5 m. The digging process is illustrated in Figure 8a.

3 – Second phase of digging of 1,5 m. Homogeneous support is set up in section 1(figure 8b).

4 – Third phase of digging of 1,5 m. Homogeneous support is set up. The invert is set up (figure 8c).

5 to 7 – Step 3 and 4 are repeated. At the end of step 7 a length of 9 m is excavated.

8 – If the arch is simulated, set up of the second umbrella arch on 18 m. Digging of 1,5 m and set up of homogeneous support. The raft is set up(figure 8d).

These steps are repeated during 17 sections of 1,5 m, i. e. until an excavated length of 25,5 m. This length permits to be far enough from the boundaries of the grid in order to limit their influence.

3.2 Two dimensional models

The 2D models are axisymmetric models. The tunnel geometry is considered to be circular. Different ways

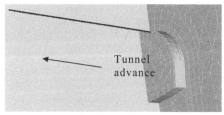

8a. Phase 2

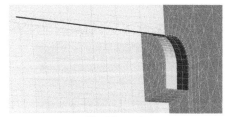

8b. Phase 3

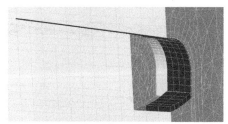

8c. Phase 4

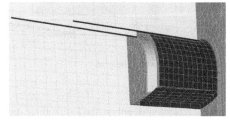

8d. Phase 8

Figure 8. Illustration of digging stages.

of representing the umbrella arch in axisymetry were studied. First, we simulated the umbrella arch with its tilt of 6°. Due to the variable section excavated, the tunnel equivalent radius varies between 6,08 m and 7,03 m.

In a second time, the tilt of the arch was not simulated in order to see if taking into account a tilt influences the numerical results in terms of displacements.

Two types of reinforcements were considered. First, we adopted a linear elastic and isotropic model. In this case, the ground is reinforced both in transversal and longitudinal direction. Secondly, we adopt a linear elastic and orthotropic model for the arch. So,

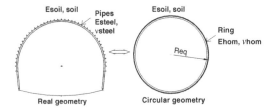

Figure 9. Homogenization principle used to calculate the equivalent characteristics.

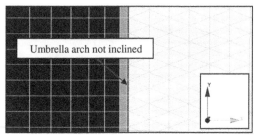

Figure 11. Detail of the umbrella arch when it is not inclined.

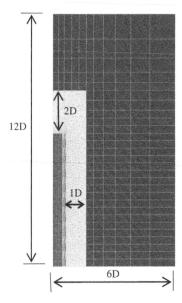

Figure 10. Dimensions of the grid.

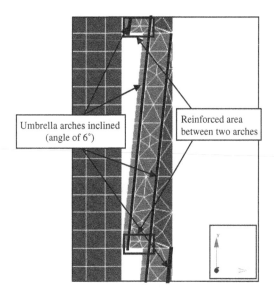

Figure 12. Detail of the grid when the umbrella arch is inclined.

the ground is reinforced only in the longitudinal direction. Only this latter is presented in this paper, which provided the closest results to those obtained by the 3D model. The umbrella arch is simulated by a circular, continuous ring with homogenized characteristics. The homogenization principle is illustrated in Figure 9. We assume that the umbrella arch acts like a crown of reinforced ground with a thickness equal to 0,10 m and with a linear elastic and orthotropic behaviour. This hypothesis is used because we wanted to represent the umbrella arch in an axisymetric numerical model. According to the case studied, the ring is inclined or not.

The slab cannot be simulated in axisymmetry. But, these models permit to visualize the movements of the tunnel face. For each representation of the arch, two cases were studied: one where the umbrella arch is not activated (reference case) and a second where it is activated.

3.2.1 The grid

The grid dimensions are 154 m (=12D) in the axial direction and 77 m (=6D) in the transverse direction in order to avoid the influence of the boundary limits. It contains around 8000 nodes and 3000 elements. The grid and its dimensions are illustrated in Figure 10.

In the model where the umbrella arch is inclined, a ground area situated between two umbrella arches is reinforced to avoid a local soil displacement. Reinforcing this area means a reinforcement of the soil cohesion which is equal to 0.7 MPa instead of 0.15. The reinforcement is activated when the support of the previous section is set up.

A detail of the umbrella arch is given in Figures 11 and 12 for the two representations of the arch. The digging phases of 1,5 m were represented with a regular grid every 0,75 meters.

3.2.2 The initial and boundary conditions

The initial stress field is anisotropic. The vertical stress is equal to 0.73 MPa which corresponds to a twenty five meters height cover. The horizontal stresses are equal to 0,51 MPa. On the all faces, the displacements are fixed in the direction perpendicular to the face.

3.2.3 Simulation of sequential excavation

The simulated phases are the following:

1 – Definition of the natural ground pressure by applying the initial stress field.

2 – If the arch is simulated, drilling of the first umbrella arch on 18 m. First phase of digging of 1,5 m.

3 – Second phase of digging of 1,5 m. Homogeneous support is set up.

4 to 7 – Step 3 is repeated. At the end of step 7 a length of 9 m is excavated.

8 – If the arch is simulated, drilling of the second umbrella arch on 18 m. Digging of 1,5 m and set up of homogeneous support.

These steps are repeated for 17 sections of 1,5 meters, i. e. until an excavated length of 25,5 m like in the 3D models.

4 RESULTS

The displacements computed using the 3D and 2D models, with or without the umbrella arch, are compared. The comparisons permit to show the influence of the arch on the displacements and to assess the quality of the adopted 2D model. The two kinds of representation of the umbrella arch are also compared in order to highlight the importance of taking into account the arch tilt in numerical models.

4.1 Comparisons between 2D and 3D results

In this part, we compare the results obtained using 3D and 2D approaches, when the arch tilt is simulated.

2D models lead to lower displacements at the crown than those computed in 3D, as we can see in Figure 13. This figure presents the evolution of the vertical displacement of the tunnel crown according to the distance from the face. So, 2D models are not safe. They do not permit to simulate correctly the behaviour of the support at the tunnel crown.

For the horizontal displacements at the tunnel wall, we observe the same phenomena ahead of the face. But, in the excavated part, there is a good agreement between 3D and 2D horizontal displacements. Moreover, the 2D approach is safe, the displacements computed in 2D being more important than in 3D. The 2D approach permits only a good assessment of the horizontal displacements at the lateral wall behind the face.

We can also see that behind the face, the umbrella arch effect on the vertical displacement is the same in 2D and in 3D. But, ahead of the face, its effect is

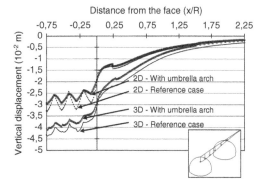

Figure 13. Vertical displacements at the crown according to the distance from the face.

Table 5. Maximal forces in a pipe at a cross section at 4.5 m behind of the face.

	Value	Units
Normal stress	−3.4	MN
Tangential stress T_y	−0.1	MN
Tangential stress T_z	0.15	MN
Bending moment M_y	0.03	MN.m
Bending moment M_z	0.02	MN.m

more significant in 3D. In both cases, the umbrella arch has little influence on the vertical displacement at the tunnel crown due to the fact that movements occur axially to the tunnel face.

For the horizontal displacements at the tunnel wall, the forepoling by umbrella arch induces a little reduction of the horizontal displacements of the lateral wall.

The maximal loads in umbrella arch, in 3D computations, at a cross section at a distance of 4.5 m behind of the face are resumed in Table 5. In this table a positive stress corresponds to a compression stress.

In Figure 14, the evolution of the extrusion of the face according to the distance from the tunnel axis is represented. The umbrella arch has the same low influence in 2D and in 3D on the extrusion of the face. The 2D approach can be used to assess the extrusion of the face because it leads to more important values than those computed in 3D. Table 6 resumes the face extrusion at the tunnel axis. The 2D approach appears safe and provides a good assessment of the face extrusion.

For the horizontal displacements along the tunnel axis, the umbrella arch has the same low influence in 2D and in 3D. 2D approach can be used to compute horizontal displacements at the face and at a distance from the face higher than one radius. In fact, in these cases, the displacements computed in 2D are more important than those obtained in 3D and thus, the 2D approach is safe.

171

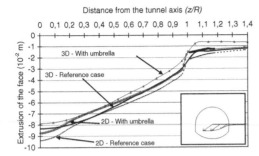

Figure 14. Evolution of the face extrusion according to the distance from the tunnel axis.

Table 6. Effect of the umbrella arch on the extrusion at the tunnel axis.

	Reference case (m)	With umbrella arch (m)	Variation (%)
2D	0.094	0.087	7
3D	0.083	0.079	5

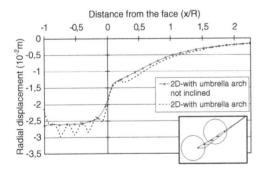

Figure 15. Evolution of the tunnel wall displacements for the two representations of the arch.

4.2 Influence of arch tilt on results

Figure 15 shows the displacements obtained in 2D at the tunnel walls in the two cases of arch representations. For the case where the arch tilt is not simulated we can see that the curve is more regular behind the face. So, the model where the tilt is not represented induces lower stresses than the model where the variation of the excavated section is simulated.

Figure 16 compares the results obtained in terms of horizontal displacements along the tunnel axis. The simplified model indicates displacements close to those obtained with the simulation of a more complex geometry. The simplification leads to a maximum difference of 10% as for extrusion of the face.

A simplified 2D model, where the arch tilt is not simulated, can be used to assess the displacements

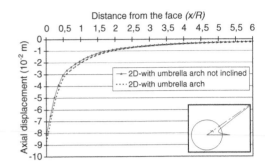

Figure 16. Evolution of the horizontal displacements along the tunnel axis in the two 2D models.

induced by the digging of a tunnel in presence of a forepoling by an umbrella arch. The displacements computed with the simplified model are lower than those obtained with a variable excavated section but the difference is very small compared to the time won: the design of the model where the tilt is represented is time consuming and the computations times are also more important.

4.3 Calculation time

The 3D calculations are time consuming. For the reference case about 12 days are necessary to complete the calculation and 14 days for the case where the umbrella arch was simulated. These times were obtained with a Pentium IV, 3.2 GHz.

5 CONCLUSIONS AND PERSPECTIVES

The influence of an umbrella arch on displacements induced by the digging of a tunnel was studied using two types of numerical models: first in a 3D model and in a second time in 2D models. The three dimensional model is relatively complex, when each inclusion is represented. Taking into account the results obtained, the simulation of the umbrella arch using a 2D simplified approach seems satisfactory. A simplified 2D model where the tilt of the arch is not simulated can be used for a first assessment of the displacements. This simplification permits to limit the model design and calculation time.

Considering these numerical simulations, it appears that an umbrella arch has low influence on displacements. This seems to be not verified by experimental results (Takechi, H et al. 2000). Thus, it appears necessary to compare the results of this study with experimental measurements. This will be realized during the construction of the tunnel near Besançon where an important monitoring is planned. By the numerical way, another method to simulate the umbrella arch in

2D could be considered. It consists on the homogenization approach developed by De Buhan & Sudret (1999). In this approach, the inclusions and the soil around are represented by a two phase elastoplastic model.

REFERENCES

Al Hallak, P. 1999. Etude expérimentale et numérique du renforcement du front de taille par boulonnage dans les tunnels en terrain meuble. *Thesis*. Ecole Nationale des Ponts et Chaussées, Paris, France.

Bae, G.J., Shin, H.S., Sicilia, C., Choi, Y.G. & Lim, J.J. 2005. Homogenization framework for three-dimensional elastoplastic finite element analysis of a grouted pipe-roofing reinforcement method for tunnelling. *International Journal for Numerical and Analytical Methods in Geomechanics* 29: pp. 1–24.

CETU, Centre d'Etudes des Tunnels. 1998. Procédés de creusement et de soutènement. *Dossier pilote des tunnels section 4*.

CETU 2003. Dossier de Consultation aux entreprises.

De Buhan, P., Sudret, B. 1999. A two-phase elasto-plastic model for unidirectionally-reinforced materials. *European Journal of Mechanics-A/Solid* 18: pp. 995–1012.

Dias, D. 1999. Renforcement du front de taille des tunnels par boulonnage – Etude numérique et application à un cas réel en site urbain. *Thesis*. Institut National des Sciences Appliquées de Lyon. Lyon, France.

Lee, J.S., Bang, C.S., Choi, I.Y. & Um, J.H. 2000. A study on the design approach of the pipe roofing reinforcement, *Journal of the Korean Society of Civil Engineers* 20(3–C): pp. 305–314.

Lunardi, P. 2000. The design and construction of tunnels using the approach based on the analysis of controlled deformation in rocks and soils. *Tunnels and Tunnelling International*: pp. 3–30.

Oreste, P.P. & Peila, D. 1997. La progettazione degli infilaggi in avanzamento nella costruzione delle gallerie. *Convegno di ingegneria geotecnica*. Perugia.

Pelizza, S. & Peila, D. 1999. Soil and rock reinforcement in tunnelling. *Tunnelling and Underground Space Technology* 8 (5): pp. 357–372.

Takechi, H., Kavakami, K., Orihashi, T. & Nakagawa, K. 2000. Some considerations on effect of the long-fore-piling-method. *AITES-ITA 2000 World Tunnel Congress*, Durban: pp. 491–498.

Tan, W.L. & Ranjith, P.G. 2003. Numerical Analysis of Pipe Roof Reinforcement in Soft Ground Tunnelling. *Proceeding of the 16th International Conference on Engineering Mechanics*. ASCE, Seattle, USA.

Uhtsu, H., Hakoishi, Y., Nago, M. & Taki, H. 1995. A prediction of ground behaviour due to tunnel excavation under shallow overburden with long-length forepolings. *South East Asian Symposium of Tunnelling And Underground Space Development Japan Tunnelling Association*. Bangkok: pp. 157–165.

Yoo, C. 2002. Finite-element analysis of tunnel face reinforced by longitudinal pipes. *Computers and Geotechnics* 29: pp. 73–94.

5. Ground freezing

Numerical Modelling of Construction Processes in Geotechnical Engineering for
Urban Environment – Triantafyllidis (ed)
© 2006 Taylor & Francis Group, London, ISBN 0 415 39748 0

An elastic-viscoplastic model for frozen soils

Roberto Cudmani
Züblin AG, Stuttgart

ABSTRACT: Basic aspects of the time- and temperature-dependent mechanical behavior of frozen soils are discussed. Based on experimental results (Orth 1985) an elastic-viscoplastic constitutive model is developed and validated by means of unconfined strain-controlled compression and creep tests. It is concluded that in spite of its simplicity the proposed model is able to simulate the creep behavior and rate-dependent strength of frozen soils realistically.

1 INTRODUCTION

The design of ground freezing measures requires a detailed consideration of the geological and hydro-geological ground properties, the geotechnical and thermal properties of the soils to be frozen and the particular boundary conditions. The design problem involves both thermal and mechanical analysis. In the thermal analysis, the time and power supply required to freeze the soil and to maintain the frozen body at a desired temperature is determined. The mechanical analysis is necessary to judge the stability and evaluate the deformations of the frozen body during the different construction stages. Although analytical solutions of ground freezing problems involving simple geometries and boundary conditions are presently available, solving real problems in complex geological, hydro-geological and geometrical conditions requires the application of numerical methods, especially the Finite Element Method (FEM). Clearly, the use of both numerical and analytical solutions requires the knowledge of the mechanical behavior of the frozen soils, which is described mathematically by constitutive models.

A frozen soil is a four-phase-mixture consisting of soil grains, ice, water and air. The fractions of these components depend on the granulometric properties of the soil, the density, the water content, the temperature and to less extent on the stress state. In the case of saturated granular soils, the shear resistance after freezing is the result of a complex interaction between the grain skeleton and the ice in its pores. On the one hand, the ice hinders dilatancy of the grain skeleton, leading to an increase of confining pressure and shear resistance. On the other hand, the grains retard the development and spreading of cracks in the ice matrix. This explains why the behavior of frozen soils

is stiffer, and significantly more ductile than that of the ice fraction. The time- and temperature-dependent behavior of frozen soils, which is principally attributed to the mechanical properties of ice, has been extensively investigated in the literature [e.g. (Andersland and Al Nouri 1970), (Ladanyi 1972), (Klein 1985), (Orth 1985), (Gudehus and Tamborek 1996)].

This contribution focuses on the mechanical behavior of frozen granular soils under quasi-monotonic loading. Main aspects of the time- and temperature-dependent behavior of frozen soils are analyzed (section 2) with the help of laboratory tests. Based on experimental results a constitutive model for frozen soils is proposed (section 3). Predictions of the mechanical behavior of frozen soils observed in the laboratory are made in order to validate the constitutive model (section 4). In a companion paper the application of the model to the analysis of a real soil freezing problem is presented (Cudmani and Nagelsdiek 2006).

2 MECHANICAL BEHAVIOR OF FROZEN SOILS

An extensive experimental investigation of the mechanical behavior of frozen soils was carried out by (Orth 1985). He performed unconfined and triaxial compression tests with frozen Karlsruhe sand (quartz sand, $d_{50} = 0.5$ mm; $max\,e = 0.85$; $min\,e = 0.57$). The samples were prepared with the pluviation method in the laboratory.

Figure 1 shows the influence of the temperature and the deformation rate on the mechanical behavior of frozen Karlsruhe sand for unconfined compression. The diagrams show the axial stress versus the axial deformation for a constant strain rate and different temperatures (a) and for constant temperature

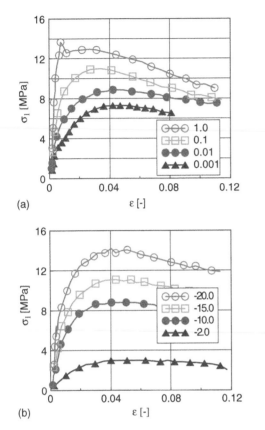

(a)

(b)

Figure 1. Axial stress versus axial strain for strain controlled unconfined compression tests. (a) Different $\dot{\varepsilon}$ and $\vartheta = -10°C$; (b) different ϑ and $\dot{\varepsilon} = 0.01\,\%/min$ (Orth 1985).

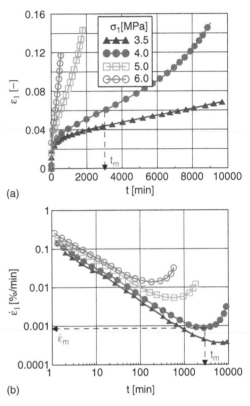

(a)

(b)

Figure 2.   One-dimensional creep tests. (a) Axial strain versus time; (b) axial strain rate versus time for different σ_1 and $\vartheta = -10°C$ (Orth 1985).

and different deformation rates (b). According to (Orth 1985) the relationship between the unconfined compression strength σ_c, the temperature ϑ and the strain rate $\dot{\varepsilon}$ can be described by

$$\sigma_c(\vartheta) = \sigma_r(\vartheta) + A(\vartheta)\ln\left(\frac{\dot{\varepsilon}}{\dot{\varepsilon}_r}\right). \tag{1}$$

The reference stress σ_r is the unconfined compression strength obtained with a reference strain rate $\dot{\varepsilon}_r$ (for instance 1%/min). Both σ_r and the proportionality constant A depend on the temperature.

Frozen soils samples creep under constant axial stress (see Fig. 2). Firstly, the creep rate decreases with time, reaching a minimum value $\dot{\varepsilon}_m$ at the time t_m. Thereafter, the strain rate increases with time. In this phase cracks develop in the ice matrix and the specimen fails. The time t_m at the turning point is called *standing time*, since up to this time the frozen soil sample shows a stable behavior. An increase of the axial stress and decrease of the temperature (not shown in

the figure) cause a decrease of the standing time. Actually, the specimen fails at two or three times t_m, but the deformation of the frozen body after t_m becomes too large for most practical applications.

(Orth 1985) found the following relationship between t_m and $\dot{\varepsilon}_m$:

$$\dot{\varepsilon}_m t_m = C \tag{2}$$

C is a dimensionless material constant. As it can be seen in Fig. 2, the deformation at the turning point $\dot{\varepsilon}_m$ increases with increasing axial stress. It is found that

$$\sigma_1(\vartheta) = \sigma_\alpha(\vartheta) + B(\vartheta)\ln\left(\frac{\dot{\varepsilon}_m}{\dot{\varepsilon}_\alpha}\right). \tag{3}$$

Herein σ_α is the axial stress leading to a reference strain rate (e.g. 1%/min) at the turning point. Both σ_α and B are temperature-dependent functions. It can be seen that the relationship between σ_1 and $\dot{\varepsilon}_m$ is formally the same as that between σ_c and $\dot{\varepsilon}$ (equation 1).

However, experimental results show that the corresponding functions in equations 1 and 3 are not

178

identical ($\sigma_a(\vartheta) \neq \sigma_r(\vartheta)$ and $B(\vartheta) \neq A(\vartheta)$). This means that the internal structure of the ice matrix in the state of maximal strength in a creep test is not the same as in a strain-controlled compression test. This is because the behavior of ice is extremely dependent on the previous history. Obviously, creep and rate-dependent strength of frozen soils underlie the same micromechanical mechanisms, the previous history alters the macroscopic response only quantitatively. Nevertheless, the compressive strengths calculated with equation 3 deviate less than 10% for deformation velocities in the range between 0.001 and 2%/min (Orth 1985). Thus, for practical purposes, it is reasonable to adopt the same function to describe both creep and shear strength. Putting the strain rate $\dot{\varepsilon}_m$ as function of ϑ and σ_1 we obtain:

$$\dot{\varepsilon}_m = \dot{\varepsilon}_\alpha \exp\left[C(\vartheta)\left(\frac{\sigma_1}{\sigma_\alpha(\vartheta)} - 1\right)\right] \quad (4)$$

Herein is $C(\vartheta) = \sigma_\alpha(\vartheta)/B(\vartheta)$. Based on results of unconfined compression tests and on theoretical considerations of the underlying creep mechanism, the functions for σ_α and $C(\vartheta)$ were proposed (Orth 1985):

$$\sigma_\alpha(\vartheta) = a_1(-\vartheta)^{a_2} \quad (5)$$

$$C(\vartheta) = \frac{K_1}{\vartheta + 273.4} + \ln\dot{\varepsilon}_\alpha \quad (6)$$

a_1, a_2 and K_1 are material constants. From equations 4, 5 and 6 the strain rate at the turning point $\dot{\varepsilon}_m$ can be calculated as a function of σ_1 and ϑ. When dealing with equations 5 and 6, care must be taken of the fact that the constants are not dimensionless.

When plotting $\dot{\varepsilon}/\dot{\varepsilon}_m$ against t/t_m for different ϑ and σ_1 Orth realized that all curves of Fig. 2 fall together. The normalized curves can be approximated with the function:

$$\dot{\varepsilon}/\dot{\varepsilon}_m = \exp(-\beta)\exp\left(\beta\frac{t}{t_m}\right)\left(\frac{t}{t_m}\right)^{-\beta} \quad (7)$$

Equation 7 fulfills the physical conditions $d\dot{\varepsilon}_m/dt = 0$ (Equation 2) and $\dot{\varepsilon}(t_m)/\dot{\varepsilon}_m = 1$. The application of equation 7 requires the determination of five material parameters: C (equation 2), a_1 and a_2 (equation 5), K_1 (equation 6) and β (equation 7). A procedure for the determination of the parameters based on unconfined creep compression tests was proposed by (Orth 1985).

Frozen soils under triaxial conditions show a qualitatively similar creep behavior as that depicted in Fig. 2 (Gudehus and Tamborek 1991). The standing time t_m increases with the mean pressure and decreases with the deviatoric stress. On the contrary, $\dot{\varepsilon}_m$ decreases with the mean pressure and increases

with the deviatoric stress. In order to consider the influence of the mean pressure on the creep behavior of frozen soils, Gudehus and Tamborek added a pressure-dependent term to equation 4 to describe creep under triaxial conditions:

$$\frac{\dot{\varepsilon}_m}{\dot{\varepsilon}_\alpha} = \exp\left[C(\vartheta)\left(\frac{\sigma_1 - \sigma_3}{\sigma_\alpha(\vartheta)} - 1\right) + b_1\left(\frac{\sigma_1 - \sigma_3}{3p}\right)^2 - b_2\right] \quad (8)$$

Unlike (Gudehus and Tamborek 1996), we adopt $b_1 = b_2$ in order to fulfill the condition $\dot{\varepsilon}_m = \dot{\varepsilon}_\alpha$ for unconfined compression with $\sigma_3 = 0$, $p = \sigma_1/3$ and $\sigma_1 = \sigma_\alpha$. The additional material constant a_1 can be determined from triaxial creep tests (Gudehus and Tamborek 1996).

According to Orth the influence of the mean pressure can be disregarded if $p < 3$ to 5 MPa which is the case for shallow constructions like tunnels and underpinning structures.

3 CONSTITUTIVE MODEL

The presented experimental results reveal an important mechanical property of frozen soils: There is no deviatoric stress that they can bear unlimitedly. In a way frozen soils behave like a viscous fluid. For this reason, the evaluation of the standing time should play a key role in the design of ground freezing. Nevertheless, most available constitutive models for frozen soils, especially those commonly used in geotechnical practice do not take this important property into account.

Assuming that the mechanical behavior of frozen soils can be described by an elastic-viscoplastic constitutive model the relationship between the stress rate tensor $\dot{\mathbf{T}}$ and strain rate tensor $\mathbf{D}$ can be written as:

$$\dot{\mathbf{T}} = \mathcal{L} : (\mathbf{D} - \mathbf{D}_v) \quad (9)$$

Herein $\mathcal{L}$ is the elastic stiffness tensor and $\mathbf{D}_v$ the viscous strain rate tensor. Based on equation 7 the following evolution equation is proposed for the viscous strain rate:

$$\mathbf{D}_v = \dot{\varepsilon}_m \frac{\tilde{\mathbf{T}}}{||\tilde{\mathbf{T}}||}\exp(-\beta)\exp\left(\beta\frac{t}{t_m}\right)\left(\frac{t}{t_m}\right)^{-\beta} \quad (10)$$

Herein $\tilde{\mathbf{T}}_{ij} = \mathbf{T}_{ij} - p\delta_{ij}$ is the deviatoric stress tensor and $||\tilde{\mathbf{T}}|| = \sqrt{\tilde{\mathbf{T}}_{ij}\tilde{\mathbf{T}}_{ij}}$ its euclidian norm. Equation 10 assumes that the tensors $\tilde{\mathbf{T}}$ and $\mathbf{D}_v$ are coaxial, i.e. creep deformations are exclusively deviatoric. The strain rate at the turning point can be calculated from equation 8 using $q = \sqrt{3/2}||\tilde{\mathbf{T}}||$ instead of $\sigma_1 - \sigma_3$:

$$\frac{\dot{\varepsilon}_m}{\dot{\varepsilon}_\alpha} = \exp\left[C(\vartheta)\left(\frac{q}{\sigma_\alpha(\vartheta)} - 1\right) + b_1\left(\frac{q}{3p}\right)^2 - 1\right] \quad (11)$$

It must be noted that equations 10 and 11 coincide with equations 7 and 8 for one-dimensional stress conditions. The constituive parameters are the same as defined in section 2.

For testing and validation purposes the constitutive equations were implemented in an element test program. For the solution of boundary-value problems the model was implemented into the commercial FE-Code TOCHNOG via user-subroutine.

Table 1. Material parameters of frozen Karlsruhe sand.

C [-]	a_1 [MPa/$^\circ C$]	a_2 [-]	β [-]	K_1 [$^\circ C$]	b_1 [-]
0.024	3.050	0.591	0.692	3817	0

4 VALIDATION OF THE CONSTITUTIVE MODEL

4.1 Simulation of laboratory tests with frozen Karlsruhe sand

In order to validate the proposed constitutive model, laboratory tests performed by Orth with Karlsruhe sand were simulated with the element test program. In the following only the simulations of unconfined compression tests are presented.

The material constants of the frozen Karlsruhe sand are listed in Table 1.

For the element test simulations a Young modulus $E = 500$ MPa and a Poisson ratio $\nu = 0.3$ were adopted.

Figure 3 shows the results of numerical strain-controlled unconfined compression tests carried out with the same conditions as in Fig. 1. As none of the simulated experiments was used to determine

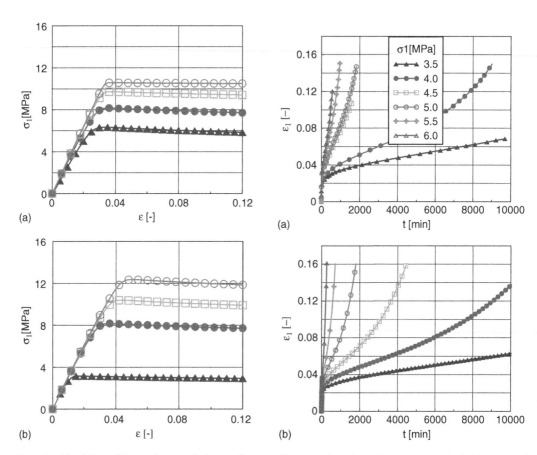

Figure 3. Simulation of the strain-controlled unconfined compression tests from Fig. 1. (a) Different $\dot{\varepsilon}$ and $\vartheta = -10^\circ C$; (b) different ϑ and $\dot{\varepsilon} = 0.01$ %/min.

Figure 4. Experimental (a) and numerical (b) results of one-dimensional creep tests for different σ_1 and $\vartheta = -5^\circ C$ (Experiments after (Orth 1985)).

the model parameters, we are actually predicting the behavior observed in the tests. As it can be seen, the model is able to simulate the measured rate- and temperature-dependent behavior quite realistically, but as expected, it is not able to predict the rate-dependence of shear strength very precisely. Nevertheless, since the calculated stress-strain behavior does not differ too much from the measured one, and the calculated strengths lies on the safe side, the prediction accuracy should suffice for (any) practical purposes.

Results of experimental and numerical creep tests are presented in Figs. 4 to 6. The diagrams show the axial deformations over the time for different axial stresses and different temperatures. The proposed constitutive model is able to describe the increase of the axial strain due to creep quite well and to predict the standing time realistically. It is noted, that only creep tests with $\vartheta = -10$ and $-20°C$ results were used to calibrate the model.

4.2 Simulation of laboratory tests with frozen Cologne sandy gravel

For a complex tunnel project in the city of Cologne different soil freezing measures are planned. In the zone to be frozen, the ground consists of quaternary sandy and gravelly soil layers. In order to evaluate the mechanical properties of the frozen soils, laboratory tests with frozen sandy gravel and gravelly sand samples were carried out (Ehl 2002). The samples were prepared in the laboratory, using a similar procedure as that employed by (Orth 1985). In the following, the proposed constitutive model will be used to simulate the behavior of the frozen sandy gravel samples. The material constants of the frozen soil are listed in Table 2.

Except the Young modulus, which was $E = 1000\,MPa$ for the strain-controlled unconfined compression tests and $E = 600\,MPa$ for the creep tests, the material

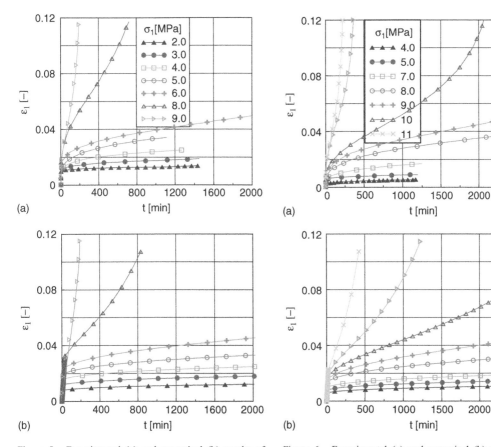

Figure 5. Experimental (a) and numerical (b) results of one-dimensional creep tests for different σ_1 and $\vartheta = -10°C$ (Experiments after (Orth 1985)).

Figure 6. Experimental (a) and numerical (b) results of one-dimensional creep tests for different σ_1 and $\vartheta = -15°C$ (Experiments after (Orth 1985)).

Table 2. Material parameters of frozen Cologne sandy gravel.

C [-]	a_1 [MPa/°C]	a_2 [-]	β [-]	K_1 [°C]	b_1 [-]
0.024	1.3	0.75	0.4	4000	0

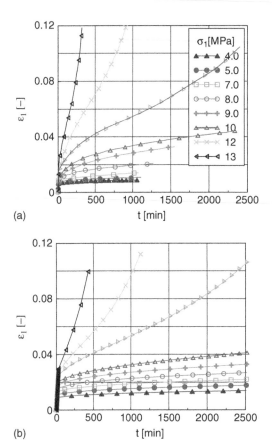

(a)

(b)

Figure 7. Experimental (a) and numerical (b) results of one-dimensional creep tests for different σ_1 and $\vartheta = -20°C$ (Experiments after (Orth 1985)).

parameters were the same in all numerical simulations. A Poisson ratio $\nu = 0.3$ was assumed.

Figure 8 compares the results of the experiments and simulations of unconfined compression tests with $\vartheta = -10$ and $-20°C$ and $\dot{\varepsilon} = 1$ %/min. Although the scattering of experimental data is quite strong, the constitutive model was able to predict the stress-strain behavior and the compressive strength quite well. The scattering of the data can be attributed to friction on

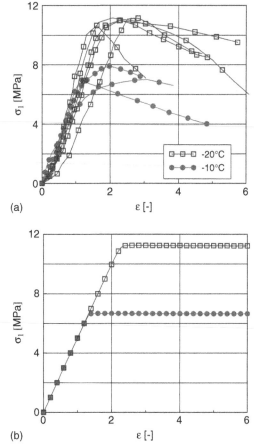

(a)

(b)

Figure 8. Experimental (a) and numerical (b) results of strain-controlled unconfined compression tests for $\dot{\varepsilon} = 1$%/min and $\vartheta = -10$ and $-20°C$.

the end platens, lack of initial homogeneity of the sample with regard to its density and temperature or/and variations of the temperature during the tests.

Results of experimental and numerical creep tests are presented in Fig. 9. The predictive capacity of the model is confirmed by these simulations. The sudden increase of strain at $t \approx 50$ h, observed in the experiment with $\sigma_1 = 3.2$ MPa, was not predicted by the model. However, from the fact that the increase of the strain rate (inclination of the curve) is only temporary we deduce that this behavior must be caused by a test anomaly.

5 CONCLUSIONS

The behavior of frozen soil results from a complex interaction between the grain skeleton and the ice

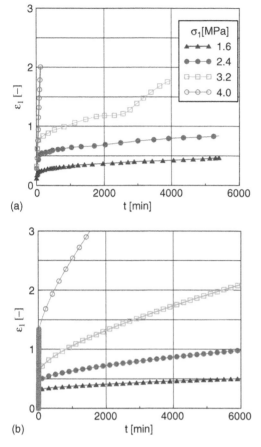

(a)

(b)

Figure 9. Experimental (a) and numerical (b) results of one-dimensional creep tests for different σ_1 and $\vartheta = -10°C$.

matrix. Nevertheless, the solution of practical problems is often based on constitutive models, which fail to describe the particular time- and temperature-dependent properties of this geotechnical composite.

Thus, creep deformations and the standing time, which play an important role in the assessment of the stability and serviceability of a frozen soil body, cannot be evaluated.

In spite of its relative simplicity, the proposed elastic-viscoplastic model enables a realistic description of the rate- and temperature-dependent response of frozen soils under monotonic and quasi-monotonic loading. Due to its simplicity the model can be easily implemented in a commercial FE-code and the few material parameters can be determined with the aid of conventional frozen soil laboratory tests.

REFERENCES

Andersland, O. and J. Al Nouri (1970). Time-dependent strength behaviour of frozen soils. *Soil Mechanics and Foundation Division, ASCE 96*(SM4), 1249–1265.

Cudmani, R. and S. Nagelsdiek (2006). FE-analysis of ground freezing for the construction of a tunnel cross connection. In *Proc. Numerical Simulation of Construction Processes in Geotechnical Engineering for Urban Environment*.

Ehl, G. (2002). Nord-Süd Stadtbahn Köln-Gefriertechnische Untersuchungen. Baugrundgutachten.

Gudehus, G. and A. Tamborek (1996). Zur Kraftübertragung Frostkörper-Stützelemente. *Bautechnik 9*, 570–581.

Klein, J. (1985). *Handbuch des Gefrierschachtbaus*. Glückauf, Essen.

Ladanyi, B. (1972). An engineering theory of creep of frozen soils. *Canadian Geotechnical Journal 13*, 63–80.

Orth, W. (1985). *Gefrorener Sand als Werkstoff-Elementversuche und Materialmodell*. Ph. D. thesis, ibf. Heft 100.

Numerical Modelling of Construction Processes in Geotechnical Engineering for
Urban Environment – Triantafyllidis (ed)
© 2006 Taylor & Francis Group, London, ISBN 0 415 39748 0

Numerical and physical modeling of artificial ground freezing

A. Sres, E. Pimentel & G. Anagnostou
Institute for Geotechnical Engineering, ETH Zurich, Switzerland

ABSTRACT: Artificial ground freezing (AGF) is a method for stabilizing otherwise unstable grounds and for preventing groundwater infiltration in tunneling, shaft and deep excavation works. Although this ground improvement method has been investigated and used extensively in the last five decades, there are still many cases where its application may be desired, but confident planning is not possible due to the lack of reliable models. This applies, for example, in the threshold cases where groundwater flows rapidly or where soil expansion may occur due to the formation of ice lenses. In this contribution, a 3D implementation of the thermo-hydraulic (TH) model has been developed. The model has been optimized in order to achieve high computational efficiency, allowing complex simulations to be performed. The numerical model has been validated by simulating existing laboratory experiments. A new laboratory model for simulating the freezing process and considering boundary conditions close to the threshold cases is also presented.

1 INTRODUCTION

In artificial ground freezing (AGF) the pore water is temporarily converted into ice, thus yielding a high-strength and low-permeability frozen soil mass. AGF is, besides grouting, a method for stabilizing otherwise unstable ground and for preventing groundwater infiltration in tunneling, shaft and deep excavation works. Grouting is in most cases the first choice, due to cost and also for technical reasons. Its applicability is limited, however, by the grain size distribution of the soil and by environmental factors. In cases where grouting is not possible as a result of these factors, artificial ground freezing may be an alternative.

The problems of artificial ground freezing are associated with high groundwater seepage velocities, expansion of the frozen soil or consolidation in the thawing phase.

In the first case, the development of the frozen body will be very time-consuming or even completely hindered. Due to the lack of comprehensive investigations, only rough reference values exist for determining the critical velocity, i.e. the threshold of the applicability of this construction method. Jessberger (1996) outlines a critical velocity of 2 m/d. Based on field data, Darcy's law for the seepage flow and thermodynamic factors, Sanger & Sayles (1979) developed a simplified method for estimating critical velocity in relation to freeze pipe radius and spacing.

The expansion of frozen soil is caused by two mechanisms: (i) the conversion of soil pore water into ice during freezing and (ii) the formation of ice lenses. The first mechanism causes a volumetric change of

the soil pores by up to 9% depending on the drainage conditions. Serious deformations can be expected as a consequence of the second mechanism, where pore water migration occurs due to negative pressures or due to water pressed out during the consolidation of frozen soil. Usually both phenomena, i.e. the volume expansion and the formation of ice lenses, take place simultaneously but they differ considerably in their magnitude and predictability. A further deformation problem can arise from the ice lenses, since they can cause later, in the thawing phase, ground settlements. The formation of ice lenses by migration of water to the ice front can occur generally in each soil type depending on the seepage flow and temperature conditions (Andersland & Ladanyi 2004).

Up until now there have been no reliable models enabling the confident planning of AGF in the case of seepage flow, or an optimization of the cooling energy so that ground deformations are kept to a minimum during freezing and thawing.

2 MODELING – STATE OF THE ART

To simulate the expansion of the frozen body we need the temperature distribution over time. The surface of the body is then given by the freezing temperature. To predict the temperature distribution in a freezing soil under conditions of seepage flow a thermo – hydraulic model (TH) is needed. Such models are well known. For example, Frivik & Comini (1982) implemented a simplified model for the case of a plane and horizontal domain with experimentally determined material parameters. Makowski (1983) developed a 2D finite element program with improved models for the material parameters of the soil – water admixture. This model is limited to simple cases, because the complex geometries usual in practical cases cannot be simulated adequately. There is therefore a need for a 3D implementation.

Models for frost expansion and the associated heave on surface have been developed for the practical dimensioning of road structures as well as for the investigation of fundamental aspects of frost phenomena. These models differ mainly in respect of the extent to which empirical observations have guided their development. Kujala (1997) divided them into empirical, hydrodynamic, rigid-ice, segregation potential and thermo-hydro-mechanical (THM) models. The most rigorous ones are the THM models. In addition to heat and mass flow, these models take into account the mechanical properties of frozen soil and apply consistently to both the 2D and 3D cases. Fremond & Mikkola (1991) developed such a model, based on the conservation laws of mass, momentum and energy and the entropy inequality. In their model, saturated soil is considered as a mixture of three constituents: soil skeleton, water and ice. The constitutive equations of a porous medium are derived by considering the local equilibrium state and choosing appropriate expressions for the free energy and the dissipation potential. A key element of their work is the description of the free energy of bounded water. The model can describe the suction resulting from pore water freezing, the transfer of pore water and heat, and the frost heave. Mikkola & Hartikainen (2001) implemented this model for the 2D case. The results agreed well with experiments. A similar model was developed by de Boer et al. (2003) without considering the free energy of the bound water. The expansion of the soil is taken as a function of the expanding freezing water. They implemented their model only for the 2D case. The comparison of the results with experiments showed, however, a poor agreement.

To verify the models, good experimental data are needed. In addition, it must be pointed out that there is a lack of experimental results for complex cases.

3 GOVERNING EQUATIONS OF THE THERMO – HYDRAULIC MODEL

The TH model considers only the two coupled fields temperature T and piezometric head H. The latter is defined as $H = p/(\rho^w g) + z$, where p, ρ^w, g and z denote the pore water pressure, the water density, the gravity acceleration and the geodetic height, respectively. The seepage process is governed by the equation of heat transport

$$C^s \dot{T} - \partial_k \left(\lambda \partial_k T \right) + C^w v_k \partial_k T = 0 , \qquad (1)$$

where λ is the thermal conductivity, C_w the volumetric heat capacity of the water and C_s the socalled *effective* heat capacity of the saturated soil. C_s includes the latent heat which occurs during phase change and is either measured in laboratory tests or computed based upon the enthalpy method (cf. Del Guidice 1978). The Darcy velocity v_k is given by Darcy's law

$$v_k = -k \cdot \partial_k H , \qquad (2)$$

where k denotes the hydraulic conductivity. The velocity field fulfils the equation of continuity, which for a quasi steady-state seepage process is reduced to

$$\partial_k v_k = 0. \qquad (3)$$

The equations of continuity (3) and heat transport (1) are coupled by the Darcy velocity v_k and the temperature dependency of the hydraulic conductivity k, which takes into account the low permeability of frozen soil.

The equations defined above have to be solved in the region Ω with time $t \geq 0$. The boundary is then given by $\Gamma = \partial \Gamma \times t$. It is subdivided into Dirichlet Γ_D and Neumann Γ_N boundaries, where $\Gamma = \Gamma_D + \Gamma_N$ applies.

The Dirichlet boundary conditions are:

$$T(x_k, t) = T_d(x_k) , \qquad (4)$$

$$H(x_k, t) = H_d(x_k), \qquad (5)$$

with the position vector x_k in Γ_D and $t \geq 0$. The Neumann boundary conditions are:

$$\lambda n_i \partial_i T(x_k, t) = q(x_k, t) , \qquad (6)$$

$$k n_i \partial_i H(x_k, t) = m(x_k, t), \qquad (7)$$

with x_k in Γ_N and $t \geq 0$. Convective heat flow can be defined as $q_i = \alpha(T - T_{fluid})n_i$ with n_i the normal of the surface, where the fluid could be air or water. In this way, boundaries to the atmosphere and rivers can be modeled.

The initial conditions are:

$$T(x_k, t = 0) = T_0(x_k) , \qquad (8)$$

$$H(x_k, t = 0) = H_0(x_k). \qquad (9)$$

The values may be constant for the whole region or they may depend on the location. In the latter case a steady state calculation is carried out in order to obtain the head and temperature fields prevailing before the freezing operation.

4 FINITE ELEMENT FORMULATION

The numerical simulation will be done by the finite element method in a given three-dimensional domain.

The spatial discretization of equation (1) and (2) has been accomplished by means of the Garlekin method. The unknown scalar fields T and H are approximated throughout the solution domain at any time t by the following relationships:

$$T(x_k, t) = \sum_{i=1}^{8} h_i(x_k) \cdot T_i(t) , \tag{10}$$

$$H(x_k, t) = \sum_{i=1}^{8} h_i(x_k) \cdot H_i(t) , \tag{11}$$

where $h_i(x_k)$ denotes the biquadratic element shape function; and $T_i(t)$ and $H_i(t)$ the nodal values. The resulting equations in matrix form are:

$$C(T)\,\dot{T} + K(T)\,T - B(T,H)\,T = Q , \tag{12}$$

$$S(H)\,\dot{H} + A(T,H)\,H = F . \tag{13}$$

As mentioned in Section 3.1, the latent heat which occurs during phase change is included in the heat capacity. This so-called effective heat capacity is either measured in laboratory tests or computed based upon the enthalpy method (cf. Del Guidice 1978).

Integration over time is managed by the following the backward Euler scheme (Reddy & Gartling 2001), which is a single-step, non-iterative method for the solution of T^{n+1}:

$$\left[\frac{C(T^*)}{Dt} + K(T^*) \right] T^{n+1} =$$

$$\frac{C(T^*)}{Dt} T^n + Q + B(T^*, H^*) T^n \tag{14}$$

with the extrapolated temperature and piezometric head

$$T^* = \frac{3}{2} T^n - \frac{1}{2} T^{n-1} , \tag{15}$$

$$H^* = \frac{3}{2} H^n - \frac{1}{2} H^{n-1} . \tag{16}$$

This scheme is used also to solve equation (13).

5 NUMERICAL IMPLEMENTATION

In the interests of computational efficiency some simplifications have been made. These include the choice of elements. We use isoparametric hexahedral elements with biquadratic shape functions. They are generated in such a way that they all have same shape, but different dimensions. The Jacobi matrix and its determinant thus become very simple, depending only on the side length of a cube.

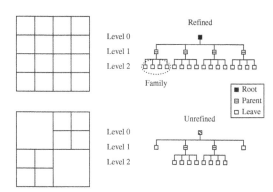

Figure 1. Illustration of the refinement and unrefinement procedure in a 2D version (Quadtree). In the first row, a macro element (ME) is shown following the refinement procedure and the resulting tree. The ME is now made up of finite elements of the same size. The second row shows the macro element after the unrefinement procedure. The first and the last family have been removed.

The geometrical description of a simulation problem is treated separately from the computational mesh (Korvink 1993). The simulation domain is subdivided into a coarse mesh of cubes (so-called macroelements). Each macro-element is the root of a mesh refinement tree. These macro-elements are further subdivided into finite elements following an appropriate tree structure. A macro-element will be divided into eight sub-cubes. Those elements became the first eight leaves in the tree and represent the first level of refinement. Eight leaves with the same root (parent) are also called a family. Then, in the next step of refinement, the elements will be divided into eight sub-cubes again. Each element becomes the root (parent) of the newly created family (leafs). This refinement procedure will be carried on until the desired level of refinement or size of elements is achieved (Fig. 1). The computational mesh is then built from the leaves, which are the elements at the ends of the branches of the tree.

Using the appropriate level of refinement a grid can be developed on which all structures, such as soil layers or freezing pipes, are defined (Fig. 2). The refinement levels also define the accuracy of the geometrical modeling. To reduce the numbers of elements, an unrefinement procedure will work in exactly the opposite way to the refinement procedure. Applying specific rules, it merges eight elements (provided that they comprise a family) into one bigger one (smaller level). This is achieved by removing the family and marking the parent as a leaf (Fig. 1). The aim is to obtain as few nodes as possible and a good distribution of the size of the elements for numerical accuracy.

To avoid asymmetric matrices, the convection term $B(T, H)$ was brought to the right side of equation (14) using the temperature of the former time step.

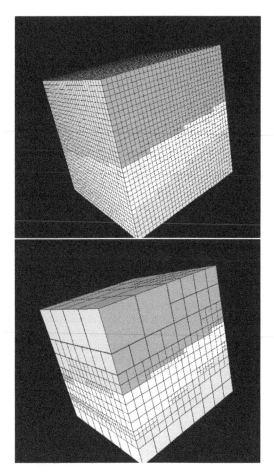

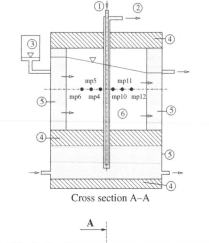

Cross section A–A

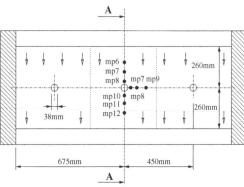

Figure 3. On the top: the sheer plan of the experimental test section: (1) brine inlet, (2) brine outlet, (3) constant pressure water feeding system, (4) extruded polystyrene, (5) expanded clay, (6) test soil. On the bottom: the plan view of the experimental apparatus, in correspondence with the measuring plane, showing the locations of the measuring thermocouples.

Figure 2. Simulation region with different soil layers. On the top with original set pattern and on the bottom after unrefinement.

The equations (12) and (13) are solved sequentially (staggered algorithm). The fields are coupled through the convection term $B(T, H)$. This procedure enables more efficient use of the computer memory. Compressed formats for matrices and vectors as data structures are used and an iterative solver based on the conjugate gradient method is applied. To achieve good performance, the global matrix is preconditioned with the incomplete Cholesky method.

6 VALIDATION OF THE NUMERICAL MODEL

In order to validate the TH model presented above, we simulated numerically the laboratory tests carried out by the Norwegian Institute of Technology, Trondheim (Berggren 1979, Frivik & Comini 1982). In these experiments, the growth of the ice wall across a row of freeze tubes, which were placed perpendicularly to the direction of seepage flow, was monitored (Fig. 3).

In these experiments, a high-capacity refrigeration system supplied brine to a row of three, 38 mm outer diameter freeze tubes at a suitably chosen flow rate and temperature. The tubes were placed vertically in the test section, with 450 mm spacing between their axes. Large flow rates of the coolant and heavy thermal insulation at the top and at the bottom of the test section were used in order to ensure two dimensional temperature fields in correspondence with the horizontal measuring plane. Heavy thermal insulation was also employed at the sides of the test section in order to reduce lateral heat flow.

Water was supplied to the test section at a constant pressure by means of a feeding bottle, which was suspended at different heights. A homogeneous bulk flow through the test soil was ensured by a circulation system made up of perforated feeding and drainage pipes,

Table 1. Physical properties of the "Hokksundsand".

T [°C]	C^s [MJ/m³K]	λ [W/mK]	$k/\rho^w g$ [m²s/Pa]
15.0	2.48	2.00	$3.16\ 10^{-8}$
−0.04	2.44	2.00	$2.46\ 10^{-8}$
−0.045	1,871	2.00	0.0
−0.06	1,079	2.41	0.0
−0.1	262	2.72	0.0
−0.3	36.5	2.99	0.0
−1.0	5.30	3.15	0.0
−5.0	2.00	3.25	0.0
−15.0	1.80	3.26	0.0
−40.0	1.60	3.27	0.0

immersed in expanded clay, and by two perforated aluminum plates with a 35 percent net flow area.

Copper-constantan thermocouples were used to measure the temperature. Figure 3 shows the location of these in the experiment, used here for verification.

The soil considered was an average texture "Hokksundsand" whose solid fraction was characterized by 35% mass quartz content and by a density of 2,700 kg/m³. Dry density values for this sand are, approximately, 1,600 kg/m³, corresponding to 41% porosity. The total mass water content for the saturated sand was thus in the order of 20%. The volumetric heat capacity versus temperature curve $C^s(T)$ were determined experimentally. The thermal conductivity versus temperature curve $\lambda(T)$ was estimated from knowledge of the soil texture, density of dry sand, degree of saturation, quartz content and unfrozen water content. Finally, the hydraulic conductivity k(T) of the unfrozen sand was determined "in situ" by means of a linear regression procedure and from measured values of flow rates and corresponding pressure gradients in the test section under steady-state isothermal conditions. Zero permeability values were assumed at below freezing temperatures. The values of $C^s(T)$, $\lambda(T)$ and k(T) are listed in Tab. 1 and used in the numerical simulation.

The computational domain takes into account the symmetry properties of the system. The side lengths of the finite elements were 30 mm. The conditions for the temperature at the side boundaries, where the water flows in and out, and on the surface of the freeze pipe are set as time-dependent Dirichlet conditions with values shown in Figure 4, which were measured together with the soil temperature. A Neumann condition with $q = 0$ applies to the remaining boundaries. For the piezometric head field, a constant head difference $\Delta H = 0.0243$ m was maintained by appropriate Dirichlet conditions. The respective initial seepage velocity amounts to 0.9 m/day. The initial temperature was set to 8.0°C. The time period of 25 hours was simulated with time steps of 600 sec.

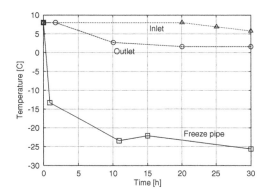

Figure 4. Surface temperature of the freeze pipe and temperature of the water inlet and outlet used for thermal boundary conditions.

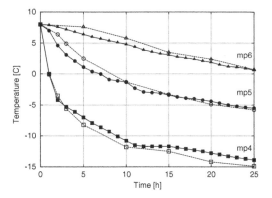

Figure 5. Measured (dashed lines) and computed (solid lines) temperatures at points mp4 to mp6 located in front of the freeze tube.

The results of the simulation with the TH model are shown in Figures 5–7. A good correlation between measured and simulated values can be observed. The small deviations are largely due to local inhomogeneites of the soil in the laboratory model.

To illustrate the performance of the numerical algorithm, the maximum number of nodes that can be simulated without memory swapping was estimated, together with the resulting computer time for a time step. An Intel PC with a P4/3.2GHz processor and 1 GB RAM memory on a SUSE Linux 9.2 operating system was used. With this configuration it is possible to simulate a TH model with up to 1,000,000 nodes, where each time step will take about 200 sec. So, for example, the TH model will take about 11.5 hours to simulate this system (10^6 nodes) over a time interval of one week, with construction time subdivided into time steps of one hour.

189

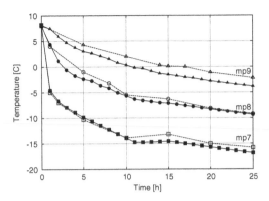

Figure 6. Measured (dashed lines) and computed (solid lines) temperatures at points mp7 to mp9 located at the side of the freeze tube.

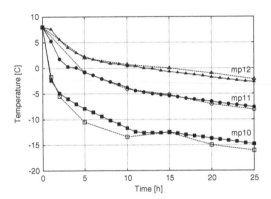

Figure 7. Measured (dashed lines) and computed (solid lines) temperatures at points mp10 to mp12 located in the rear the freeze tube.

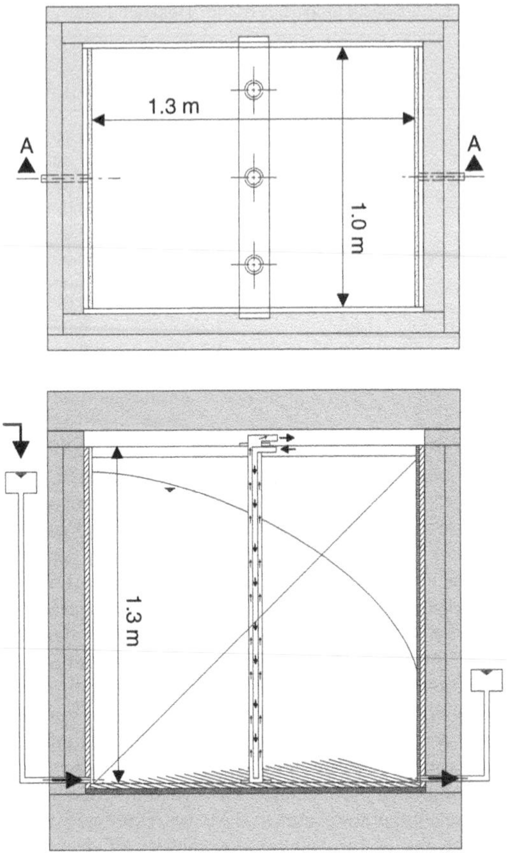

Figure 8. On the top: top view of the laboratory model without isolation cover. On the bottom: cross section A–A.

7 PLANNED LABORATORY TESTS

The laboratory experiment discussed above provided valuable insights into frozen body development, but it does have some limitations. For example, the soil deformations were not monitored, only relatively low head gradients are possible and the model dimensions of $H \times L = 0.6 \times 0.6 \, m^2$ are too small for simulating complex cases.

With the specific aim of verifying the THM model, a new laboratory test has been developed, allowing experiments to be performed with the relevant parameters. The dimensions and components of the model have been designed so as to achieve a large degree of flexibility in defining the boundary conditions and for obtaining comprehensive information on soil behavior.

The laboratory model includes a watertight box with effective dimensions $H \times L \times W = 1.2 \times 1.3 \times 1.0 \, m^3$ (Fig. 8). The box has an all around thermal insulation of at least 14 cm foam and a water inlet and outlet at the bottom of the two frontal faces. The freezing pipes consist of an interior PVC pipe and a copper pipe with an outer diameter of $D = 41 \, mm$. The pipes will be arranged in rows of three. Depending on the boundary conditions the amount, position and alignment of the pipe rows can be varied. The testing program includes tests with three vertical rows and tests with one row in either a vertical or horizontal position. The locations of the rows in the model are not fixed and can be varied to take account of hydraulic conditions.

The tests will be performed in an air conditioned room with a constant temperature of 20°C. A cooling unit with a power of 1.5 kW is able to cool the pipe rows to below −30°C. The monitoring of the tests will include sensors distributed in the model for measurement of temperature, pore water pressure and suction and soil deformation. At least 30 temperature sensors, 8 tensiometers and 16 accelerometers will be placed in the model. Additionally, the amount of water which in- and out-flows the model will be recorded over time.

Different soil mixtures varying between sandy silt up to silty sand will be used as material for performing the tests. The soil portions will be mixed in order to achieve a predefined grain size distribution and permeability.

8 CONCLUSIONS

The problems associated with the application of artificial ground freezing are high groundwater seepage velocities, expansion of the frozen soil or consolidation in the thawing phase. Up until now, there have been no reliable models enabling the confident planning of artificial ground freezing in borderline cases.

A new TH model for 3D has been numerically implemented with optimized computational algorithms. Initial verification shows close agreement with experimental results. Current research is focusing on the enhancement of the model through mechanical coupling and on experimental testing of the THM model.

REFERENCES

Andersland, O. & Ladanyi, B. 2004. *Frozen ground engineering – second edition*, Hoboken, New Jersey: John Wiley & Sons Inc

Berggren, A. 1979. *Artificial Freezing of Seepage Flow (B.Sc. thesis in Norwegian)*. Trondheim: Division of Refrigeration Engineering NTH

de Boer, R., Bluhm, J., Wüling, M. & Ricken, T. 2003. Phasenübergänge in porösen Medien. *Forschungsbericht aus dem Fachbereich Bauwesen 98, Universität Duisburg-Essen*. Rotterdam: Balkema

Del Guidice, G. 1978. Finite element simulation of freezing processes in soils. *International Journal for Numerical and Analytical Methods in Geomechanics* 2:223–235

Fremond, M. & Mikkola, M. 1991. Thermodynamical modelling of freezing soil. *International Symposium on Ground Freezing 91. Yu Xiang and Wang Changsheng (Eds.)*, Rotterdam: Balkema

Frivik, P.E. & Comini, G. 1982. Seepage and Heat Flow in Soil Freezing. *Journal of Heat Transfer* 140:323–328

Hartikainen, J. & Mikkola, M. 1997. General thermomechanical model of freezing soil with numerical application. *Ground Freezing 97*. Rotterdam: Balkema

Jessberger, H.L. 1996. *Grundbau Taschenbuch*. Berlin: Ernst & Sohn, Verlag für Architektur und techn. Wiss

Korvink, J.G. 1993. *An implementation of the adaptive finite element method for semiconductor sensor simulation (PhD Thesis)*. Zurich: Institute of Structural Engineering ETH

Kujala, K. 1997. Estimation of frost heave and thaw weakening by statistical analyses and physical models. *Proceedings of the International Symposium on Ground Freezing and Frost Action in Soils*. Rotterdam: Balkema

Makowski, E. 1983. *Modellierung der künstlichen Bodenvereisung im grundwasserdurchströmten Untergrund mit der Methode der finiten Elemente*. Bochum: Lehrstuhl für Grundbau und Bodenmechanik

Mikkola, M. & Hartikainen, J. 2001. Mathematical model of soil freezing and its numerical implementation. *Int. J. Numer. Meth. Engng.* 52:543–557

Reddy, J.N. & Gartling, D.K. 2001. *The Finite Element method in Heat Transfer and Fluid Dynamics, second Edition*. Boca Raton(USA): CRC Press LLC

Sanger, F.J. & Sayles F.H. 1979. Thermal and rheological computations for artificially frozen ground construction, *Ground Freezing developments in geotechnical engineering, Eng. Geol.* 13, vol. **26**, Amsterdam: Elsevier

Numerical Modelling of Construction Processes in Geotechnical Engineering for
Urban Environment – Triantafyllidis (ed)
© 2006 Taylor & Francis Group, London, ISBN 0 415 39748 0

Freezing of jet-grouted soil – comparison of numerical calculations and field data

Martin Kelm & Jürgen Raschendorfer
Ed. Züblin AG, Stuttgart, Germany

ABSTRACT: A shaft of 22 jet-grouted columns was built for a large-scale field test in order to determine the behaviour of jet-grouted material subject to freezing. 19 freezing pipes and several temperature monitoring points were installed. The jet-grouted columns were cooled down to −39°C through freezing pipes for 54 days. A numerical simulation of the entire field test was carried out using the finite-element method. The numerically obtained results were compared to the recorded temperatures. The comparison shows a very good agreement.

1 INTRODUCTION

Ground freezing has been used for over 120 years as an aid to mine shaft sinking. For about 30 to 40 years, it is also used extensively in tunnelling, underpinning and construction of large underground cavities. Especially in heterogenous subsoil and under complex geometrical boundary conditions ground freezing has proven to be an economic solution. Furthermore, ground freezing is an ecologically friendly soil improvement, as after the construction process the frozen soil body disappears completely. Modern computational capacities and sophisticated finite-element codes enable complex numerical simulations including heat transfer.

For the NordSüd Stadtbahn Köln subway project frozen jet-grouted construction elements are planned.

Since jet-grouted soil has different physical properties than undisturbed soil with pore water, one of the questions was what would happen to the jet-grouted material during freezing. Furthermore, it is not clear whether freezing of jet-grouted soil can be described with the same types of thermal parameters as used for undisturbed soil. In order to shed light onto these questions, a large-scale field test was carried out.

2 LARGE-SCALE FIELD TEST

A shaft of jet-grouted columns with an external diameter of about 7.6 m and a depth of about 15.8 m was built, see Figure 1. The wall consists of 22 columns which were jet-grouted using various techniques (such as "Duplex", "Triplex" etc.). The diameter of the columns amounts to about 1.6 m. The upper part of the columns are located in fill material, whereas the lower parts of the columns lie in Quarternary sandy gravel. The shaft base consists of 21 columns with a height of about

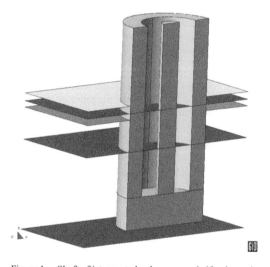

Figure 1. Shaft of jet-grouted columns, one half, schematic illustration.

2.5 m. In the middle of the shaft a single column was jetted.

A total number of 19 freezing pipes were installed (L1 to L19), see Figure 2. 18 pipes were drilled into the shaft wall and one into the single column (L19). The latter pipe is isolated in the upper part.

Three lines of monitoring points were installed, MQ1, MQ2 and MQ3, each consisting of at least three monitoring points, cf. Figure 2. At each monitoring point the temperature is recorded in various depths.

Before starting the large-scale field test by switching on the freezing system, the in-situ soil temperature was documented. Once the frozen body reached its designated extent, the shaft was excavated. Due to

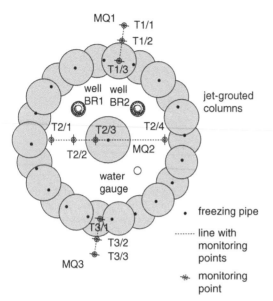

Figure 2. Cross-section through shaft, monitoring points and freezing pipes.

a leakage all freezing pipes were switched off after 39 days. Two days later, only 11 pipes were switched on again. After 54 days, the remaining pipes were turned off. The temperature was recorded during all stages.

3 NUMERICAL MODEL

A numerical model was developed to simulate the freezing process during the field test. Numerically gained results are compared to recorded data.

The calculations were carried out using a professional 2D/3D version of the finite element code Tochnog. Tochnog features a wide range of modelling capabilities for geotechnical problems including a partial differential equation for heat transfer.

3.1 Theoretical aspects

The differential equation for heat transfer used in Tochnog reads:

$$\varrho C\left(\dot{T} + \beta_i \frac{\partial T}{\partial x_i}\right) =$$

$$k\left(\frac{\partial^2 T}{\partial x_1^2} + \frac{\partial^2 T}{\partial x_2^2} + \frac{\partial^2 T}{\partial x_3^2}\right) - aT + f \qquad (1)$$

where

ϱ	[kg/m^3]	density
T	[K]	temperature
k	[W/mK]	heat conductivity
C	[J/kgK]	specific heat capacity

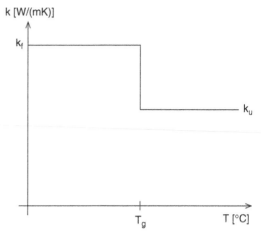

Figure 3. Heat conductivity k-function used in the numerical simulations.

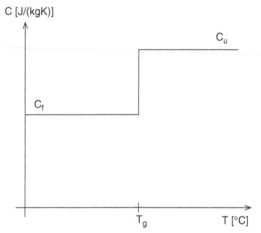

Figure 4. Specific heat capacity C.

β_i	[m/s]	groundwater flow velocity
a	[W/Km^3]	absorption
f	[W/m^3]	heat flux

All parameters are functions of temperature T. Density is assumed constant. As ground water flow is rather low, convective terms are neglected, i.e. $\beta_i = 0$. The energy demand during soil freezing consist of two parts (Sanger and Sayles 1978). The amount of heat:

1. which has to be withdrawn from the soil in order to freeze the soil
2. to maintain the soil body frozen within its environment.

The presented simulations only consider the first energy demand.

194

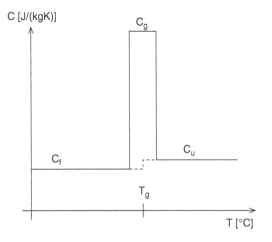

Figure 5. Modelling of crystallisation heat using a modified heat capacity distribution.

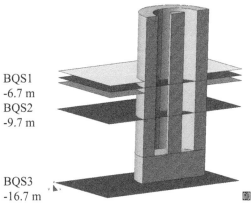

Figure 6. Modelled cross-sections (dark gray planes), groundwater table (light gray plane) and boundary between fill material and quarternary gravel (middle gray plane).

Heat conductivity k and heat capacity C depend on the temperature. Figures 3 and 4 illustrate their functions used in the numerical simulations.

Index u denotes the parameters of **u**nfrozen soil and index f of **f**rozen soil. T_g indicates the freezing point. The heat conductivity is larger for frozen soil. Whereas the heat capacity is larger for unfrozen soil.

The effects of heat released during crystallisation is simulated by increasing the heat capacity in the vicinity of the freezing point (Klein 1985), see Figure 5.

Heat transfer from the relatively warm air in the shaft after excavation to the surface of the jet-grouted columns is included using the following equation:

$$q_c = \alpha_c(T - T_c) \qquad (2)$$

where

q_c	[kg/m^3]	heatflux
α_c	[W/(m^{2}K)]	convection coefficient
T	[K]	temperature
T_c	[K]	temperature of environment

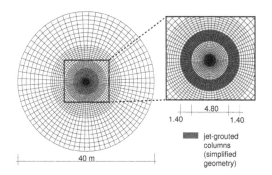

Figure 7. Finite element mesh, BQS1 and BQS2.

switched off. Two days later, only 11 pipes are turned on. All remaining pipes are switched off after 54 days according to the field test.

The numerical model simulates a diameter of 40 m and consists of 13592 linear 4-noded elements with 14025 nodes, see Figure 7.

3.2 Geometry – cross-sections

2D-models are used for the simulation of the soil freezing phase. Three cross-sections are modelled, as shown in Figure 6. The first cross-section (BQS1) lies at a depth of −6.7 m, just below the groundwater level. The second cross-section (BQS2) is located in the layer of Quarternary sandy gravel at a depth of −9.7 m. The third cross-section (BQS3) lies below the shaft at a depth of −16.7 m.

The freezing pipes are idealized as dots with a predefined temperature. A temperature curve starting with the initial soil temperature and decreasing to −39°C is used. After 39 days all freezing pipes are

3.3 Boundary and initial conditions

Due to jet-grouting (hydration) the temperature of the soil increases. Thus, the initial temperatures in the numerical model are chosen according to recorded data. As a simplification the model is divided into areas with the same initial temperature, as illustrated in Figure 8.

Table 1 lists the initial temperatures of all areas depicted in Figure 8 for each cross-section.

The initial temperature outside the listed areas is set to 15°C (BQS1) and 12°C (BQS2 and BQS3), respectively. At the outer model boundary these temperatures are fixed.

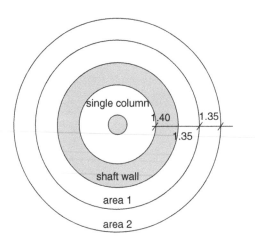

Figure 8. Areas with different initial temperatures. For each area the temperature is constant at the beginning of simulation.

Table 1. Initial temperatures.

	BQS1 [°C]	BQS2 [°C]	BQS3 [°C]
Single column	40	30	–
Inside of shaft	36	30	–
Shaft wall	32	22	–
Below shaft	–	–	20
Area 1	28	25	16
Area 2	20	18	14

Table 2. Thermal parameters used in numerical calculations, e: void ratio.

		Fill	Quarternary gravel	Jet-grout
ϱ	[t/m³]	1.86	2.22	2.45
e	[–]	0.85	0.37	0.14
k_u	[W/mK]	1.41	2.59	1.72
k_f	[W/mK]	2.64	3.77	2.06
C_u	[J/kgK]	1665	1244	919
C_f	[J/kgK]	1148	990	816

The temperature in the excavated shaft (air) is set to a constant value of 5°C.

3.4 Thermal parameters

All thermal parameters used in the numerical simulations are summarized in Table 2. It is assumed that the freezing of jet-grouted material can be described with the same types of parameters as soil.

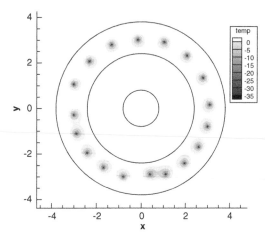

Figure 9. Temperature distribution after 7 days for cross-section BQS1, only temperatures below 0°C are depicted.

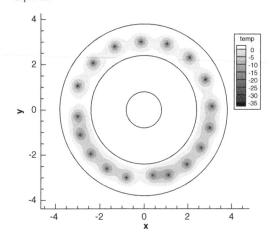

Figure 10. Temperature distribution after 14 days for cross-section BQS1.

The convection coefficient between the surface of the jet-grouted columns and the air inside the shaft after excavation is set to 6 W/(m²K) (Schneider 2004).

4 RESULTS AND COMPARISON

The calculated development of the frozen body is displayed in Figures 9–20 for cross-section BQS1. The results are similar for both other cross-sections (BQS2 and BQS3). Contour plots of the temperature distribution are depicted at intervals of seven days. Only values below 0°C are shown with different shades of gray. The white parts of the plots have temperatures above 0°C. Thus, the freeze/thaw boundary is equal to the boundary between white and the lightest gray. Additionally, the simplified wall of jet-grouted columns and the

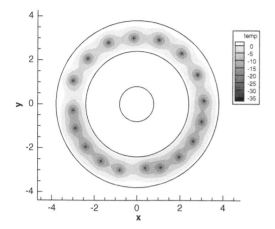

Figure 11. Temperature distribution after 21 days for cross-section BQS1, the frozen body is closed.

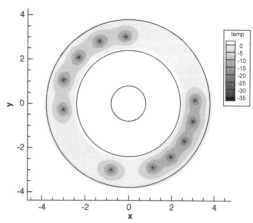

Figure 13. Temperature distribution after 42 days for cross-section BQS1, some pipes are switched off.

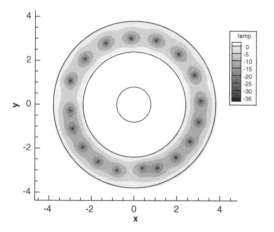

Figure 12. Temperature distribution after 35 days for cross-section BQS1, the frozen body has the same contour as the simplified jet-grouted shaft.

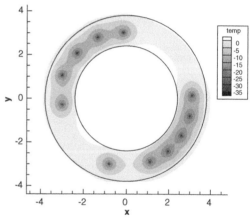

Figure 14. Temperature distribution after 49 days for cross-section BQS1, the shaft is excavated.

single column in the center of the shaft are drawn in the plots.

The first contour plot (Fig. 9) shows the temperature field seven days after switching on all freezing pipes. The freezing pipes have a temperature of −39°C and can be clearly distinguished as black dots. Around each freezing pipe, a small area is frozen. The frozen bodies only touch at two points due to the small distance between those pipes.

In cross-section BQS1 the freezing pipe in the single column at the center of the shaft is isolated and therefore is not modelled. As a consequence, no frozen body develops.

In Figure 10 the temperature distribution after 14 days is displayed. The frozen bodies around the freezing pipes touch at all points except one.

The frozen body closes at some point in time between 14 and 21 days, see Figure 11. After 35 days (Fig. 12) the frozen body has almost the same extent as the wall of jet-grouted columns.

Figure 13 displays the temperature distribution after 42 days, counting from the beginning of the field test simulation. After 39 days, all pipes were switched off for two days due to a leakage. Around all pipes the temperature is higher than in Figure 12, especially around the switched-off pipes. Around these, the temperature only lies between 0°C and −5°C.

Between $t = 42$ d and $t = 49$ d the shaft, including the single column in the middle, is excavated, see Figure 14. However, the ring of frozen jet-grouted soil is getting thinner around the switched-off pipes due to thawing at the surface of the jet-grouted columns.

197

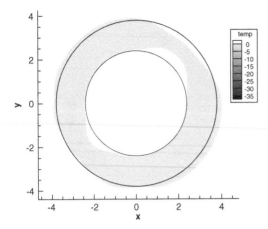

Figure 15. Temperature distribution after 56 days for cross-section BQS1, all pipes are switched off.

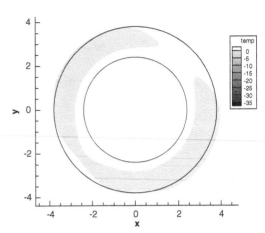

Figure 17. Temperature distribution after 70 days for cross-section BQS1.

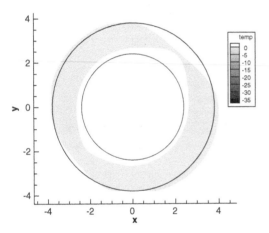

Figure 16. Temperature distribution after 63 days for cross-section BQS1.

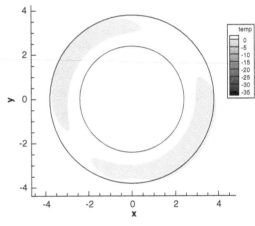

Figure 18. Temperature distribution after 77 days for cross-section BQS1.

Figures 15–20 show the thawing of the frozen body after switching off the remaining freezing pipes. The temperatures drop very fast to values between 0°C and −5°C after all pipes are turned off at $t = 54$ d. The jet-grouted soil thaws much faster at the inner surface of the shaft wall which is exposed to air than at the outer surface surrounded by soil.

The frozen ring of jet-grouted soil is broken up after about 63 days (Fig. 16) and the frozen state has almost completely disappeared after 91 days, see Figure 20.

Additionally, the numerical results are evaluated along selected monitoring points and lines, as illustrated in Figure 2. The points allow for a continuous monitoring of the soil temperature history. The monitoring points of the numerical model are identically to those used in the large-scale field test. As a consequence, the numerically gained temperature histories

can be directly compared to the recorded histories. Figure 21 to Figure 24 display the temperature histories of four monitoring points in cross-section BSQ1.

Point T1/1 lies at a distance of about 1 m outside the jet-grouted columns. Figure 21 compares the temperature recorded during the field test and the numerical data. The numerically gained curve is plotted with a dashed line and the recorded curve is plotted with a continuous line with small squares. The squares mark the discrete recording points in time. Both curves almost coincide. At the beginning the temperature is relatively high (ca. 28°C). It slowly decreases to a value of about 7°C. The soil is not frozen.

Monitoring point T1/2 is located at about 0.2 m to the jet-grouted shaft wall. The temperature history can be found in Figure 22. Again both curves show the agreeing values. Starting with a temperature of

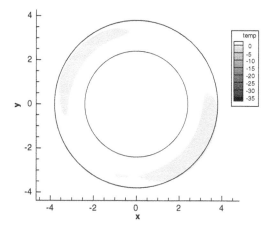

Figure 19. Temperature distribution after 84 days for cross-section BQS1.

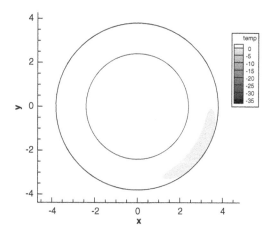

Figure 20. Temperature distribution after 91 days for cross-section BQS1.

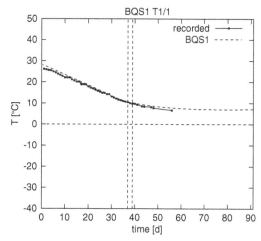

Figure 21. Temperature history for point T1/1.

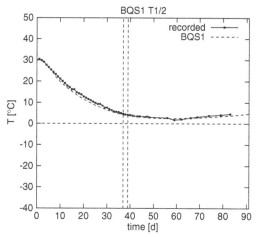

Figure 22. Temperature history for point T1/2.

about 32°C it drops to a minimum of about 3°C. After switching off the freezing pipes the temperature slowly increases. The soil is also not frozen at this monitoring point.

Point T1/3 and T3/1 are both located within a jet-grouted column in the middle between two freezing pipes. Up to and including the freezing point both curves are almost identical. Below the freezing point, the numerical model underestimates the temperature. Nevertheless, the temperature drop due to the turn-off of the freezing pipes at $t = 39$ d for two days and the slow temperature decrease around the freezing point after finally switching off the remaining pipes agree well with the recorded data.

The authors like to point out that no parameter fitting was done. The original set of thermal parameters

determined by laboratory tests before the field test were used for the model.

5 CONCLUSIONS

The numerically obtained results and the recorded values show a very good agreement despite a complex field test history including deactivated freezing pipes, excavation of the shaft etc.

The ground freezing process of soil (here: fill material and Quarternary gravel) can be very well numerically predicted using the described methods.

The assumption that jet-grouted soil can be described with the same types of thermal parameters as undisturbed soil is on the safe side, as higher

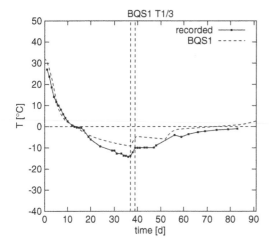

Figure 23. Temperature history for point T1/3.

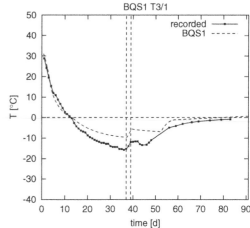

Figure 24. Temperature history for point T3/1.

temperatures in the jet-grouted columns are calculated in the numerical simulations.

Groundwater flow is not included in the presented numerical simulations. However, groundwater flow can influence the freezing process in such a way that the predicted frozen body does not develop. Therefore groundwater flow has to be taken into account in cases of high flow velocities (larger than about 2 m/d (Klein 1985)).

Both the numerical simulations and the field data prove that single switched-off freezing pipes do not lead to a failure of the whole system (loss of water-tightness). Even turning off all freezing pipes does not cause an abrupt failure as the jet-grouted soil thaws slowly.

Based on these results, several thermal calculations are carried out to design various frozen bodies.

REFERENCES

Klein, J. (1985). *Handbuch des Gefrierschachtbau im Bergbau.* Verlag Glückauf, Essen (Edt.).
Sanger, F. and F. Sayles (1978). Thermal and rheological computations for artificially frozen ground construction. *Eng. Geol. 13,* p. 311–337.
Schneider, K. (2004). *Bautabellen für Ingenieure* (16th ed.). Werner Verlag.

Numerical Modelling of Construction Processes in Geotechnical Engineering for Urban Environment – Triantafyllidis (ed)
© 2006 Taylor & Francis Group, London, ISBN 0 415 39748 0

FE-analysis of ground freezing for the construction of a tunnel cross connection

Roberto Cudmani & Siegfried Nagelsdiek

Ed. Züblin AG, Stuttgart

ABSTRACT: The Finite Element Method (FEM) is used in combination with an elastic-viscoplastic constitutive model for the analysis of a ground freezing problem related to the construction of a tunnel cross connection in the city of Cologne. As the considered problem is essentially three-dimensional, but a 3D-calculation was not feasible, an approximate solution is obtained by coupling a simplified 3D- with a 2D-model.

It is concluded that the FEM is an appropriated and powerful technique for the analysis of ground freezing problems. However, the time-dependent response of the frozen body, especially the evaluation of the maximum allowable standing time, can be only properly predicted if the real constitutive behavior of the frozen soil is taken into account in the calculation.

1 INTRODUCTION

For safety reasons modern tunnels consists of two separated pipes, which are connected at different distances by cross connections. These serve as pedestrian passage and are escape routes in emergency cases like vehicle collision and fire. When the tunnels are driven in soils below the ground water table a series of auxiliary construction measures are required to open the cross sections once the main pipes were bored. They must guarantee watertightness and stability during the excavation of the cross sections. Among the different available techniques, soil freezing is one the most preferred in urban areas. It can be used in almost any type of soil containing pore water and its effect on the natural ground and water properties are temporary and harmless for the environment.

The design of soil freezing measure involves an estimation of the frozen soil geometry (pre-dimensioning) and the analysis of the mechanical behavior of the frozen soil body when subjected to the expected external loads. These steps must be repeated if the dimensions of the frozen soil must be changed to fulfill the design requirements or for optimization purposes. For simple geometries and boundary conditions, the solution of ground freezing problems can be done by means of analytical models. In this approach, the behavior of the frozen soil is assumed to be elastic or elastic-ideal plastic (klein 1985). For the solution of complex ground freezing problems, numerical methods, especially the Finite Element Method (FEM), are applied. Although the FEM allows for a more realistic simulation of the material behavior, numerical solutions in practice are often obtained with the same constitutive models as in the analytical approaches.

The aim of this contribution is to show the application of the FEM in combination with a new constitutive model for frozen soils to the analysis of a complex ground freezing problem. With this object the construction of a tunnel cross connection in the North-South subway line in the city of Cologne is simulated.

2 THE *NORTH-SOUTH SUBWAY* IN COLOGNE

The *North-South subway* in Cologne, which is being constructed at the time, will connect the central station of Cologne in the north with the south part of the city (Fig. 1). The planed subway line has a total length of 4 km and includes two parallel tunnels with a length of 2.7 km which will be constructed by means of Tunnel Boring Machines (TBM).

The two tunnels cross the historical city centre of Cologne with plenty of archeological objects from the ancient Roman time and the middle ages (Leondaris and Escher 2005). Six underground stations are going to be constructed along the line. For this purpose, several cross connections of the two tunnels will be built in the stations "Kartäuserhof", "Severinstrasse" and "Rathaus". The cross connections will be excavated with conventional methods from a narrow braced excavation between the two tunnels. Ground freezing will provide watertightness and static support of the surrounding soil during the excavation of the cross sections.

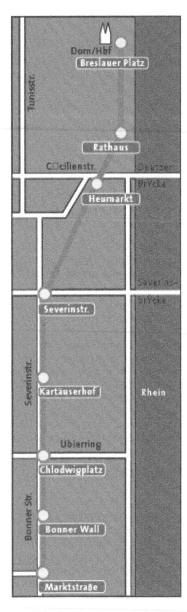

Figure 1. North-South subway line in the city of Cologne.

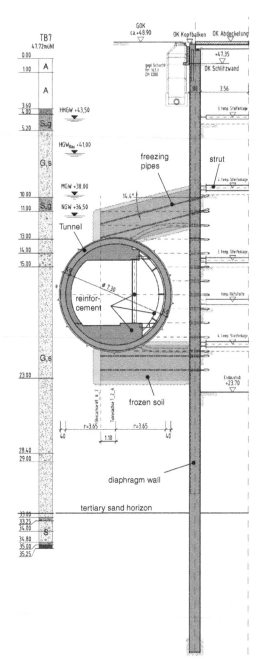

Figure 2. Cross section of the analyzed cross connection at the station Severinstrasse.

3 PROBLEM AND GROUND DESCRIPTION

The subsoil in Cologne consists of sandy and gravelly quaternary and tertiary sediments from the Rhine river. The ground surface is located 48.9 m over the zero level at the considered cross section. A sandy, silty fill, having a thickness of 5 m, lies on quaternary sand and gravel layers which extends down to a depth of 35 m. Underneath, tertiary fine sands are found.

The hydrological situation is characterized by a quaternary ground water aquifer, whose lower boundary is the horizon of the less permeable tertiary fine sand layers. Large seasonal fluctuation of the ground water level can occur since the ground water table

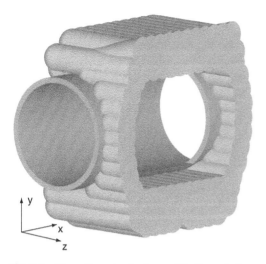

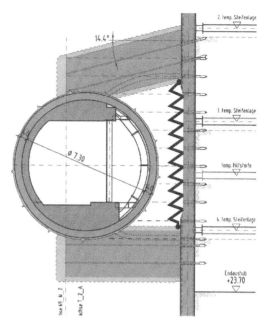

Figure 3. Three-dimensional scheme of the frozen body.

communicates with the water level of the Rhine river. For this reason, lowest (NGW) and highest (HGW) ground water levels of 43.5 m and 36.5 m over the zero level must be considered in the design. During the construction time a highest ground water table (HGW$_{Bau}$) of 41.0 m must be taken into account.

The analyzed section is shown in Fig. 2 (due to symmetry only half of the section need to be modelled). In the drawing, we recognize the tunnel, the narrow braced excavation as well as the roof and the bottom of the frozen soil body. Fig. 3 shows the complete frozen soil body schematically. As it can be seen, the roof and the bottom of the frozen soil body are not isolated from each other but connected vertically by lateral walls.

4 CONSTITUTIVE MODEL FOR THE FROZEN SOIL

An elastic-viscoplastic constitutive model presented in a companion paper (Cudmani 2006) is used for the simulation of the frozen soil behavior. According to this model, the relationship between the stress rate tensor $\dot{\mathbf{T}}$ and the strain rate tensor $\mathbf{D}$ is given by:

$$\dot{\mathbf{T}} = \mathcal{L} : (\mathbf{D} - \mathbf{D}_v(t, \vartheta, \mathbf{T})) \qquad (1)$$

Herein is $\mathcal{L}$ the elastic stiffness tensor and $\mathbf{D}_v$ is the viscous strain rate tensor, which depends on the time, the temperature and the stress state. Along with the time- and temperature-dependent response, equation 1 allows the evaluation of the maximum standing time t_m. Under constant loading frozen soils creep with decreasing strain rate for $t < t_m$. Beyond t_m the creep rate increases almost exponentially with time until the material fails. The application of equation 1 requires

Figure 4. Detail of the 2D-Problem with substitute spring.

the determination of five material parameters. A procedure for their evaluation based on unconfined creep compression tests was proposed by (Orth 1985).

For the FE-calculations, the model was implemented in the commercial FE-Code TOCHNOG via user-subroutine.

5 FE-MODELS

As it can be deduced from Figs. 2 and 3, the considered problem is essentially three-dimensional due to the frozen soil geometry. However, since a 3D-simulation was not feasible with the available computing capacity, there was no other alternative than to assume 2D-plain strain conditions. Clearly, the difficulty with the 2D-model is to consider the frozen soil body properly. If the vertical support of the roof and the bottom provided by the lateral walls is not taken into account, the calculation could lead to the conclusion that the ground freezing does not provide stability. The roof could easily move downward and the bottom upward when the cross connection is opened.

In order to consider the three-dimensionality of the frozen soil body, an analysis based on the combination of a 3D-model and a 2D-model is proposed. In the 2D-plain strain model a vertical section perpendicular to the tunnel axis, through the middle of the cross connection is considered (Fig. 2). The lateral wall is substituted by a spring connecting the roof with the bottom of the frozen soil body (Fig. 4). The 3D-model is used to estimate the stiffness of the substitute spring.

203

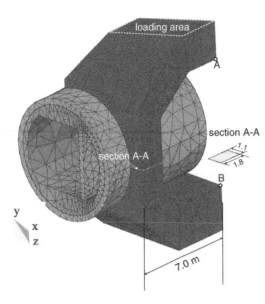

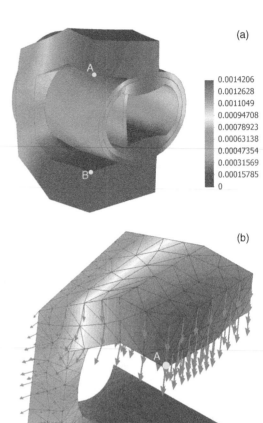

Figure 5. 3D-Model.

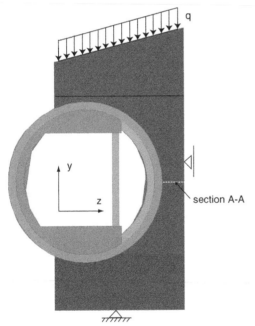

Figure 6. Side view of the 3D-Model.

Figure 7. Contour of displacements for $q = 500$ kPa: a) Complete model, b) frozen soil body.

A smooth contact is assumed between the frozen soil and the diaphragm wall (not considered in the model) as well as between the frozen soil and the tunnel. The displacements of the frozen soil body at the contact with the diaphragm wall are restrained in the horizontal direction. The bottom of the frozen soil body is constrained in all three directions. The soil is not considered in the 3D-model. Thus, the bedding of the tunnel and the frozen soil body are disregarded. The tunnel and the inner reinforcement are modelled as elastic bodies.

The behavior of the frozen soil body is modelled with an elastic-ideal plastic constitutive law with

5.1 3D-Model

The 3D-model includes the frozen soil body and a tunnel segment with the inner reinforcement. Due to symmetry only the half of the body need to be considered (Figs. 5 and 6).

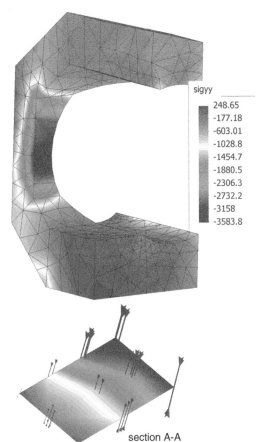

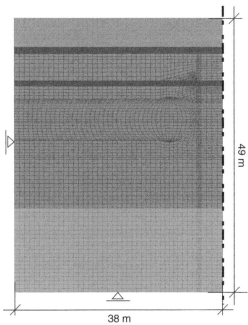

sigyy

248.65
-177.18
-603.01
-1028.8
-1454.7
-1880.5
-2306.3
-2732.2
-3158
-3583.8

section A-A

Figure 8.   Contour of vertical stresses for $q = 500$ kPa.

49 m

38 m

Figure 9.   2D-FE-model at the beginning of the calculation.

Figs. 7 and 8 show the displacements and vertical stresses for a load $q = 500$ kPa, which approximately corresponds to the total overburden pressure acting on the top of frozen soil body. Using the data in the contour plots, a relative displacement $\Delta u \approx 0.0014$ m and a force $F_l \approx (3583 - 248)/2 \cdot 1.8 \cdot 1.1 = 3301$ kN are calculated. Assuming a linear increase of F_l and Δu with the external load, a substitute stiffness $k_w = 3301/0.0014/2.8 = 842$ MN/m/m is obtained.

Mohr-Coulomb yield condition. The material parameters of the frozen soil are $E = 290$ MPa, $\varphi = 23.8°$ and $c = 0.48$ MPa. They were determined for a temperature of the frozen zone of $\vartheta = -10°C$ and a standing time of 3 months by (Ehl 2002). Drained conditions are assumed in the calculation. The model consists of 9900 linear tetrahedra elements and 1552 contact spring elements to model the contact between the tunnel and the inner reinforcement as well as between the frozen soil and the frozen soil.

In the numerical simulation a vertical load q is applied on the top face of the body (loading area in Fig. 5). This load causes a relative displacement of the roof with respect to the bottom of the body $\Delta u = u_A - u_B$. The stiffness of the substitute spring is calculated as $k_w = F_l/\Delta u/B$ (units MN/m/m). F_l is the force transmitted from the roof to the bottom through the lateral wall and B is the length of the loaded area in z-direction. F_l is the force acting on the cross section A-A.

5.2   2D-Model

Unlike the 3D-model, the 2D-simulation takes into account the actual soil profile, the ground water level and the construction stages. The dimensions of the 2D-Model and the boundary conditions are shown in Fig. 9. The following material and element types were used to model the different components of the model:

1. Soil: elastic-ideal plastic material behavior with Mohr-Coulomb yield condition, three- and four-nodes continuum elements. The use of a more complex constitutive law is not justified as the aim of the calculation was to analyze the behavior of the frozen soil body.
2. Frozen soil: elastic-viscoplastic material behavior, three- and four-nodes continuum elements.
3. Tunnel, inner reinforcement, diaphragm wall: Linear elastic material behavior, three- and four-nodes continuum elements.
4. struts: two-nodes bar elements.

Table 1. Material parameters for the soil layers.

Layer	E [MPa]	φ [°]	c [kPa]	ψ [°]
Fill	3.7	32.5	0	5.0
Quaternary sand	23	32.5	0	5.0
Quaternary gravel	52	35	0	5.0
Tertiary sand	74	35	0	5.0

Table 2. Construction stages considered in the 2D-calculation.

Step	Time	Construction stage
1	–	Initial stress state
2	–	Load of neighbor building
3	–	Installation of the diaphragm wall
4	–	Tunnel excavation
5	–	Construction of the tunnel shell
6	–	Construction of inner reinforcement
7	–	1. Excavation
8	–	1. Strut
9	–	2. Excavation
10	–	2. Strut
11	–	3. Excavation
12	–	3. Strut
13	–	4. Excavation
14	–	4. Strut
15	–	5. Excavation
16	–	5. Strut
17	0–30	Soil freezing
18	30–40	Excavation of the cross connection
19	40–45	Opening the tunnel shell
20	45–210	Creep

Figure 10. FE-model at end of the calculation step 5.

5. contact between frozen soil and wall/tunnel: smooth (frictionless) contact, two-nodes contact springs. Contact loss is allowed for.
6. contact between tunnel and soil/inner reinforcement: friction and contact loss are allowed for, two-nodes contact springs.

The parameters for the frozen soil are presented and validated in a companion paper (Cudmani 2006). The main material parameters for the soil layers are listed in Tab. 1.

The construction stages considered in the calculation are listed in Tab. 2.

The construction of the diaphragm wall and the soil freezing are simulated as an exchange of material models, i.e. deformations due to wall installation and soil freezing are disregarded. From step 1 to 16 the response of the model is independent on time. The variable time is merely a loading factor. On the contrary, after activating the elastic-viscoplastic model in step 17, the variable time is equal to the real time.

Figs. 10 to 14 show the 2D-FE-model at different steps of the calculation.

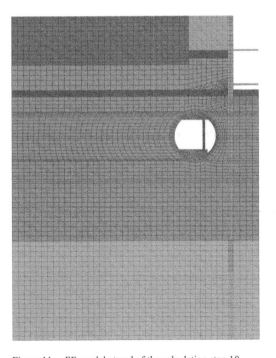

Figure 11. FE-model at end of the calculation step 10.

Figure 12.   FE-model at end of the calculation step 16.

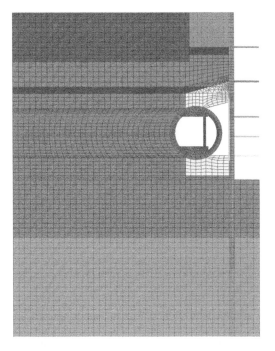

Figure 13.   FE-model at end of the construction time.

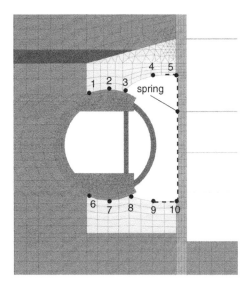

Figure 14.   Detail of the FE-model at the end of the construction time.

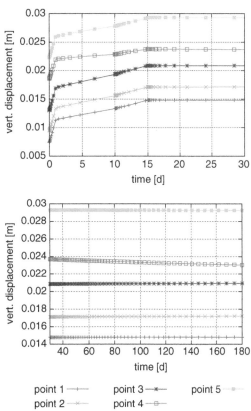

point 1 —+—   point 3 —*—   point 5 —▪—
point 2 —×—   point 4 —□—

Figure 15.   Evolution of vertical displacement in different points of the roof of the frozen soil body.

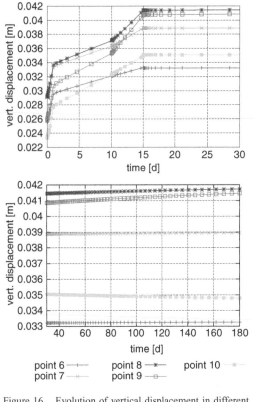

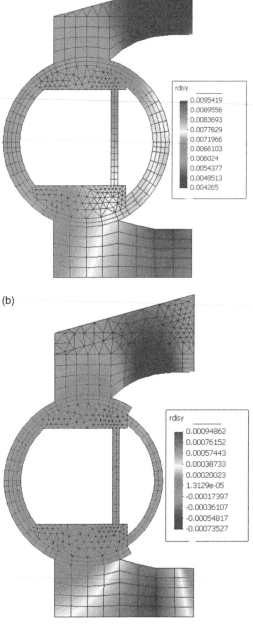

(a)

(b)

Figure 16. Evolution of vertical displacement in different points of the bottom of the frozen soil body.

In the following, calculation results are presented for the design ground water level HGW$_{Bau}$ = 41.0 m. A mean temperature of the frozen soil $\vartheta = -10°$C and drained conditions of the soil layers are assumed. The discussion is focused on the behavior of the frozen soil body and not on the general behavior of the system.

The vertical displacements over time in different points of the frozen soil body are presented in Figs. 15 and 16. The position of the points is shown in Fig. 14. In the diagrams, $t = 0$ corresponds to the time at the end of soil freezing (step 17), since during soil freezing (creep) deformations are negligible. Obviously, the displacements in $t = 0$ are not zero due to the previous construction stages. It can be seen that both the roof and the bottom of the frozen soil body move upward during the opening of the cross connection because the net vertical force acting on the frozen soil body is positive. The net vertical force results from the addition of the earth pressure, the buoyancy force and the weight of the frozen soil body. Responsible for the positive net force is the lowering of the ground water level in the excavation zone, which is carried out during the excavation in the step 18. The displacements during the creep phase (step 20) are relatively small

Figure 17. Contour of relative vertical displacements in the steps 18 (a) and 20 (b).

compared with the displacements due to opening of the cross connection (steps 18 and 19) and the creep rate decreases with the time, i.e. the frozen soil body shows a stable behavior.

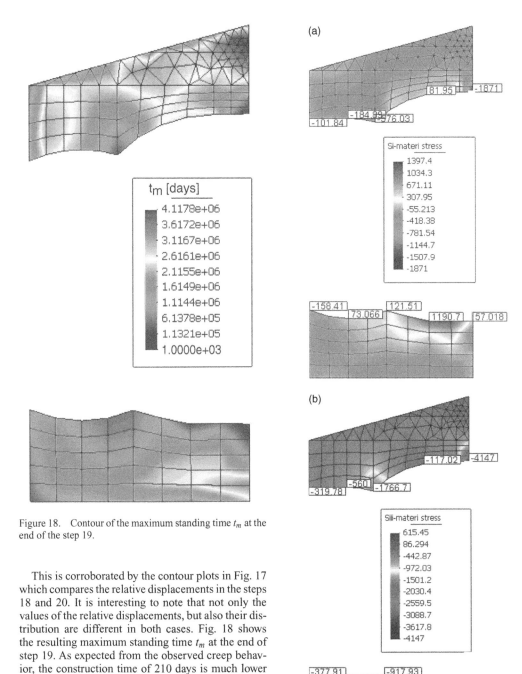

Figure 18. Contour of the maximum standing time t_m at the end of the step 19.

This is corroborated by the contour plots in Fig. 17 which compares the relative displacements in the steps 18 and 20. It is interesting to note that not only the values of the relative displacements, but also their distribution are different in both cases. Fig. 18 shows the resulting maximum standing time t_m at the end of step 19. As expected from the observed creep behavior, the construction time of 210 days is much lower than t_m. The contour plots of maximum and minimum principal stresses at the end of step 20 are presented in Fig. 19. As it can be deduced from the plots, the deviatoric stresses $q \approx (\sigma_1 - \sigma_3)/2$ in the frozen body are small compared to the unconfined compression strength ($\sigma_c \approx 7$ MPa for $\vartheta = -10°$C). According to the constitutive model for the frozen soil, this is the reason for the stable behavior of the frozen soil body.

Figure 19. Contours of (a) maximum and (b) minimum principal at the end of the construction time.

209

6 CONCLUSIONS

A complex ground freezing problem related to the construction of a tunnel cross is modelled with the FEM. Since a three-dimensional FE-analysis of the problem was not feasible, a numerical solution is obtained by coupling the mechanical responses of a simplified 3D-model and a complex 2D-model. In spite of this simplification, the numerical simulation allows for an assessment of the stability and the serviceability of the frozen soil body after opening the cross connection. However, the key quantities to judge the stability, which are the deformation rate and the maximum standing time, can be only evaluated if an appropriated constitutive model is used for the frozen soil.

REFERENCES

Cudmani, R. (2006). An elastic-viscoplastic model for frozen soils. In *Proc. Numerical Simulation of Construction Processes in Geotechnical Engineering for Urban Environment*.

Ehl, G. (2002). Nord-Süd Stadtbahn Köln-Gefriertechnische Untersuchungen. Baugrundgutachten.

Klein, J. (1985). *Handbuch des Gefrierschachtbaus*. Glückauf, Essen.

Leondaris, H. and Escher, M. (2005). Tunnelbau am rhein – nord-süd stadbahn köln. *Felsbau 23, 5*, 129–132.

Orth, W. (1985). *Gefrorener Sand als Werkstoff-Elementversuche und Materialmodell*. Ph. D. thesis, Veröffentlichungen des Instituts für Bodenmechanik und Felsmechanik/Universität Karlsruhe. Heft 100.

Numerical Modelling of Construction Processes in Geotechnical Engineering for
Urban Environment – Triantafyllidis (ed)
© 2006 Taylor & Francis Group, London, ISBN 0 415 39748 0

Ground freezing: an efficient method to control the settlements of buildings

C. Kellner & N. Vogt
Zentrum Geotechnik, Technische Universität München, Germany

W. Orth
Ingenieurbüro für Bodenmechanik und Grundbau, Karlsruhe, Germany

J.-M. Konrad
Université Laval, Québec, Canada

ABSTRACT: The construction of the new soccer stadium in Munich–Fröttmaning made it necessary to extend the platforms of the subway station U6 "Marienplatz" below the historical town hall of Munich, as up to 32 000 passengers per hour use this station. For this, two new tunnels for pedestrians were driven parallel to the existing subway tunnels. They were planned as shotcrete construction and driven under atmospheric conditions. In order to reduce the ground displacements, the tunnels were driven below caps of artificially frozen soil. As frozen soil has higher stiffness and bearing capacity compared to its unfrozen condition, the displacements could be reduced by use of this frozen soil cap. However, the well known – but difficult to quantify – phenomenon of frost heave could lead in unscheduled additional displacements, which might reduce or even nullify the positive effects of the frozen cap. Therefore the magnitude of the frost heave displacements had to be predicted prior to undertake the construction. The following steps were taken to analyse the problem at hand:

– Laboratory tests to establish bearing-capacity, creep-behaviour, and frost heave response
– Finite-Element calculations
– Installation of a monitoring system (automatic levelling system) on the construction site
– Strain gauge measurements in the shotcrete lining of the tunnel.

1 INTRODUCTON

The underground station "Marienplatz" is a central traffic junction in the centre of Munich, where two metro-lines U6 and U3, leading from north to south, are connected with 7 lines leading from east to west. The subway U6 is the only line, where passengers can get from the city centre to the new soccer stadium "Allianz-Arena". To increase the capacity for the changing passengers the department of public construction, city of Munich, planned to enlarge the platforms by driving two new tunnels parallel to the existing platforms. The old and new tunnels were connected by 11 short breakthroughs.

Figure 1 shows a plan view including the relevant buildings. The existing subway lines and the underground station are marked in black colour, the new tunnels in dark gray. The buildings are printed in bright grey. In a cross section the new tunnels are located below the historical town hall (Figure 2).

As the settlements of the historical town hall, which is a listed building, during the construction of the existing tunnels in the 1960s had reached about 30 mm, the expected additional displacements had

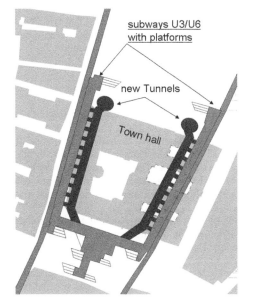

Figure 1. Plan view.

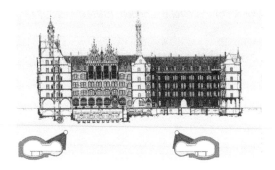

Figure 2.   Cross section with town hall and tunnels.

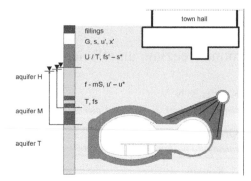

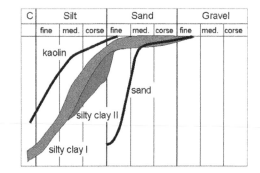

Figure 3.   Presentation of the subsoil.

C	Silt			Sand			Gravel		
	fine	med.	corse	fine	med.	corse	fine	med.	corse
kaolin									
						sand			
			silty clay II						
silty clay I									

Figure 4.   Grain size distribution.

to be limited strictly. For this purpose compensation grouting was advertised for bids. It was planned to make horizontal drillings out of two access shafts under the town hall in the shape of a horizontal fan. Dewatering was planned with horizontal drains parallel to the tunnels.

The Max Bögl GmbH worked out a specific proposal, where artificial ground freezing was used to guarantee the stability and safety of the tunnel roof and to reduce the deformations. Two pilot headings were driven laterally above the new tunnels instead of the horizontal drillings for the compensation grouting. The freezing pipes were implemented in the pilot tube. Dewatering was done by pumping from well points instead of horizontal drains. Additional accompanying scientific measures should ensure tolerable deformations. For this both the magnitude of the frost heave displacements and the settlements had to be predicted prior to undertake the construction. Therefore frost heave tests and FE- computations were carried out. During the construction process these results were permanently compared to measurements in order to control and ensure tolerable deformations.

2   GROUND CONDITIONS

Due to the subsoil investigations for the building of the existing tunnels in the 1960s and the mapping of the working face during that time, the ground conditions were well known. They are shown schematically in Figure 3. On top there are fine to coarse quaternary gravels and fills. They are followed by tertiary layers, which consist of alternating sequences of fine grained soils in a semi-solid to solid consistency and fine to medium sands of high density. According to the geological development the thickness of the single layers varies strongly. The sandy layers carry artesian groundwater.

According to updated subsoil investigations and a pumping test the tertiary sands could be ranged into three aquifers. The first aquifer H is situated below the first fine grained top layer. It consists of a sandy layer with a thickness of several meters. From 14 m to

26 m below ground level a second fine grained layer follows. In this layer inclusions of almost non-cohesive silty-sandy material with varying thickness are embedded frequently (aquifer M). In this Aquifer also the tunnel roof is located at about 17 m below ground level. Underlying another sandy layer follows with a thickness reaching sometimes more than 10 meters (aquifer T). Figure 4 shows the grain size distribution of the tertiary silty clays and sands. The soils referred to as sand I and clay I later in the text were taken during the excavation of the access shaft at a level of the planned frozen cap, whereas clay II was taken during the excavation of the tunnels. In addition, the diagram shows the grain size distribution of a very homogeneous industrial kaolin, a so called "china clay". As this material is known to be highly frost susceptible, additional laboratory tests were performed with it. Table 1 shows the soil properties, which have been later used also for the finite-element-calculations.

3   THE PROPERTIES OF FROZEN SOIL-LABORATORY TESTS

3.1   *Bearing capacity*

The bearing behaviour of frozen soil differs from that of unfrozen soil and concrete. Frozen soil is a

Table 1. Soil properties

	φ [o]	c [kN/m²]	Einitial loading [kN/m²]	Reapplication of load [kN/m²]
Quaternary Gravels	37,5	4*	120.000	180.000
Tertiary Sands	37,5	4*	120.000	180.000
Tertiary silts and clays	22,5	30	90.000	120.000

* apparent cohesion

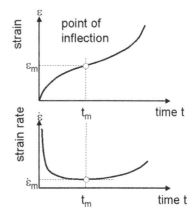

Figure 5. Typical creep curve of frozen soil.

viscoplastic material, this means it shows significant time dependent deformations under constant load. The creep rate increases with the stress level and decreases with lowering the temperature. On the other hand in constant strain rate tests the yield stress increases with the strain rate and with lowering the temperature.

The reason for the viscoplastic behaviour is the ice-matrix in frozen soil, that gives the main portion of strength but is a viscoplastic material itself. This is caused by the comparatively low activation energy within the ice crystal, that causes thermally-activated displacements of single molecules in the ice crystal if deviatoric stress is applied.

3.1.1 Deviatoric stress

During creep under deviatoric stress the creep rate in frozen soil first decreases up to a minimum $\dot{\varepsilon}_m$ at a certain time t_m (point of inflection of the creep curve) and than increases again until failure occurs finally. As the transition of decreasing to increasing creep rate is very smooth, in literature usually three creep stages are defined: The primary creep with decreasing creep rate, secondary creep with constant creep rate and tertiary creep with increasing creep rate. Theoretical investigations as well as high precision tests however indicate that a secondary creep stage with constant creep rate does not exist. Furthermore even under very low deviatoric stress the stage with increasing creep rate is always reached. So frozen soil is rather a fluid of high viscosity.

In Figure 5 the typical creep curve of frozen soil is given. The creep curves at various stresses and temperatures have nearly the same shape when they are normalised by a linear transformation of the time axis (Figure 6) because the strain at the point of inflection is nearly constant under different test conditions. Consequently the different creep curves can be represented by a unique mastercurve when the time axes is transformed linearly. The dependence on temperature and stress then can be described by one characteristic point of the curve.

It is convenient to use the minimum creep rate $\dot{\varepsilon}_m$ or alternatively the creep time t_m at the point of inflection.

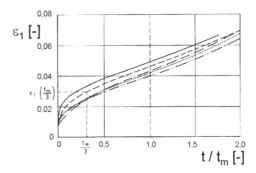

Figure 6. Normlized creep curves of frozen sand at different temperatures and stress levels.

Those can easily be found from the strain rate versus time curves (Figure 6). An extended study of micromechanical processes occurring in crystalline bodies under stress led to the formula (Orth 1986, 1988):

$$\dot{\varepsilon}_m (\sigma, T) = \dot{\varepsilon}_\alpha \exp \left[\left(\frac{K_1}{T} + \ln \dot{\varepsilon}_\alpha \right) \left(\frac{\sigma}{\sigma_\alpha (T)} - 1 \right) \right] \quad (1)$$

where:

– $\dot{\varepsilon}_m$ = minimum strain rate during creep (Figure 5),
– σ = the applied deviatoric stress, in uniaxial stress states proportional to uniaxial stress σ_1,
– T = absolute temperature [K],
– $\dot{\varepsilon}_\alpha$ = reference strain rate, commonly $\dot{\varepsilon}_\alpha = 1$ %/min
– $\sigma_\alpha(T)$: that stress under which at temperature T there is $\dot{\varepsilon}_m = \dot{\varepsilon}_\alpha$ (T),
– K_1: a characteristic temperature [K] that was found to be constant if T < 268.4 K ($\cong$ −5°C),

From the similarity of the creep curves and the strain-rate curves, another important result can be derived, which was also found in research on pure ice (Mellor 1982).

$$\dot{\varepsilon}_m \cdot t_m = C = \text{constant, and} \qquad (2)$$

$$t_m = C / \dot{\varepsilon}_m \qquad (3)$$

Knowing the minimum creep rate $\dot{\varepsilon}_m$, the time t_m (when the approximately constant strain ε_m is reached) can be calculated and vice versa.

In Figure 7 the time t_m versus σ at different temperatures is plotted for a clay. Theoretically this figure can be derived from a set of approximately six uniaxial creep tests on frozen soil. For practical purpose one should have redundancy by some more tests to evaluate the scatter of the soil behaviour. The derivation of creep tests is described at (Orth 1986, 1988).

3.1.2 Triaxial stress states

The influence of triaxial stress states on the behaviour of frozen soil depends on the grain skeleton (grain size, grain shape, type of mineral, density) as well as on the degree of water saturation. However, for many practical applications of soil freezing, triaxial stress can be neglected for two reasons: Firstly, significant grain friction in frozen soils is mobilized at comparatively high strain (more than 10% axial strain). This is unacceptable in most cases.

Secondly, the triaxial stress 10 or 20 metres below ground surface is too low to cause considerable frictional strength.

Calculations showed that the frictional strength 10 m below the surface with an angle of friction of $\varphi = 35°$ is only about 10% of the uniaxial stress that can be applied at $-10°C$ and $t_m = 6$ month. Of course, friction can become important at high triaxial stress states (e. g. in deep frozen shafts) if there is coarse soil. For many practical problems, however, it is sufficient to use uniaxial tests and to design using deviatoric stress only.

3.2 Creep behaviour

The calculations of creep deformations in frozen soil requires some special procedures as most of the commonly used Finite-Element- programs cannot represent the viscoplasticity of frozen soil. Therefore a loading time for each section of the frozen body was derived from the driving rate in the tunnel. For this loading time and for the given temperatures in the frozen body obtained from thermal calculations and measurements the creep deformation can be calculated versus stress. The division of stress by strain gives the Young's modulus. As the creep deformations increase with increasing stress the Young's modulus decreases with increasing stress.

In Figure 8 the Young's modulus for loading time of sixty hours and the temperature of $-10°C$ is plotted

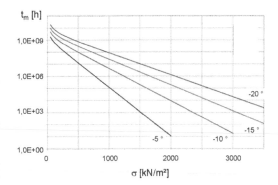

Figure 7. Time t_m versus σ at different temperatures.

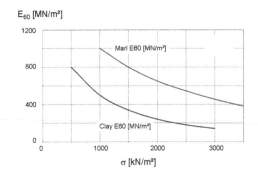

Figure 8. Youngs Modulus E60 (60 hours loading time, $-10°C$) versus stress level.

versus stress level both for clay and for marl. In the FEM-calculations the so derived Young's moduli were introduced. This was done stepwise in an iteration until the calculated stress level was in correspondence to the Young's modulus from the creep tests.

This way of calculation led to reasonable results what can be seen by the settlement measurements carried out during construction. Furthermore the creep of frozen soil was represented by the record of normal forces in the reinforcement of the roof, that indicates a rapid transition of the overburden pressure from the frozen body to the shotcrete shell within a few days (Figure 34). As the surface deformations caused by creep of frozen soil were within the given limits but on the other hand the frozen body transferred its load to the shotcrete shell within the calculated time it was confirmed that the creep behaviour of frozen soil can be represented realistically by the given calculation procedure. An essential condition for this is a reliable knowledge of the creep deformation behaviour of the frozen soil both in quality and quantity.

3.3 Frost heave response

Experiences and laboratory tests, which are described in literature (Jelinek 1968) already long time ago,

214

showed that silts cause compared to other types of soil the biggest heave displacements, when they are exposed to frost impact. A well excepted explanation states, that silts on the one hand (compared to clay) are still permeable enough to enable a water transport to the frost front and on the other hand (compared to gravel) have small enough small voids to enable a suction of water. As already shown in Figure 3, the artificially iced soil was situated in the alternating sequences of aquifer M, so that the combination of the frost susceptible silts and clays and the water carrying sands let expect comparatively high heave rates. Therefore the frost heave response and especially the magnitude of the frost heave displacements had to be predicted prior to undertake the construction.

Frost heave in general is caused by the volume change from water to ice and furthermore by the growth of ice lenses close to the frost front. It appears that for a realistic prediction the examination of both heave effects at realistic boundary conditions is necessary. The bulk load on the frozen soil and the water supply have great influence on the resulting frost heave rates. In the laboratory these effects are usually analysed with an experimental setup, which is shown in Figure 9 as a schematic view. The sample, which is to be examined, is put in a rigid container. The top of the sample is loaded. During the time the sample freezes uniaxially from the top water intake at the bottom is possible.

Former heave tests (Konrad 1982) showed in general, that the pore water is pressed out of the sample in the beginning of the frost action, when the frost front is penetrating rather rapidly into the sample. At a certain point the expulsion of water is reversed and water is drawn back to the frost front.

The laboratory tests aimed at recording separately the heave displacements resulting from volume change and ice lens formation. The first priority of the heave tests for recording the effects of volume change was a rapid penetration of the frost front to avoid an initiation of an ice lens growth. In contrast, in the tests for the examination of the ice-lens growth a steady state of heat flow and as a consequence a fixed position of the frost front was aspired.

The tests were performed with bulk loads corresponding to $6 \, kN/m^2$ to about $400 \, kN/m^2$ to meet different objective targets. Tests with lower bulk loads lead to comparatively high heave rates and allowed above all the visual evaluation of the ice lens. Tests with higher bulk loads were performed to estimate the magnitude of frost heave, which had to be expected at the construction site.

3.3.1 Experimental setup

The samples were produced in a modified proctor test with an initial water content of $w = 20\%$. The dimension of the samples corresponded to the CBR-tests

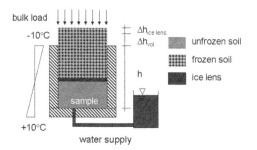

Figure 9. Schematic view of a sample.

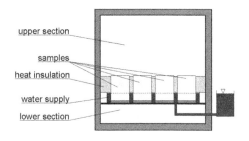

Figure 10. Schematic view of the climatic chamber.

with one restriction in so far, that the sample was slightly conical. The bottom of the sample had a diameter of 145 mm, whereas the top had a diameter of 150 mm. This shape was chosen to minimise the friction effects at the side walls during the heave tests. After the production the samples were repotted in a plastics form of the same size and shape. The plastics material had almost the same thermal conductivity as soil.

The sample is put on a dripstone and together with it in a loading device. It is loaded uniaxially on the top by an air pressurized system. The loads are controlled by a load cell. The movements of the top of the sample are measured with dial extensometers.

The whole equipment is put in a climatic chamber, where a water supply is enabled at the bottom of the sample and the sidewalls are covered with heat insulation. Figure 10 shows a principal sketch of the climatic chamber with two temperature sections and Figure 11 shows the experimental setup.

3.3.2 Volumetric heave

3.3.2.1 Fully saturated soil and undrained conditions

An upper boundary for the maximum possible heave due to the volume change can be derived under the assumption of undrained conditions, fully saturated soil and a completely freezing of the porewater.

215

Figure 11. Apparatus for load application.

The water content of a sample is defined as the ratio of the mass of water and dry matter:

$$w = \frac{m_w}{m_d} = \frac{\rho_w \cdot V_w}{\rho_d \cdot V}, \text{ thus}$$

$$V_w = w \cdot \rho_d \cdot \frac{V}{\rho_w} \quad (4)$$

Due to freezing this volume increases about 9%.

$$V_{w,1} = 1{,}09 \cdot V_w \quad (5)$$

If lateral strains are constraint, this volume change leads to an displacement at the surface of:

$$Heave = 0{,}09 \cdot w \cdot \rho_d / \rho_w \cdot h \quad (6)$$

The heave displacements, which are derived in this way for saturated soils and undrained conditions are an upper boundary for those heave displacements, which can be expected under drained conditions and/or unsaturated soil. When freezing a soil with a water content of $w = 20\%$ and a dry density of $\rho_d = 1{,}8 \text{ g/cm}^3$ the maximum possible heave displacements are about 3,2% of the frozen zone.

3.3.2.2 Unsaturated soil and drained conditions

Table 2 gives an overview of the test results of volumetric heave at a freezing temperature of $-20°C$ and different bulk loads for some unsaturated samples under drained conditions. The values are average values of all tests, which have been carried out at the same bulk load.

It can be seen, that the values of the volumetric heave decrease with an increasing load. At small bulk loads the sands and clays show more or less the same values. At higher values of about 250 kN/m^2 there were practically no heave displacements in the sand, whereas the clay showed volumetric heaves between 0,5% and 1,0%.

The Kaolin showed clearly decreasing heave displacements at higher bulk loads. However, it has to be mentioned restrictively, that the Kaolin showed upheavals due to swelling during the time the samples were stored. The values given in Table 2 for the kaolin partly are based on different water contents at the beginning of the frost penetration and are therefore only partly comparable

3.3.3 Ice lens growth

Figure 12 shows a test result of tertiary clay at a load of 6 kN/m^2. The freezing temperature was chosen to $-10°C$ and the water supply had a temperature of $+10°C$. A short time after freezing set in the samples showed nearly linear frost heave displacements, so that the speed was approximately constant. With the time the curves flattened slightly, but didn't stagnate at a test duration of more than 5 days.

Table 3 shows measured values for the velocity of ice lens growth, which were taken in a time slot from 12 to 24 hours after freezing set in. This time slot was chosen, because the heave by volume increase had come mostly to an end after 12 hours.

Table 2. Volumetric heave depending on the bulk load.

Load [kN/m²]	Clay I [%]	Sand [%]	Kaolin [%]
6	1,57	1,77	(10,07)
100	1,11	–	(7,27)
250	0,70	0,05	(4,25)
400	–	–	(3,29)

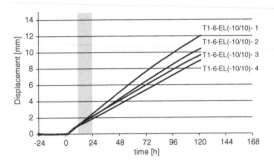

Figure 12. Ice lens growth of 4 equal samples of clay I at a bulk load of 6 kN/m².

Table 3. Heave rates due to formation of ice lenses, depending on the bulk load.

load [kN/m²]	Clay I [mm/d]	Clay II [mm/d]	Sand [mm/d]	Kaolin [mm/d]
6	2,08	3,06	1,54	16,39
100	1,93	1,56	–	12,25
250	0,45	0,87	0,01	7,46
400	0,07	0,56	–	4,93

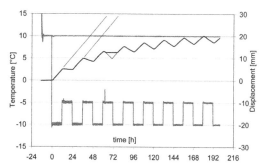

Figure 15. Ice lens growth in clay I at a bulk load of 6 kN/m².

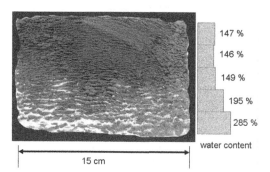

Figure 13. Samples of sand and kaolin after freezing.

Figure 14. Ice lens in kaolin.

It can be seen, that the velocity of the ice lens growth decreases with higher loads. The tertiary clays showed a higher ice lens growth then the sands. The heave rates of the Kaolin exceeded those of the clays by far. Figure 13 shows one sample of the tertiary sand and one of the kaolin after one heave test. The different heave displacements can be seen clearly. Figure 14 shows a cross section of the frozen zone with an increase in the number of ice lenses towards the frost front.

3.3.4 Influence of an oscillating freezing temperature

The freezing system at the construction site was driven in intervals. After the frozen cap had been built up completely, the power for upholding the necessary

Figure 16. Ice lens banding in the frozen zone of one sample.

frozen volume was piped in time intervals into the ground. To study the quantitative effect of this oscillating control of the heat flux a test was made, where the freezing temperature in the climatic chamber was varied between −5°C and −10°C in time intervals of 12 hours. The experimental setup and testing was like described above.

Figure 15 shows a linear increasing heave displacement over time when frost starts. As soon as the freezing temperature in the upper chamber was lowered to −5°C, the linear curve gets a break and the sample showed a slight settlement. When the freezing temperature again was lowered to −10°C again linear heave displacements started. The later cycles led to a similar progression of the heave displacements, where at the velocities lowered slightly down with time. The horizontal part of the curve in the third cycle is due to a measurement error. The dashed line is an estimated heave displacement during that time. Figure 16 shows a cross section of the frozen zone after the test had been stopped. Single distinct ice-lenses are clearly recognizable. It can be assumed, that every cycle produced a single ice lens.

3.4 Results

The degree of heave displacements due to volume change obviously depends on the initial water content of the soil. An upper boundary for the displacements, which can occur, can be calculated on the assumption of a complete saturated soil, undrained conditions and completely freezing of the porewater.

217

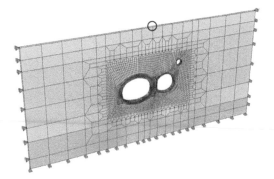

Figure 17. Finite-element model.

Figure 18. Construction steps of the old tunnels.

Figure 19. Construction of a side wall.

The heave rates, which will occur in fact will be lower, as a part of the porewater will flow away from the frost front during freezing.

When reaching nearly steady state conditions in heat flow, the expulsion of water is reversed and water is drawn back to the frost front. Ice lens formation starts.

If the freezing system is driven in an oscillating way, the frost front will not be stable. During the time the frost front retreats, heave rates will be comparably lower and the total heave will be minimised.

4 CALCULATION OF THE GROUND DISPLACEMENTS

4.1 Finite-element model

The system, which was used for the calculation of the ground displacements is shown in Figure 17. A 2-dimensional model under plain-strain conditions and using "hardening soil" as a constitutive law was generated in the software package "Sofistik". The single soil parameters were already shown in Table 1. Generally, the heading of a tunnel leads to longitudinal and transversal change of forces above the working face, and as a consequence to a 3-dimensional stress and strain distribution in the ground. A plain strain model can only reproduce the changes of the transversal forces. To count for effects in longitudinal directions, the model was manipulated. Those elements which represent parts of the soil, which are already excavated in reality still keep parts of their element stresses and stiffness to simulate the longitudinal load distribution. The principle is shown in detail in (Ostermeier 1991) and is commonly known as "alpha – procedure".

Due to the nonlinearities of the system it was necessary to model the single construction steps of the existing and the new tunnels, to come to a at least rough estimation of the ground displacements, which had to be expected.

4.2 Existing tunnels: simulation of the construction

The existing tunnels for the subway line U3/U6 below the town hall had been built from 1966 to 1970 by the help of dewatering. In a first construction step the side headings had been driven. After the lining of corrugated sheet iron had been assembled, the side headings had been completely filled up with concrete. As a consequence the thickness of the sidewalls reaches 2 m and more. The tunnel roof was driven under a bladed shield construction and the already finished side headings took the function of a support for the tunnel roof. After that, the bottom of the tunnel was driven and the sealing as well as the facing were built in. The Figure 18 shows the single construction steps and Figure 19 the construction of a side wall with the lining of corrugated sheet iron.

Figure 20 shows the results of the computational simulation of the driving. The grey curves show the settlements, which have been measured during the construction of the tunnels. The blue line shows the calculated displacements. The maximum displacement, which has been calculated with about

Table 4.	Load cases for the simulation of the old tunnel.
Nr.	Load case
I	**Primary stress distribution**
1	Activating gravity and surface loads
II	**Heading of the existing tunnels**
2	Dewatering
3	Tension release in side wall I
4	Excavation of side wall I
5	Concreting of side wall I Tension release in side wall II
6	Excavation of side wall II
7	Concreting of side wall II and tension release in the tunnel roof
8	Excavation of the tunnel roof
9	Concreting of the tunnel roof and tension release in the tunnel floor
10	Excavation of the tunnel floor
11	Concreting of the tunnel floor
12	Concreting of the lining
13	End of dewatering

Table 5.	Load cases for the simulation of the new tunnel.
Nr.	Load case
III	**Heading of the primary tube**
14	Tension release
15	Excavation
16	Lining
IV	**Soil freezing**
17	Soil freezing in the iced cap
V	**Heading of the new tunnel**
18	Dewatering
19	Tension release in the tunnel roof
20	Excavation of the tunnel roof and tension release in the bottom
21	Excavation of the tunnel bottom
22	Concreting outer lining
23	Concreting inner lining
24	End of freezing, thawing of the iced cap
25	End of dewatering

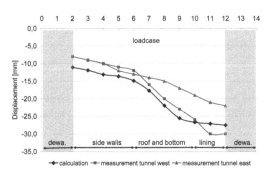

Figure 20. Settlements due to driving of the existing tunnels.

27 mm lied between the measured displacements of about 30 mm at the west tunnel and about 22 mm of the eastern tunnel. The calculated displacements due to dewatering of 11 mm overestimated the measured displacements of up to 8 mm slightly.

All in all the calculated displacements matched the measured displacements, so that a forecast of the displacements, which had to be expected during the building of the new tunnels was possible.

4.3 New pedestrian tunnels: simulation of the construction steps

Basically the simulation of the construction steps of the new tunnel was done in the same way as it was done for the existing tunnels before. Table 5 shows the single load cases, which were used to simulate the construction process. The displacements were set to zero at the beginning of load case Nr.14.

At the beginning of the freezing of the cap (loadcase Nr. 17) the soil properties of the cap were changed to those of the frozen soil. As a new aspect the frost heave

displacements had to be taken into consideration, too. From the laboratory tests described above, both values for the expanse of the iced soil due to volumetric heave and heave rates due to ice lens growth were derived. The displacements, which had to be expected at the boundary of the iced soil could roughly be estimated from these values. A linear section across an arbitrarily formed and closed frozen body, which is completely covered by soil will always meet twice the boundary of the body and once the body itself. A general statement for the displacements, which have to be expected at the boundary of the body is:

Displacement =
Thickness of the cap · "Volumetric Heave" +
2 · Comissioning time · "heave rate" (7)

This statement includes a number of simplifications and assumptions, which will be described below:

– in general: As soon as the volume change of the porewater due to freezing starts, the volume of the iced soil increases and the surrounding soil gets compressed. (similar to the swelling of soil) This means, that the values for both frost heave characteristics "Volume change" and "Icelens growth" depend on the acting surcharge and as a consequence on time
– volumetric heave: Theoretical Solutions of the frost penetration (e.g. Neumann) and experiments show, that the velocity of the frost front decreases with time. This means, that it takes the longer to freeze layers of the same thickness the bigger the distance to the heat sink is. As a consequence the porewater in that layers has more time to run in unfrozen areas and not to cause heave effects. Hence, the heave characteristic "Volumetric heave" is a function in time. As a consequence the test results shown in

table 2 are "moments" of an nonlinear, instationary process.
- ice lens growth: Experimental results (Konrad 1981, 1982) show, that the velocity of the ice lens growth depends linearly on the temperature gradient. The experimental results shown in table 3 are all based at a temperature gradient from $-10°C$ to $+10°C$ over a distance of 100 mm. It would be necessary to put through the experiments at a temperature gradient, which is representative for the construction site. This could be derived by thermal calculations and/or adjusting the heave prediction on the measured temperature gradients at frozen cap.

In accordance with the situation at the construction site an initial load at the later frozen part of the soil of about $340 kN/m^2$ was assumed. This load would be increased due to the additional load resulting from volume change and ice lens growth. For a subsequent analysis cautiously an average load of about $400 kN/m^2$ was assumed.

Taking into consideration a thickness of the frozen body of about 3 m and a commissioning time of the iced soil of about 70 days, the displacement at the boundary of the iced soil could be estimated to:

Displacement =
3000 mm · 0,4 % + 2 · 0,5 mm/d · 70 d = 82 mm.

To study the effects of the displacements of the frozen roof in about 15 m to 17 m below ground level, the corresponding elements of the finite element model were subjected to a temperature load case, which lead to comparable displacements. Figure 21 shows the elements of the frozen cap and Figure 22 the shape of the vertical displacements on the surface and along vertical cuts at the side of the frozen roof. It can be seen, that frost heave occurred asymmetric with larger values around the pilot heading. For a frozen zone, which was completely situated in tertiary clay, heave displacements on ground level of about 8 mm were calculated.

For the following reasons it could be expected, that the heave displacements at the construction site wouldn't meet the calculated values:

- The calculated displacements are based on the assumption of a steady-state heat extraction. At the construction site the freezing process was driven in a oscillating way, this means in time intervals. Experimental tests with a oscillating frost temperature showed, that the heave displacements are lower compared to a steady state heat extraction.
- The driving of the tunnel and the commissioning time of the frozen zone happen partly at the same time. Settlements due to the driving of the tunnel are superposed with the heave process. Settlements and heave displacements are compensating each other.

4.4 Results

Figure 23 shows the magnitude of the calculated ground displacements compared to the displacements which later were measured at the construction site. These measurements are described later in the text. The dashed part in the curve of the calculated displacements stands for the generated frost heave displacements described above. All in all there can be found a good accordance of calculation and measurement.

5 CONSTRUCTION STEPS

The building project started with the production of the bored pile walls of the access shafts. From an intermediate excavation level the pressing of the preliminary tubes started, each about 100 m long. For this a hood shield with a compressed-air chamber

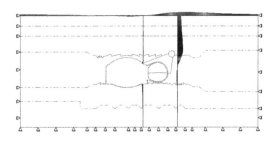

Figure 22. Frost heave displacements.

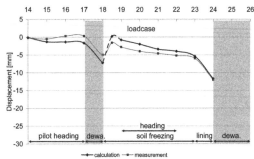

Figure 23. Settlements due to driving of the new tunnels.

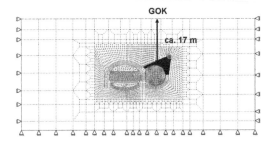

Figure 21. Frozen cap.

was used and hydraulic transport was used for the excavated material. As the ground water level wasn't completely below the bottom of the tube at that time, the compressed-air chamber was driven with an air pressure between 0,2 bar and 0,5 bar depending on the waterlogged line at the face. The outer diameter of the tube was 2,40 m and the thickness of the wall 0,20 m. Bentonit was used during pressing to reduce the friction and finally substituted by a formation stabilizer.

About 350 freezing holes were carried out by pressure drilling out of the preliminary tube. They had a diameter of 88,9 mm and an overall length of about 3800 m. The freezing holes were arranged in the shape of a vertical fan, so that the soil above the tunnel roof built a closed frozen mass. Brine freezing with $CaCl_2$ at a temperature of $-38°C$ was used to create the frozen soil and the freezing system had a power of 2×275 kW. In order to minimise the commissioning time of the frozen mass, the freezing body above each tunnel was divided up into three parts "North", "Middle" and "South", which were driven separately. The following Figures 24 to 29 show:

- the oval access shaft
- the putting in of the hood shield
- the pressing of a tube out of the access shaft
- the assembling of the freezing system
- the freezing system during operation
- the assembling of a support during tunnel driving.

The driving of the new tunnels, each with a full section of about 50 m² was done with a shotcrete – method as a full-face excavation with advance crown. The frozen soil was worked off by use of a cutter scraper. The average tunnel construction process was 2 m a day. The shotcrete lining had a thickness of the wall of 30 cm. As the new tunnel had to be connected to the existing tunnel, the lining of the old one and about 13 cm of the tunnel facing were worked off. After the lining was completed, 11 connection points between the tunnels were opened and the facing was built in.

Timetable of construction steps of geotechnical interest: Tunnel Weinstraße

Figure 25. View of the hood shield.

Figure 26. Pressing of a tube .

Figure 24. View of the access shaft.

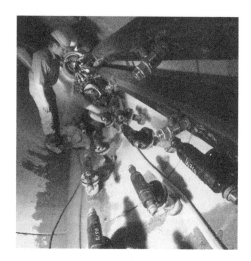

Figure 27. Assembling of the freezing pipes.

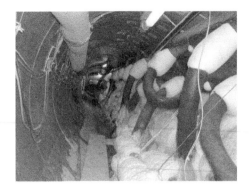

Figure 28. View of the pressed tube and the pipelines for ground freezing.

Figure 29. Assembling of a support during tunnel driving.

- pressing of the preliminary tubes
  07.12.2003–14.12.2003
- freezing section "north":
  10.05.2004–25.08.2004
- freezing section "middle":
  01.06.2004–20.09.2004
- freezing section "south":
  22.06.2004–21.09.2004
- heading of the new tunnel:
  24.06.2004–13.09.2004.

The construction of the tunnel "Dienerstraße" was done with an offset of about 3 month.

5.1 Measurements of settlements and heave displacements

The settlements during the construction time were measured by a geometrical levelling on the ground surface and additionally by an automatic levelling system in the second basement of the town hall. The measuring network of the geometrical levelling had about 130 survey points, which were mounted on the cladding of the town hall and on surrounding buildings. The automatic levelling system had 10 survey points and could be connected to the geometrical measuring network.

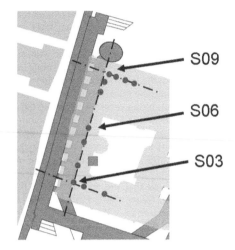

● measurement station
■ reference station

Figure 30. Position of the measurement points of the automatic levelling system.

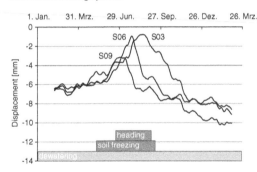

Figure 31. Records of ground displacements along the tunnel.

Measurements were carried out every 15 minutes. The main axis ran parallel to the tunnels. The main purpose was to measure differences between the single survey points with high accuracy. Figure 30 shows the position of the survey points. We will have a stronger look at the points S03, S06 and S09 now.

Figure 31 shows the records of the measured settlements and heaves of the automatic levelling system. As the automatic levelling system covered only a relatively small area it was not able to measure settlements due to dewatering, because all survey points were affected. From the geometrical levelling values for the settlement due to dewatering of about 6 mm arised. Hence, the curves in the figure were shifted of 6 mm before freezing and heading were carried out.

By starting the freezing of the soil the expected heave displacements set in, which reached about 5 mm. After the heading of the tunnel crossed under the survey points, settlements occurred. The heading reached the survey points S09, S06 and S03 in descending order, which lead to the undulated shape of the deflection curves. The settlements continued after the freezing of the soil was stopped. The speed of the settlements lowered down steadily, so that 3 month after the soil freezing was stopped, the settlements reached absolute values between 10 and 12 mm.

Figure 23 already showed the comparison between measured and calculated results. The measured results were allocated to the corresponding load cases of the finite-element calculation. Summarizing it can be noticed, that the measured ground displacements had the same order of magnitude as the calculated ground displacements.

5.2 Strain gauge measurements

The development of the normal forces in the shotcrete lining was measured with strain gauges at two steel girders in the lining. The girders were appliqued and protected at the Zentrum Geotechnik and then transported to the construction site.

The cross section of the girders is trapezoidal with a reinforcement steel Ø 25 mm at each corner and a joint in the tunnel roof. The measuring points were arranged above the crown, at each side of the joint in the tunnel roof and next to the support at the existing tunnel, one at each upper chord and lower chord of the girders. Figure 32 shows the position of the measuring points, whereas the points on the upper chord and lower chord are represented by one mark.

The strain gauge measurements show a rather rapid increase of the normal forces in the girders, after they had been mounted. After about 10 days the normal forces show only comparatively small changes of the normal forces. Even after the freezing system was turned off, the normal forces didn't increase strongly.

5.3 Results

The measurement of settlements on the ground surface showed:

– total settlements of approx. 12 mm
– settlements due to dewatering of approx. 4 to 5 mm
– total heave displacements of approx. 3 to 4 mm.

Strain gauge measurements:

– The rapid increase of the normal forces after they were mounted can be explained by a redistribution of forces due to the heading of the tunnel. The shotcrete lining acts as a temporary support for the frozen mass in the direction of the tunnel.
– As there are no greater increases in the normal forces of the girders after the freezing was turned off, it can be assumed, that the internal forces inside the frozen body already sooner were diverted on the shotcrete lining due to creeping effects.

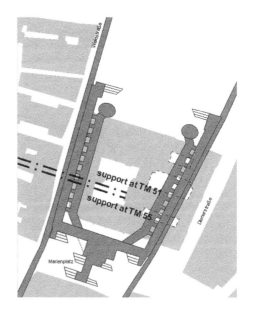

Figure 32. Position of the girders with strain gauges.

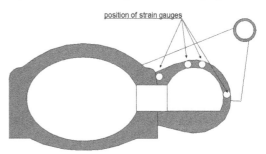

Figure 33. Position of the strain gauges at the girders.

– Under the assumption of a plain strain state in the lining and a Young's modulus of 15.000 MN/m² there can be calculated upper boundaries for the normal forces in the shotcrete lining of about 1400 kN/lfdm. This shows a good accordance with the results of the finite-element calculations.

6 SUMMARY

In 2004, a tunnelling project below the historical town hall in the centre of Munich was carried out by order of the department of public construction, city of Munich. The contractor Max Bögl GmbH developed a specific proposal to drive the tunnel below a cap of artificially frozen soil in order to reduce the ground displacements. As frozen soil has higher stiffness and bearing capacity compared to its unfrozen condition, the displacements could be reduced by use of this frozen soil cap.

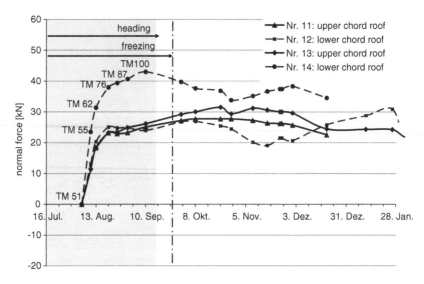

Figure 34. Record of the normal forces in the reinforcement of the roof.

To predict the ground displacements during freezing and tunnelling both laboratory tests and Finite-Element-Calculations were carried out. It appeared that for a realistic prediction the examination of heave effects caused by ice lens growth and volumetric effects at realistic boundary conditions (bulk load and water supply) was necessary. However, the results of both laboratory tests and calculations depend on a number of influencing variables, which had to be chosen and adapted carefully to the specific situation. Based on the experiments a prediction of the ground displacements at the tunnel-construction site was made, which later showed a good accordance with the measurements.

The authors would like to thank the department of public construction, city of Munich, and Max Bögl GmbH for the professional cooperation and the permission to publish the measurements and the calculations.

REFERENCES

Bosch, B. 2005. Abgesicherte Beurteilung von Frosthebungen bei geringen Auflasten. *Diplomarbeit Nr. 631 Zentrum Geotechnik (unpublished)*

Eicher, L. 2004. Bahnsteigerweiterung am U-Bahnhof Marienplatz in München. *Tiefbau, amtliches Mitteilungsblatt der Tiefbau-Berufsgenossenschaft*: Heft 12, Dezember 2004: 784–789

Eicher, L. & Bayer F. & Vogt, N. 2005. Baugrundvereisung zur Verfestigung und Firstsicherung beim Bahnsteig-Erweiterungstunnel U3/U6 in München. *Schriftenreihe des Zentrum Geotechnik*: Heft 37: 163–172

Fillibeck, J. & Kellner, C. & Rieken, W. & Scharrer, S. 2005. Bahnsteigerweiterung der U6 unter dem neuen Rathaus in München - Spritzbetonvortrieb mit Vereisung. *Bautechnik*: 82 (2005), Heft 7: 416–425

Heitzer, K. 1981. Ein neues Verfahren zur Bestimmung der Frostempfindlichkeit von Böden und Boden – Chemikal – Gemischen und deren Tragfähigkeit nach dem Auftauen. *Dissertation, Lehrstuhl und Prüfamt für Grundbau, Bodenmechanik und Felsmechanik der TU München*

Jelinek & Jessberger & Lackinger 1968, Frostwirkungen im Straßenbau. *2. Donau-Europäische Konferenz Wien(Proc.)*: 106–138

Konrad, J. M. & Morgenstern, N. R. 1980, A mechanistic theory of ice lens formation in fine-grained soils. *Canadian Geotechnical Journal*: vol.17, no.4: 473–486

Konrad, J. M. & Morgenstern, N. R. 1981. The segregation potential of a freezing soil, *Canadian Geotechnical Journal*: vol.18, no.4: 482–491

Konrad, J. M. & Morgenstern, N. R. 1982. Effects of applied pressure on freezing soils. *Canadian Geotechnical Journal*: vol.19, no.4: 494–505

Konrad, J. M. 1993, Frost heave in soils: concepts and engineering. *Canadian Geotechnical Journal, Sixteenth Canadian Geotechnical Colloquium*: vol. 31: 223–245

Mellor, M. & Cole, D.J. 1982. Deformation and failure of ice under constant stress and constant strain rate, *Cold Regions Science and Technology*: 5: 201–219

Orth, W. 1986. Gefrorener Sand als Werkstoff, Elementversuche und Materialmodell. *Veröffentlichungen des Institutes für Bodenmechanik und Felsmechanik der Technischen Hochschule Fridericiana in Karlsruhe*: Heft 28

Orth, W. 1988. A creep formula for practical application based on crystal mechanics, *Proceedings of the fifth International Symposium on Ground freezing*: Nottingham, England: A.A.Balkema

Ostermeier, B. 1991. Ein Beitrag zur Erfassung des Vortriebsgeschehens beim Bau von Tunneln im Lockergestein mit Spritzbetonsicherung – Ebene und räumliche Berechnungen –. *Berichte aus dem Konstruktiven Ingenieurbau, Technische Universität München*: Band 1/91

Numerical Modelling of Construction Processes in Geotechnical Engineering for
Urban Environment – Triantafyllidis (ed)
© 2006 Taylor & Francis Group, London, ISBN 0 415 39748 0

Underground line U5 'Unter den Linden' Berlin, Germany Structural and thermal FE-calculations for ground freezing design

B. Brun & H. Haß
CDM Consult GmbH, Bochum, Germany

ABSTRACT: In the aftermath of the fall of the Berlin Wall and Germany's reunification the German government moved from Bonn to Berlin. Not only ministry workers and their families moved to Germany's capital, it also has become more and more attractive for tourists. The existing local public traffic network cannot cope with the increasing number of passengers and needs to be extended. One of the main projects is the extension of the already existing line 'U5'. Because of the sensitive metropolitan situation, the construction of two stations and one switching station will be executed under the protection of frozen soil bodies by using the sequential excavation method (SEM), also known as New Austrian Tunneling Method (NATM). The frozen soil bodies' task is to cut off the water and provide structural support. The planning and design of the subway line U5 includes the largest volume of frozen soil body world wide to date. All structural and thermal calculations for ground freezing design have been performed using the finite element method (FEM). The FE-calculations consider all phases of construction and interrelations, e. g. nearby buildings. The calculations are used for the structural design and static analyses of the frozen soil body as well as for the calculation of the deformation during construction. The thermal calculations are performed for the calculation of the build-up time of the frozen soil body, the determination of the average temperature of the frozen soil and to check the frozen soil parameters, depending on the average temperature in the structural calculations.

1 INTRODUCTION

1.1 *Project description*

To accommodate increasing numbers of passengers using the public underground transportation systems the traffic network in Berlin has to be extended. One of the main projects is the extension of the already existing underground line 'U5' from the eastern city center of Berlin to the future main station in the West. The project U5 'Unter den Linden' is a portion of this extension of the U5. It is called 'Unter den Linden' after the boulevard the tunnel alignment follows. As a first part of the project, the U55-Shuttle at Brandenburger Tor, is currently under construction.

The client of this project is the Senate Administration for Building and Traffic on behalf of the Berlin-Traffic-Company (Berliner Verkehrsbetriebe, BVG).

This section has been planned and designed by a joint venture, consisting of the engineering companies PSP (Philipp, Schütz + Partner, Munich), ZKP (Zerna, Köpper, Partner, Bochum), DMT (Deutsche Montan Technologie, Essen), and CDM (CDM, Bochum). In this joint venture CDM is responsible for the planning and the design of ground freezing and jet grouting.

1.2 *Tunnel alignment*

The total length of the underground line between the stations 'Berliner Rathaus' (city hall) and 'Brandenburger Tor' is approx. 1.7 km. Along this alignment, three underground structures are proposed. The whole section contains the stations 'Spreeinsel' and 'Brandenburger Tor' as well as one switching station (Fig. 1).

Starting at 'Berliner Rathaus' east of the 'Spreeinsel' station with a tunnel boring machine equipped with a pressure-balanced slurry shield, the route passes underneath the Spree river, the so-called 'Palast der Republik', the former East-German parliament building, and reaches the 'Spreeinsel' station almost underneath the Spree-channel.

From the 'Spreeinsel' station the route runs to the switching station. In this area, the slurry shield will intersect with an old tunnel of the underground line U6.

The gradient of the U5-tubes then subsides and the alignment intersects the tunnels of railroad lines S1/S2. Finally, the TBM-tunnels reach the 'Brandenburger Tor' station. In this area, the new section of U5 will be connected to the adjacent section, which has already been completed (Fig. 2).

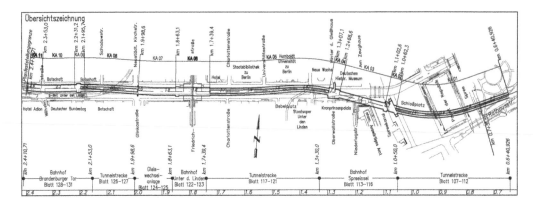

Figure 1. U5 'Unter den Linden', tunnel alignment.

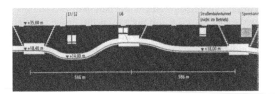

Figure 2. Schematic longitudinal section of U5.

1.3 Geology

Preliminary investigations showed fills overlying mostly sand and gravel (quaternary layers) with stones and boulders within the tunnel excavation. In addition, marl and localized peat deposits could be encountered.

The groundwater level fluctuates between 2 m and 5.5 m below ground surface. The flow velocity of the groundwater is not significant.

1.4 Project constraints

Extensive constraints were set by the client that had to be considered during the design phase. There should be no impact on the traffic and the neighborhood, so that all surface work has to be kept to the technically possible minimum.

Furthermore, the planning has to utilize construction methods that cause only slight deformations of adjacent structures. Therefore, an overall lowering of the groundwater table would not be approved.

2 GENERAL SEQUENCE OF CONSTRUCTION

Launching and receiving shafts are located within the alignment, providing access for the tunnel boring machines, ground freezing operations and other tasks. After the completion of all shafts, two tunnel tubes will be excavated by using a pressure-balanced slurry

shield with a shield diameter of 6.7 m. The total length of the shield driving will be about 3,000 meters with a maximum depth of approx. 25 m below ground surface. The maximum estimated water pressures at the tunnel face will be about 1.8 bar (26.5 psi).

After the completion of the two shield driven tubes the stations will be excavated by underground mining. The installed shield tubes will be removed and the station will be constructed by using the sequential excavation method (SEM), also called 'shotcrete method'. All these work steps will be performed within a frozen soil body.

3 GROUND FREEZING

3.1 General

Two different freezing methods are proposed: brine freezing for the construction of the stations, some cut-off crosswalls as well as temporary sealing of connections, and freezing with liquid nitrogen LN_2 for the removal of an old sheet pile within the slurry shield track.

Because of the construction of the stations (using the shotcrete method) under the protection of a frozen soil body, some technical safety and constructional demands arise.

The frozen soil bodies have to cut off the water and provide structural support. These boundary conditions will only be met, if there is a totally closed frozen soil body that shows the required thickness.

3.2 Drilling

Furthermore, there should be no abandonment of borings due to boulders. Afterwards the freeze pipes must be installed in an exact position.

Drillings for the freeze pipes will require a high level of vertical and horizontal accuracy and have to

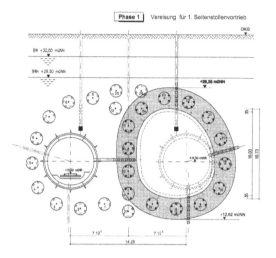

Figure 3.   Phase 1, Frozen soil body around one of the tubes.

be performed against water pressures up to 1.8 bars. Given the required level of accuracy and the potential of boulders at the tunnel face, microtunneling is the preferred method for drilling the freeze pipe holes. For the most part of the tunnel length welded steel jacking pipes with an outside diameter of 1.5 m are proposed as freeze pipe tunnel liner. The diameter was chosen to allow for a suitable access and work space needed for the installation of freeze pipes.

3.3   Particular features

The following describes particular features of the 'Spreeinsel' station (length: 170 m), the 'Switching Station' (length: 120 m) and the 'Brandenburger Tor' station (length: 190 m).

The overall volume of frozen soil depends on the different lengths of the soil bodies and amount to approx. 30,000 m³. The required thickness of each frozen soil body will be 2.5 m. Between 27 and 31 microtunnels are needed to install nearly 25,000 m of freeze pipes.

A special feature of the 'Spreeinsel' station is its construction below the Spree-Channel. Here, the uppermost microtunnel is located only slightly below the bottom of the channel and is blanketed by less than 0.1 m. An overburden of at least two diameter thickness is required to avoid the microtunnel machine driving into the channel. Due to this, additional ballast at the bottom of the channel is required, which has to be removed after the completion of the station.

In the area of the construction site of the 'Switching Station' the frozen soil body lies partially above the groundwater level. In this area it will be necessary to increase the degree of saturation of the surrounding soil by artificial watering. It is not necessary to

increase the saturation over the whole length as the tunnel gradient declines towards the west.

For this section the planning provides four horizontal drillings on a distance of approx. 40 m. These boreholes will be used for watering. Experience from other CDM projects showed that surface infiltration is not sufficient for a controlled build-up of the frozen soil body. Therefore, CDM suggested constructing microtunnels for a barrier system along both sides of the location route.

The length of the construction site of the 'Brandenburger Tor' station is about approx. 190 m. Special conditions of this structure are a partially curved station, a partial widening of its cross section as well as a building pit with connection to the top of the frozen soil body.

3.4   Construction phases

The construction phases used for the construction of all stations are described below.

In Phase 1 the frozen soil body around one of the tubes, driven by the slurry shield TBM, will be frozen. Under protection of the frozen soil body, the soil around the tube will be excavated and the already existing shield tubes will be removed (Fig. 3).

After completion of the shotcrete lining, the final lining will be constructed. During the construction of one of the side tunnels the soil around the second will be frozen (Phase 2). After completion of the first side tunnel the identical construction of the second one will start. It is not possible to drive both sides at the same time, because one of the tubes is required for access and material transport. After completion of the final linings in the side tunnels both freezing circles are activated (Fig. 4, Phase 3) in order to prepare Phase 4 – the connection of both side tunnels.

To connect both side tunnels to complete the structure it is necessary to switch off the freezing pipes in the center. At the same time the outer freezing circle has to be maintained. After excavation and completion of the shotcrete lining in the central section the final lining can be constructed and both sides will be connected to form one structure.

4   DESIGN OF GROUND FREEZING

4.1   General

The frozen soil body is designed for cutting off the groundwater and providing structural support of the tunnel excavation in the non-cohesive quaternary layers.

The temperature distribution in the frozen soil is inhomogeneous, colder near the freeze pipes and less cold in-between, especially at the boundary of the frozen soil body. This boundary is at the $-2°C$ isothermal line, whereas at the core of the frozen soil body

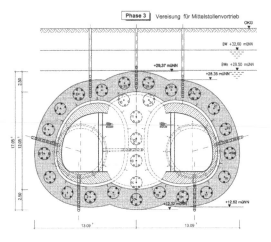

Figure 4.   Phase 3 freezing pipes in the center switched-off.

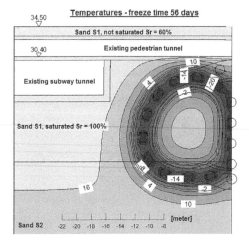

Figure 5.   Results of thermal FE-calculation, temperature distribution at 'Brandenburger Tor' station.

the temperature is −18°C. For the design calculations CDM assumed an average temperature within the frozen soil body of −10°C.

For the thermal calculations the Finite-Element-Method was used. This method allows the estimation of the time dependent development of the frozen soil bodies, the temperature distribution within the frozen soil bodies at particular times, the freezing time that is required to build up the required thickness of the frozen soil body, the heat flux as well as the required capacity of the freeze plant. The thermal calculations are performed with the software TEMP/W.

The Finite-Element-Method was used for the structural calculations as well. CDM used this method for the design of the frozen soil bodies as well as the calculation of the total deformations during construction. The structural calculations are performed with the software PLAXIS.

4.2   Thermal design

The thermal design can be based on analytical methods for the freeze body development and the heat flux using analytical methods in the preliminary design phase. In most final designs thermal calculations by FEM are required to verify results of the pre-design and to optimize the freeze pipe spacing and arrangement, the freeze plant capacity and the overall layout of the freezing system. Using this method, the actual conditions (freeze pipe spacing and location, interrelation of adjacent freeze pipes, dependencies of freeze body development between different layers etc.) can be optimized. The actual locations of thermocoupling can be incorporated in these calculations and the time dependent temperatures can be used for direct comparison with the related thermocouple readings during freezing operation.

With FEM the following design conditions can be considered:

– Different soil layers,
– Different initial temperatures in the soil layers,
– Varying freeze pipe spacing and varying freeze pipe temperatures (if required),
– Different and temporarily varying freeze plant capacities for brine freezing,
– Simulation of intermittent freezing during maintenance,
– Additional heat sources or thermal boundary conditions, if existing.

The results of the thermal calculations should include:

– The time dependent development of the frozen soil body,
– Determination of the average frozen soil temperature to prove the time dependent frozen soil parameters (e. g. strength, stiffness etc.) used for the structural design,
– Determination of the time required to freeze the designed frozen soil body,
– Optimization of freeze pipe arrangement and freezing operation,
– Time dependent temperature distribution in the soil,
– Determination of energy consumption,
– Design of capacity of the refrigeration plant (brine freezing).

Figure 5 shows an example of the temperature distribution in the frozen soil body at the 'Brandenburger Tor' station after a freezing time of 56 days as result of a thermal FE-calculation.

Figure 6 shows an example for the evaluation of the time dependent development of the frozen soil body and the determination of the average frozen soil

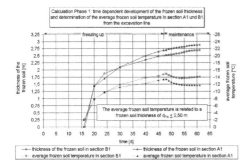

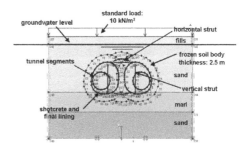

Figure 6. Results of thermal FE-calculation, time dependent development of the frozen soil body and the determination of the average frozen soil temperature.

Figure 7. Structural FE-calculation, geometrical model.

temperature in two different sections as the result of a thermal FE-calculation. The FE-calculation considers the freezing up and the maintenance of the frozen soil body.

Referring to Figure 6 the frozen soil body between the freeze pipes closes up after 16 days. The required thickness of the frozen soil body of 2.5 m will be achieved after 47 days. At the end of the freezing up of the required frozen soil body an average frozen soil temperature of $T \leq -10°C$ will be achieved.

During the maintenance of the frozen soil body the increase of thickness of the frozen soil is minimal. The average frozen soil temperature rises during the maintenance of the frozen soil body, but remains below $-10°C$.

4.3 Structural design

Structural design is required for all construction phases in which the frozen soil body serves both as structural element (acting against soil and water pressure) and for cutting off groundwater. For many practical applications, it is sufficient to check the stresses in the structural design of freeze bodies. The required structural design data for the allowable stress, Young's modulus, and shear parameters will be determined using the projected frozen soil stand-up time.

Therefore, the basis for a sound structural design of a load-bearing frozen soil structure is extensive knowledge of the time and temperature dependent strength and deformation properties of the material. Frequently, the complex time-dependent stress-strain characteristics are simplified for the structural design; however, in some cases the entire stress-strain-history from the start of loading has to be considered. Detailed time-dependent solutions can be reached by using numerical methods. Strength properties of the frozen soil body change according to the temperature distribution throughout its cross section. Whereas the strength is highest in the center of the frozen soil

body, strength decreases towards its boundaries (0 °C – isothermal line for fresh water). Due to this fact, it is possible that the stress at the boundaries exceeds the bearing capacity of the soil body. In this case, a plastification and subsequent stress redistribution will take place. The resulting stress distribution is then similar to that of the temperature distribution.

Final structural analysis of the freeze body can be performed using FEM. Frozen and unfrozen soil stress-strain behavior can be simulated with a non-linear model. The advantage of FEM is that it realistically accounts for both the frozen and the unfrozen soil whereas the analytical approach uncouples the frozen soil body from the surrounding unfrozen soil, and the external loads of earth and water pressure plus any surcharge loads are applied to the freeze body.

The calculated deformations should include the effects of:

– Excavation (stress redistribution),
– Creep influences of the frozen soil,
– Thawing.

Depending on its shape (e. g. straight freeze wall instead of a curved shape) the frozen soil body may only take limited loads, so that the design of additional supporting measures is required. Additional measures may consist of shotcrete application, anchors, beams or similar.

The following figures show printouts of the structural FE-calculations for the underground railroad line U5. The construction phases of all stations are almost the same.

Figure 7 shows the general geometry of the stations. The model should closely represent the real situation. Since construction extends over long distances along the tunnel axis, a plain strain model is applicable.

The ground conditions are based on the evaluation of the geotechnical site explorations. Horizontally layered geometries with a horizontal ground surface are used. A permanent working load of $10\,kN/m^2$ is applied on the ground surface.

The structural elements like the tunnel segments, the shotcrete and the final lining as well as the required

Table 1. Parameters of unfrozen and frozen soil at 'Spreeinsel' station.

Bodenart	E [kN/m^2]	ν [–]	φ [°]	Ψ [°]	c [kN/m^2]	γ [kN/m^2]	γ [kN/m^2]	k_0 [–]	zul. σ_D [kN/m^2]	zul. σ_Z [kN/m^2]	Standzeit [Monate]
Aufschüttung II	$1{,}6 \times 10^5$	0,3	33	3	0	16	9,5	0,455	–	–	–
Talsande/ obere Sande	$1{,}6 \times 10^5$	0,3	34	4	0	17,7	10,45	0,441	–	–	–
Schmelzwassersande	$2{,}3 \times 10^5$	0,3	36	6	0	17,4	10,4	0,65	–	–	–
Geschiebemergel	$1{,}8 \times 10^5$	0,3	29,5	0	21	21,95	12,3	0,65	–	–	–
Talsande/obere Sande gafroren, T = −10°C	$2{,}45 \times 10^5$	0,3	27,5	0	410	17,7	10,45	–	1350	250	3
Schmelzwasser Sande gafroren, T = −10°C	$3{,}2 \times 10^5$	0,3	30	0	530	17,4	10,4	–	1750	350	3
Geschiebemergel gafroren, T = −10°C	$1{,}75 \times 10^5$	0,3	17,5	0	420	21,95	12,3	–	1150	200	3

Table 2. Material Parameters of the tunnel segments at 'Spreeinsel' station.

Bauwerk	Dicke d [m]	angesetzte Dicke d* [m]	E_b [kN/m^2]	ν	EA [KN/m]	EI [kN/m^2]	γ_b [kN/m^3]	$w = \gamma_b \times d^*$ [kN/m^2]
Tübbinge	0,35	0,35	$3{,}4 \times 10^7$	0,2	$1{,}19 \times 10^7$	$1{,}215 \times 10^5$	25	8,75
Außenschale 35	0,35	0,3	$1{,}95 \times 10^7$	0,2	$5{,}85 \times 10^6$	$4{,}3875 \times 10^4$	25	7,5
Außenschale 40	0,40	0,35	$1{,}95 \times 10^7$	0,2	$6{,}825 \times 10^6$	$6{,}955 \times 10^4$	25	8,75
Außenschale 45	0,45	0,4	$1{,}95 \times 10^7$	0,2	$7{,}8 \times 10^6$	$1{,}04 \times 10^5$	25	10

horizontal and vertical struts are modeled by beam elements. The tunnel lining was modeled as a curved beam element. The lining properties can be adapted during the different phases of construction. The thickness of the frozen soil body is assumed to be 2.5 m.

The Mohr-Coulomb model is used as an approximation of the soil behavior. The creep behavior of the frozen soil body has been calculated by using comparative material properties. An example of the used soil parameters at 'Spreeinsel' station is shown in Table 1.

For the stiff massive structures of the tunnel segments as well as the horizontal and vertical struts the linear elastic model is used. An example of the used material parameters of the tunnel segments at 'Spreeinsel' station is shown in Table 2.

After the input of the geometrical model the FE-mesh is generated. It is expected that stress concentrations will occur in the frozen soil body and so a local refinement of the mesh generation is used here.

The construction phases (see Paragraph 3.4) are exactly modeled in the structural FE-calculation. Each calculation phase of the structural FE-calculation is described as follows. The first calculation phase is the initial phase. In the initial conditions the self-weight of water of 10 kN/m^3 has been entered. The initial water pressures are generated on the basis of a horizontal general phreatic line. All structural components are deactivated. The initial stress field is generated.

In the second calculation phase the tunnel segments are activated on both sides. The soil clusters inside the tunnel are deactivated. The water pressure is deactivated inside the two tunnel tubes. In calculation Phase 3 the soil body on the right side will be frozen. The properties of the frozen soil are adapted on the right side. Under protection of the frozen soil body, the soil will be excavated by using the SEM and the already existing shield tubes will be removed. The soil inside the frozen soil body is deactivated.

After the excavation of the soil under protection of the frozen soil body on the right side the shotcrete lining in the right tunnel will be completed in the fourth construction phase. The shotcrete lining is activated.

After completion of the shotcrete lining in calculation Phase 5 the final lining will be constructed. The lining properties are adapted during this phase of construction. The vertical strut is activated. During construction of the right side, the soil of the left side will be frozen. After completion of the right side, the construction of the left side will start.

The left side will be constructed using the same method as the right side (see Phases 6, 7 and 8).

After the completion of both, the left and right final linings, both freezing circles will be activated in order to prepare for the last calculation phase – the connection of both sides. In order to connect both sides of the construction it is necessary to discontinue the

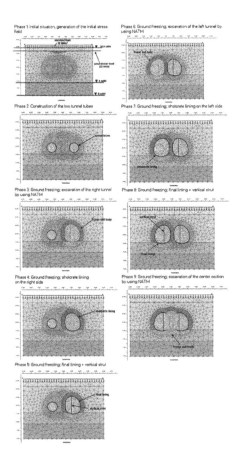

Figure 8. Structural FE-calculation, calculation phases.

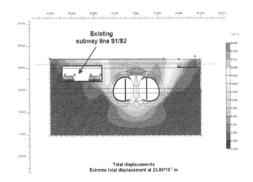

Figure 9. Results of structural FE-calculation, total diplacements at 'Brandenburger Tor' station.

freezing in the center section whereas at the same time the outer freezing circle of the whole cross section has to be maintained. The soil properties are adapted for the whole freezing body, except in the center section. The soil between the two sides will be excavated using SEM and the existing microtunnels with the freezing

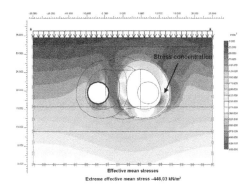

Figure 10. Results of structural FE-calculation, main effective stresses at 'Spreeinsel' station.

pipes will then be removed. This construction phase is the last phase important for the design of the frozen soil body.

After completion of the shotcrete lining in the center section the final lining can be constructed and both sides will become a monolithic construction. Each calculation phase of the structural FE-calculation is shown in Figure 8.

The structural FE-calculations are necessary to consider all phases of construction and their interactions. CDM uses the structural FE-calculations for the design of the frozen soil body. For each construction phase, especially the deformed mesh, the total displacements and the effective stresses are evaluated.

The following figures show examples for the results of the structural FE-calculations. Figure 9 shows deformations after one construction phase at 'Brandenburger Tor' station. The tunnel of the suburban railway line S1/S2 lies in close proximity to the proposed 'Brandenburger Tor' subway station. The different deformations are shown by varying colors. Please note that the calculated deformations are only an estimate. The input parameters, selected for the structural FE-calculation, are mainly fixed for the static analysis. It is anticipated that the actual deformations will be smaller than the calculated deformations.

Figure 10 shows main effective stresses after one calculation phase of the 'Spreeinsel' station. The different stresses are also shown by varying colors. It can be clearly seen that there is a stress concentration on the frozen soil body on the right side, which is not critical.

In the structural FE-calculation of the freeze body the stability is established by the calculation of the stresses and deformations on an elastic, ideal plastic two-dimensional model of the freeze body and the surrounding soil. An example of the load bearing proof of the freeze body is provided for one calculation phase and the soil layer sand in Figure 11.

The design of the freeze body thickness is performed by the proof of stresses in the freeze body.

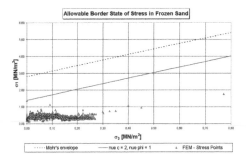

Figure 11. Results of structural FE-calculation, graph of main effective stresses.

In each construction phase the stresses of those elements modelling the freeze body are determined. For these relevant stresses it is proven that the allowable stress exceeds the existing stress. By the use of elastic, ideal plastic material principles for the FE-calculations of the freeze body the compliance with the boundary condition will be enforced according to the numerical accuracy.

The indicated allowable compressive stress relates to the uniaxial stress. According to the Mohr-Coulomb's criteria higher principal compressive stresses are allowable for triaxial stress conditions, provided that the lateral pressure is adequate.

An example of the principal stresses in the freeze body, based on FE-calculations, are represented as triangles in the σ_1–σ_3 diagram in Figure 11.

The allowable stress conditions, which relate to an approach of the combination of partial safety factors for η_φ and η_c (usually $\eta_\varphi = 1.0/\eta_c = 2$), can also be represented in a σ_1–σ_3 diagram as a straight line for the Mohr-Coulomb-criteria. The diagram also provides the principal stresses, which are based on the failure values of shear strength ($\eta_\varphi = 1.0/\eta_c = 1.0$), in form of an axis that is indicated as principal stress axis.

The evaluation of the FE-calculation proves the compliance with the ultimate compressive strength for the frozen soil. Therefore, the stability of the freeze body is given in all investigated phases.

5 CONCLUSIONS

Because of the metropolitan situation of the extension of the existing metro-line 'U5' in Berlin, a number of requirements and boundary conditions had to be considered during the planning and design phase. These boundary conditions are responsible for the decision to plan the execution of the freezing method during the construction of the stations and the switching station.

When the planning of this project will be finished, the total length of all frozen soil bodies will be about 490 m, the total volume is approx. 84,200 m^3.

For the final design calculations of the ground freezing it is suggested to use the Finite-Element-Method. The thermal and structural calculations have been performed by using FEM to consider all phases of construction and their interaction.

With FEM, the following important design conditions can be considered for the thermal calculations:

– Different soil layers,
– Different initial temperatures in the soil layers,
– Varying freeze pipe spacing and varying freeze pipe temperatures (if required),
– Different and temporarily varying freeze plant capacities for brine freezing,
– Simulation of intermittent freezing during maintenance,
– Additional heat sources or thermal boundary conditions, if existing.

The final structural analysis of the freeze body has been performed using FEM. Frozen and unfrozen soil stress-strain behavior can be simulated using a non-linear model. The advantage of FEM is that it realistically accounts for both the frozen and the unfrozen soil whereas the analytical approach uncouples the frozen soil body from the surrounding unfrozen soil, and the external loads of earth and water pressure plus any surcharge loads are applied to the freeze body. The structural calculations include the effects of:

– Excavation (stress redistribution),
– Creep influences of the frozen soil,
– Thawing.

As to date, the U5-project is by far the largest tunnel project worldwide planned and designed for use of the ground freezing method.

REFERENCES

Baugrund Berlin: Baugrundgutachten für die U-Bahnlinie U5-Abschnitt Pariser Platz – Bahnhof Berliner Rathaus (Stadtbezirk Mitte von Berlin), 24.01.1996
Haß, H. & Schultz, M. Tunneling Through Soft Ground Using Ground Freezing. in: Hutton, J. D., Rogstad, W. D., RETC Conference 2005 Proceedings, Littleton CO, USA, 2005
Haß, H. & Seegers, J. 2000 b. Design of ground freezing as temporary support during the construction of underground structures of the subway line 'U5' in Berlin. In Ground Freezing 2000 ISGF, Proceedings
Hoffmann, B. 2000. Underground Railroad Line U5 'Unter den Linden', Berlin, Structural Calculations. In 7th Plaxis user meeting, Proceedings, 9–10 November 2000, Karlsruhe, Germany
Jagow-Klaff, R. & Jessberger, H. L. 2001. In Smoltzcyk, Ulrich (Ed.). Grundbau-Taschenbuch 6. Auflage, Teil 2, Verlag Ernst & Sohn
Jordan, P. 1993. Gefrierverfahren im Tief- und Tunnelbau. In Straßen und Tiefbau 10/93

6. Piling

Numerical Modelling of Construction Processes in Geotechnical Engineering for Urban Environment – Triantafyllidis (ed)
© 2006 Taylor & Francis Group, London, ISBN 0 415 39748 0

Numerical modelling of pile jacking, driving and vibratory driving

K.-P. Mahutka, F. König & J. Grabe
Hamburg University of Technology, Geotechnics and Construction Management, Hamburg, Germany

ABSTRACT: This paper presents a numerical model which simulates the penetration of a displacement pile into a homogenous soil body by three different installation methods: A static procedure in which the pile is jacked monotonically into the soil and two dynamic procedures in which the pile is driven and vibrated, respectively. The influence of the installation procedure on the soil state around the pile is examined, especially the change in void ratio, the stress state and the occurring surface settlements. Finally, the three different installation methods and the effects on the surrounding soil state are compared. An attempt is made to explain the differences in stress and void ratio distribution after the installation and the likely effects on the bearing capacity of the piles.

1 INTRODUCTION

The bearing capacity of a displacement pile depends on a number of factors including ground conditions, pile dimensions and – often neglected – the installation method. Usually influences of the installation method refer only to the driveability and the penetration speed, ignoring the actual impact on the bearing capacity. Nonetheless, it is well known that similar piles in the same soil can show a significant difference in bearing capacity depending on whether they where jacked, driven or vibrated into the ground (Hartung, 1994).

The reason for this observation lies in the fact that during pile installation the surrounding soil is disturbed heavily. Different installation methods may change the soil conditions, like stress state and relative density in different ways which leads to varying bearing capacities.

To gain a better understanding of the processes in the ground during pile installation, a finite element model is presented which simulates the penetration process of a prefabricated pile in a non-cohesive soil. Different installation methods have been simulated including pile jacking, pile driving and vibratory pile driving to observe the influence of the construction method on the changes in soil state around the pile.

Moreover, the simulation is valuable for further finite-element-simulations including displacement piles in order to simulate more realistic initial stress and density conditions in the near-field of the piles instead of K_0-conditions.

2 PILE CONSTRUCTION METHODS

Displacement piles are mainly inserted into the ground by the construction methods jacking, driving and vibratory driving. Pile jacking is a vibrationless method which uses static hydraulic forces to install piles. Little concerns about damages of adjacent structures due to vibrations have to be made, which makes this installation technique attractive in urban environments. The possibility of settlements or heave has still to be taken into account. Jacked piles are reported to behave much stiffer than comparable driven piles (Deeks et al., 2005). The main drawback of this method is that large counterweights are necessary to install the piles (for example existing structures) and that the method is therefore restricted to micro-piles.

These problems can be overcome with dynamic installation methods like pile driving and vibratory pile driving. Vibratory pile driving in comparison to pile driving causes vibrancies with lower amplitudes and is a low noise installation method. In general, the bearing capacity of vibratory driven piles is lower than the one of driven piles. Vibratory pile driving is based on a harmonic excitation of the pile and the surrounding soil. Thereby a rearrangement of the soil takes place and the friction between soil and pile is reduced. The harmonic excitation is generated by centrifugal masses in a vibrator. Investigation on small scale models for the movement of vibratory pile driving has been done by Rodger & Litteljohn (1980). They distinguish slow and fast vibratory pile driving, which are merging fluently and depend on system parameters like relative density, geometry of driving material and vibration parameters. The formation of the tip resistance was investigated in field studies and numerical investigations by Dierssen (1994) and Cudmani (2001). Due to harmonic excitation, oscillations are transmitted in the ground. Therefore it is possible that unallowable large vibrations or even settlements on adjacent buildings occur. This was investigated by Grabe & Mahutka (2005) with the model presented hereafter.

Using pile driving, a mass is dropped from a certain height on the top of the pile. It is possible to use drop hammers which accelerate the mass in order to insert more driving energy into the pile. In general, the pile driving systems are divided into slow and fast pile driving, depending on the dropping frequency. Slow pile driving is particularly used in soft soils.

3 FINITE ELEMENT MODEL

The finite element method is used to model the three installation processes pile jacking, driving and vibratory driving. The solution of the boundary value problem is computed numerically with the commercial code ABAQUS. In the following subsections the main modelling techniques are explained.

3.1 *Modelling of the penetration process*

The soil is discretized with axisymmetric continuum elements. The length and depth of the investigated area is about 10.0 m for all models, see Figure 1. Figure 2 shows the modelling of the pile penetration. The pile is modelled as a rigid axisymmetric surface which is laterally supported. The pile tip is rounded because of better numerical stability. In a distance of 1.0 mm to the axis of symmetry a rigid surface is modelled which forms an axisymmetric tube. At the beginning of the simulation, the soil is in frictionless contact with this surface. During penetration of the pile, the soil separates from the tube as soon as it gets into contact with the pile tip, according to a zipper, see detail in Figure 2.

A kinematic contact formulation in combination with the Coulomb friction model is used for the contact between pile and soil. This contact formulation allows the modelling of large displacements between soil and pile.

3.2 *Constitutive law*

This study investigates the penetration process in non-cohesive, granular materials. To describe the non-linear and an elastic behaviour of granular materials such as sand and gravel, the constitutive equation of hypoplasticity is used, see Gudehus (1996) and von Wolffersdorff (1996). Hypoplasticity is well suited for describing the mechanical behaviour of soils including properties like dilatancy, contractancy, and different stiffnesses for loading and unloading. Hypoplasticity is formulated in rates. The stiffness depends on the current stress state T and the void ratio e. In addition to the two state variables stress and void ratio Niemunis & Herle (1998) proposed the so-called inter-granular strain δ. With this additional state variable it is possible to model the accumulation effects and

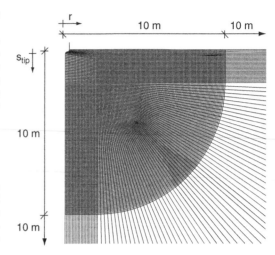

Figure 1. Axisymmetric finite element mesh for pile driving.

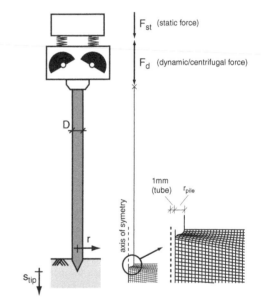

Figure 2. Modelling the penetration.

the hysteric material behaviour under cyclic loading which is of particular importance for the dynamic installation processes. For the presented calculations the material parameters of Karlsruhe sand are used, see Herle (1996). Pore pressures have not been taken into account in this analysis.

3.3 *Adaptive meshing*

During the penetration process the soil close to the pile is pushed aside and sheared. The soil elements

236

around the pile exhibit large deformations and distortions, which are significant for the pile jacking process but even higher for the two dynamic installation processes. The deformations of the elements require special attention. Therefore, the mesh is updated every time increment in the regions of large deformations. This technique is called adaptive meshing. Thus, for every time increment an optimal mesh is generated. Numerical instabilities due to large element distortion are avoided by the adaptive meshing technique.

3.4 Infinite elements

The boundaries of the model are meshed with infinite elements. This type of element allows ground waves to disappear nearly unhindered out of the model. This is very important for the modelling of the two dynamic installation processes pile driving and vibratory pile driving. Reflections, which occur by using fixed supports at the boundaries, are reduced to a minimum. For the investigation of the far-field a coupling of finite and boundary elements would have to be used, see Hagen (2005). In this study, only the plastic deformations in the near field are considered.

3.5 Integration schemes

The calculation of the non-linear systems is done in two different ways. An implicit integration scheme is used to calculate the pile jacking process. It has already been approved by modelling the penetration process of a standard CPT-cone, see Grabe & König (2004).

For the two dynamic installation processes an explicit time integration scheme is used. The central difference rule is used for the solution of the non-linear system of differential equations. The unknown solution for the next time step can be found directly by knowing the solution of the previous time step, so no iteration is needed. The explicit time integration scheme has already been approved in different fields of modelling compaction processes, see Kelm & Grabe (2004) and Mahutka & Grabe (2005). It has especially the advantage of a stable contact formulation.

4 PILE JACKING

4.1 Parameters

For the jacked pile a diameter of $D = 30$ cm is chosen, which lies within the common range of jacked pile diameters. Piles with bigger diameters are rarely jacked into the ground because of the high reaction forces necessary to penetrate the pile especially in non-cohesive soils. The pile has a weight of 1.4 t, which corresponds to a concrete pile with a length of 8 m. In contrast to the other penetrating techniques the jacked pile is penetrating displacement-controlled into the

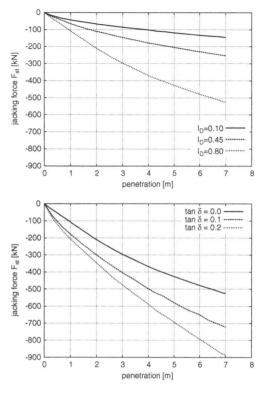

Figure 3. Reaction forces during pile jacking with increasing depth for different soil densities (top) and different contact friction ratios (bottom).

ground by keeping a constant penetration velocity. The pile is jacked 7 m into the ground.

The contact friction between pile and soil is varied between $\tan \delta = 0.0$ and 0.2 to examine the influence of the contact friction on the driving resistance while keeping the relative density of the sand constant at $I_D = 0.8$. Then the void ratio of the sand is varied between a very loose state with a relative density of $I_D = 0.1$ and a dense state with a relative density of about $I_D = 0.8$ while the contact friction is kept constant at $\tan \delta = 0.0$.

4.2 Results

The reaction force of the pile during penetration is shown in Figure 3 for different soil densities (top) and different contact friction ratios (bottom). The force increases rapidly in the first meters of penetration and keeps increasing at a lower rate the deeper the pile penetrates. It can be stated that the soil density has a significant influence on the penetration resistance. The reaction force increases disproportionately high with increasing soil density. In a dense sand with a relative density of $I_D = 0.8$ the penetration resistance

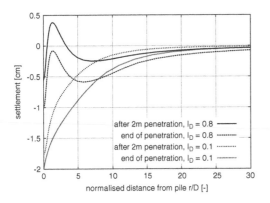

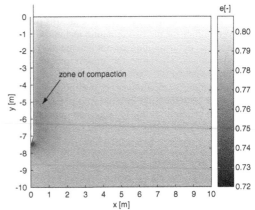

Figure 4. Vertical surface displacements during pile jacking for different penetration depths and different soil densities.

is more than three times higher than in the loose sand with a density of $I_D = 0.1$. The contact friction between pile and soil has a very strong influence on the penetration resistance as well. The influence gets bigger the deeper the pile penetrates. For a contact friction of $\tan \delta = 0.2$, the increase of penetration resistance with depth is more linear than for $\tan \delta = 0.0$ because of the increasing frictional resistance of the pile shaft. For the installation of jacked piles this suggests that the contact friction should be kept on a low level, for example by flushing.

During pile penetration, the soil is pushed aside and sheared heavily which causes significant volume changes in the soil. These effects may result in either heave or settlements at the ground surface. The vertical movements of the surface during pile penetration are shown in Figure 4 for a loose and a dense sand. For a loose sand, the penetration process causes settlements at the ground surface of around 2 cm close to the pile. During penetration, the soil beneath the pile toe is pushed downwards and densified which causes the soil above to move downwards as well. The further the pile penetrates, the greater is the surface area that is affected by settlements. After a penetration of 7 m the surface settlement in a distance of twenty pile diameters ($20D$) is still 1 mm. For a dense sand the surface movements look different compared to the loose sand. As the soil is already in a dense state before pile installation, the soil that is pushed aside during penetration causes a heave of the surface within a radius of $3D$ around the pile with a maximum heave of 4 mm. As the pile penetrates further, the heave finally turns into settlements. The surface movements are of great interest, as jacked piles are most commonly used for underpinning existing structures. For adjacent parts of the existing structure surface settlement or heave should be considered.

The change in void ratio around the pile after installation is shown in Figure 5 for both loose and dense

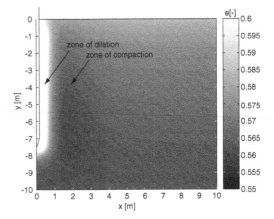

Figure 5. Calculated void ratio distribution after pile jacking for initially loose (top) and initially dense (bottom) state.

state. For the pile in loose sand, the void ratio around the pile toe decreases as the soil is densified by the penetrating pile. After the pile toe has passed the densified zone, the soil relaxes but a densified area of around $4–5D$ remains. The densification remains in a loose to medium dense state. For the pile in dense sand, the void ratio within a radius of $2D$ has increased after penetration. This can be explained by the strong shear forces that act on the soil close to the pile during the installation process which causes dilation in the dense sand. Therefore, close to the pile the compaction of the soil is superimposed by the shearing process while in a greater distance from the pile the compaction is predominant, which can be seen by a slight decrease in void ratio $5–10D$ away from the pile.

In Figure 6 the distribution of radial stresses around the pile is shown for both, loose and dense soil state. Generally, it can be stated that the radial stresses increase during penetration and exceed the K_0-state. The radial stresses are higher for the initially dense sand than for the loose sand, which can also be seen in

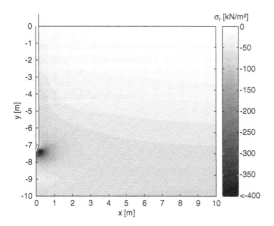

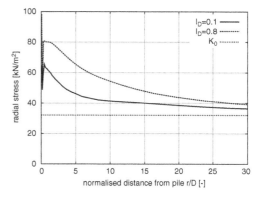

Figure 7. Radial stresses after pile jacking in a depth of 4 meters for different initial soil densities.

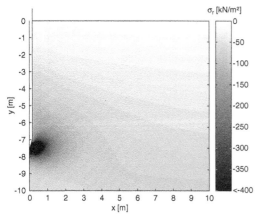

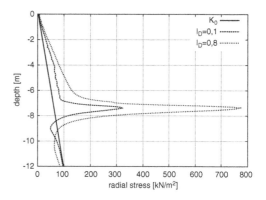

Figure 8. Lateral earth pressure after pile jacking along a vertical cross section at a distance of 10 cm from the pile shaft.

Figure 6. Radial stresses after pile jacking for initially loose (top) and initially dense (bottom) state.

Figure 7. It shows the radial stresses for a cross section in a depth of 4 meters. For the loose sand, the lateral earth pressure at the pile shaft is around twice the initial value which results in a lateral earth pressure coefficient of $K_0 = 1.0$. For the dense state K_0 is around 1.25. The increase of the radial stresses spreads out further for the dense sand than for the loose sand. The higher lateral stresses for the dense sand can be explained by the dilation of the soil which causes a stronger interlocking of the soil grains.

Figure 8 shows the radial stress in a vertical cross section in a distance of 10 cm from the pile shaft. Very high horizontal stresses can be observed at the pile toe. They remain in the ground even after the jacking process is terminated. For the pile in loose sand, the stresses have a peak value of around 320 kN/m² which is around 5 times the initial stress. For the dense sand, the stresses are even higher and reach a peak value of 13 times the initial stress value. Below the stress peak the horizontal stresses drop down to a value below K_0

which can be explained by the high radial strains that are caused by the radial spreading of the stresses below the pile toe.

5 VIBRATORY PILE DRIVING

5.1 Parameters

The diameter of the vibratory driven pile is chosen to $d = 50$ cm and the length is $l = 8.0$ m. The whole mass of $m = 3.93$ t is concentrated in a mass element. This corresponds to a concrete pile with the dimensions mentioned above. The static force due to the weight of the vibrator is $F_{st} = 20$ kN. The excitation frequency is set to $f = 25$ Hz and the centrifugal force is $F_d = 190$ kN. 20 seconds of vibratory pile driving are modelled, resulting in different maximum depths depending on the soil density.

The contact friction between pile and soil is kept constant at $\tan \delta = 0.176$. The void ratio of the sand is

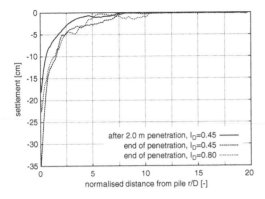

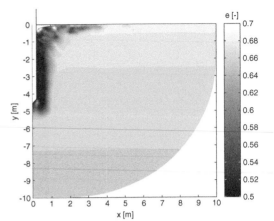

Figure 9. Vertical surface displacements during vibratory pile driving.

Figure 10. Calculated void ratio distribution after vibratory pile driving in a medium dense soil (top) and a dense soil (bottom).

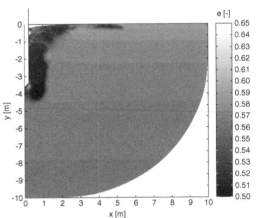

varied between a medium dense state with a relative density of $I_D = 0.45$ and a dense state with a relative density of about $I_D = 0.8$.

The influences of different excitation frequencies, centrifugal forces and layered soil on settlements and ground vibrations can be found in Grabe & Mahutka (2005).

5.2 Results

The vertical settlements of the ground surface during vibratory pile driving are shown in Figure 9 for different penetration depths and relative densities. With increasing penetration of the pile the settlements become larger. The maximum settlement for the medium dense soil directly at the pile after 20 s and a penetration depth of 4.6 m is 34.9 cm. Settlements occur up to a distance of 4 m from the pile. Even with increasing penetration the settlements at the ground surface are restricted to a radius of 8D. This can be explained with anelastic wave propagation, see Kelm (2004). Due to shock fronts, the energy that is needed for rearranging the grains and for compaction is lost with increasing distance from the pile. The maximum penetration depth for the dense soil is 3.6 m, which is lower than the penetration depth in the medium dense soil. Therefore the maximum settlement directly at the pile is lower, too. But in a distance about 25 cm from the pile the settlement curve is nearly identical to the curve of the medium dense soil. For the dense state, the surface settlements spread out farther up to 10D.

Directly around the pile the soil is compacted due to extrusion and cyclic shearing. Figure 10 shows a contour plot of the void ratio distribution after vibratory pile driving for both relative densities. At the surface a funnel of settlement occurs. The influence of compaction is limited to an area of about 2D for the medium

dense soil and about 2.5D for the dense soil. Only near at the ground surface the compaction is farther reaching, due to surface waves. The change in void ratio for the medium dense soil in different distances from the axis of symmetry is shown in Figure 11. The compaction directly at the pile and in a distance of 1D is nearly equal and rather uniform. In a distance of 4D from the pile the soil is just compacted to about 1 m depth. This is as mentioned before due to surface waves.

In Figure 12 the calculated distribution of radial stresses around the pile is shown for both relative densities. Generally, it can be stated that after penetration the radial stresses around the pile up 4D decrease below the initial K_0-state. This can also be seen in Figure 13, which shows the radial stresses for a cross section in a depth of 3 meters in comparison to the initial radial stress state. The radius of the area in which

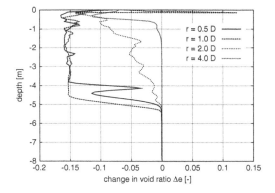

Figure 11. Calculated change in void ratio for different distances to axis of symmetry after vibratory pile driving (medium dense soil).

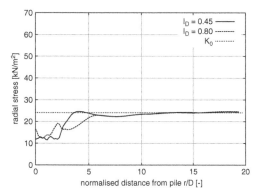

Figure 13. Calculated radial stresses after vibratory pile driving in a depth of 3 m.

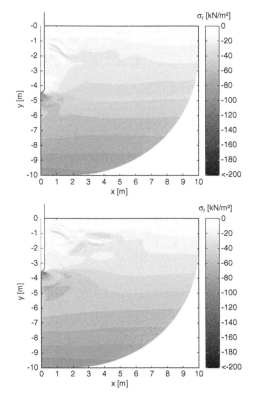

Figure 12. Calculated radial stresses after vibratory pile driving for a medium dense soil (top) and a dense soil (bottom).

stresses and void ratio change significantly during penetration is about two times bigger for the stress change (4D) than for the void ratio change (2D). Figure 13 shows the radial stresses in a vertical axis along the pile shaft for the two relative densities in comparison to the initial radial stress state of the medium dense soil due to the static weight of the pile and the vibrator. Directly at the pile tip high stress peaks occur. These radial stress peaks have about the same value than the initial static stresses at the pile tip. The stress peak in the dense soil is a bit larger than the one in the medium dense soil. Above the pile tip the lateral stresses decrease below the initial K_0-stress state in both cases. Therefore, the shear stresses and the friction between soil and pile shaft is reduced. With increasing penetration depth the radial stresses above the pile tip stay nearly constant at around $\sigma_r \approx 10\,\mathrm{kN/m^2}$.

The decrease of the radial stresses can be explained by the cyclic shearing which leads to a strong densification of the soil around the pile. As a result, soil from greater distances has to move towards the pile to fill the free space. A funnel results at the ground surface. This continuous movement causes the radial stresses to decrease approximately down to the active earth pressure.

6 PILE DRIVING

6.1 Parameters

The dimensions and the mass of the pile are the same than for the vibratory driven pile. The relative density of the soil is again $I_D = 0.45$ and $I_D = 0.8$.

The pile is driven into the ground by a drop hammer with a mass of $m = 10\,\mathrm{t}$. The falling height is $h = 1\,\mathrm{m}$ and the frequency is chosen to 60 hits per minute. A total of 30 hits is modelled. For a more efficient calculation just one discrete hit is modelled. With this calculation, the time dependent distribution of the contact force between the drop hammer and the pile, respectively, the driving cushion is determined. This contact force is applied directly on the pile as a nodal force F_d for the actual modelling of the pile driving process. The modelling of the discrete hit is carried out for a pile penetration depth of $t = 2.0$ m. A linear elastic material behaviour is taken into account for the pile and the drop hammer. The modulus of elasticity

of the pile corresponds to a concrete C30/37 and the modulus of the drop hammer to steel. Furthermore, the modulus of elasticity for the top two element rows of the pile is reduced to the stiffness of wood. With this reduction the effect of the driving cushion is taken into account. Figure 15 shows the time dependent distribution of the contact force between pile and drop hammer as well as the idealised contact force distribution for the calculation.

6.2 Results

Figure 16 shows the vertical settlements of the ground surface during pile driving for different penetration depths and relative densities. At the beginning of penetration, the medium dense soil moves upwards in a distance of 1–3D from the pile. This can also be observed for the dense soil. The maximum heave is 2 cm. Settlements occur only directly at the pile. With increasing penetration the settlements increase and the

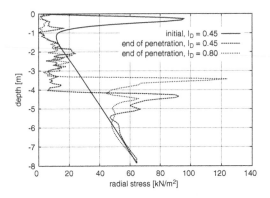

Figure 14. Calculated radial stresses along the pile after vibratory pile driving.

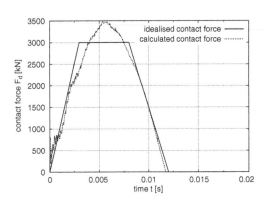

Figure 15. Calculated and idealised contact force between pile and drop hammer.

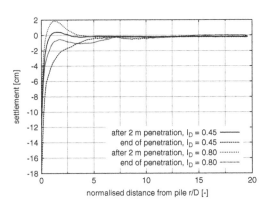

Figure 16. Vertical surface displacements during pile driving for different penetration depths and relative densities.

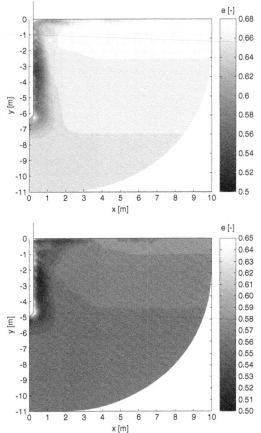

Figure 17. Calculated void ratio distribution after pile driving in a medium dense soil (top) and a dense soil (bottom).

ground surface heave turns into settlements for both relative densities. After 30 hits the settlements for the medium dense soil look like a funnel. In contrast to this, the settlements for the dense soil still show a peak in a distance of $2D$ from the pile. The influence of settlements due to pile driving is about $6D$ for both cases.

The soil close to the pile is compacted due to lateral displacement of the soil. Figure 17 shows the calculated void ratio distribution after 30 hits for both relative densities. The medium dense soil is compacted in a region of approximately 2–$2.2D$. The influence of compaction is a bit larger for the dense soil, reaching 2.2–$3D$. The soil is pushed aside due to the penetration of the pile. Additional compaction due to cyclic shearing, which could be observed for the vibratory pile driving, is lower in this case. Therefore, the displaced soil volume has to be reflected in a corresponding change in porosity Δn:

$$V_{pile} = \pi \cdot r^2 \cdot l = \pi \cdot 0.25^2 \cdot 6.5 = 1.28 \text{ m}^3$$

$$V_{void} \approx \pi(r_1^2 - r_2^2)l\Delta n$$

$$= \pi(1.25^2 - 0.25^2)7.0 \cdot 0.04 = 1.32 \text{ m}^3$$

This estimation for the medium dense soil shows that the compaction comes mainly from the displacement of the soil. The volume change of the soil is slightly larger than the volume of the penetrated pile. The additional compaction results from some cyclic shearing due to the drop impulse.

The change in void ratio for different distances from the axis of symmetry is shown in Figure 18 for the medium dense soil. A uniform void ratio distribution over depth is not reached. The largest compaction can be found in about half the penetration depth of the pile. In a distance of $4D$ from the pile the soil is compacted only in the upper 50 cm due to surface waves.

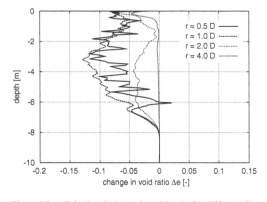

Figure 18. Calculated change in void ratio for different distances to axis of symmetry after pile driving (medium dense soil).

Figure 19 shows the radial stresses as a contour plot after 30 hits for both relative densities. It can be noticed, that the radial stresses decrease in the near area of the pile shaft. Figure 20 shows the radial stresses directly along the pile after 30 hits. Additionally, the initial radial stress state due to static load of the pile is illustrated. During the driving process, high stress peaks occur at the pile tip. The stresses are more than two times larger than the peak stresses due to static loading. For the initially dense soil the peak radial stresses are even three times larger. The stress increase reaches down to $3D$ beneath the pile tip for the medium dense soil and spreads out in radial direction for about $4D$. For the dense soil the radial stresses reach even further down and spread out further up to $5D$ (compare Figure 19). At the pile shaft the radial stresses decrease below the initial K_0 state. This can also be seen in Figure 21, which shows the radial stresses for a cross section in 4 m depth for both relative densities. Close to the pile, the radial stresses are about $\sigma_r = 13 \text{ kN/m}^2$ in both cases. In greater distance instead the stresses

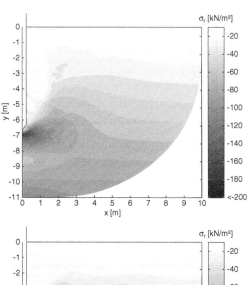

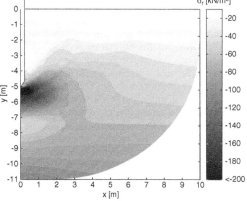

Figure 19. Calculated radial stresses after pile driving for a medium dense soil (top) and a dense soil (bottom).

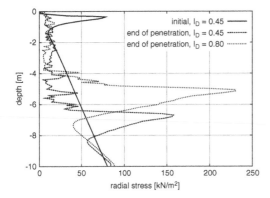

Figure 20. Calculated radial stresses along the pile after pile driving.

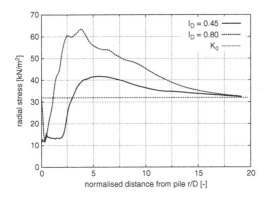

Figure 21. Calculated radial stresses after pile driving in a depth of 4 m.

increase above the initial K_0 state, as like at the pile tip lateral stress interlockings occur. After reaching a maximum value, the radial stresses decrease with increasing distance from the pile to the initial K_0 state. Again the peak value for the dense soil is larger than for the medium dense soil.

The responds of the soil state (i.e. void ratio and stress) during pile driving can be seen as a mixture of the responds during pile jacking and vibratory pile driving. In the near surrounding of the pile the soil is sheared, as mentioned above. This shearing results from the drop impulse and is not continuous and harmonic like the shearing due to vibratory pile driving. In greater distance, the soil is compacted only by compression just like in pile jacking. Therefore the radial stresses increase above the initial K_0 state.

7 COMPARISON

A quantitative comparison of the results of the three presented installation methods has to be handled with

caution because of the different boundary conditions. First, the pile diameter of the jacked pile is smaller than the diameters of the driven pile and the vibratory driven pile. Second, the energy used to install the piles are different in all three cases. Nonetheless, it is possible to compare the three methods qualitatively.

- Surface settlements:
  The largest surface settlements around the pile occur for the vibratory driven pile, even though the penetration depth is smaller than for the other two cases. This is caused by strong dynamic shearing between pile and soil. Even in a dense state a funnel of settlements can be observed. In the case of pile driving surface heave close to the pile is visible after the first few meters of penetration for the medium dense and the dense soil. Further penetration leads to increasing settlements which are smaller than for vibratory driven piles. Pile jacking causes the smallest settlements. Qualitatively, the shape of the ground surface settlements is identical to the settlements of pile driving. Especially the same characteristic for the dense soil can be seen in Figures 4 and 16.
- Change in void ratio:
  Vibratory pile driving causes the highest reduction in void ratio and accordingly the highest densification of the surrounding soil. Along the pile shaft, the minimum void ratio is reached by the end of vibration for the medium dense soil as well as for the dense soil. In case of pile driving the densification is large as well, but the void ratio by the end of the penetration process does not reach its minimum value. The densification along the pile shaft is not as uniform as for the vibratory pile driving. For both relative densities the soil is compacted. For the jacked piles instead the change in void ratio depends significantly on the initial relative density. In the case of initially loose soil the soil is densified as well, but much less than for pile driving or vibratory pile driving. In case of initially dense soil on the other hand, an increase in void ratio close to the shaft can be observed with a densification in further distance from the pile. The soil receives a change in void ratio mainly up to 2–2.5D for vibratory pile driving, 2.2–3D for pile driving and 5D for pile jacking.
- Radial stresses at the pile shaft:
  Pile jacking leads to an increase in radial stresses at the pile shaft which is higher for dense soils than for loose soils. The stress increase reaches up to a distance of more than 30D from the pile axis where the increase is still around 10% of the increase right at the pile shaft. The two dynamic driving methods instead cause a reduction of the radial stresses close to the shaft. The reduction is limited to the soil within a radius of 2–4D around the pile and is higher for the vibratory driven pile than for the driven pile. Outside this area of stress reduction a stress increase

can be observed for the driven pile while for the vibratory driven pile the stresses stay around the initial K_0-value. The change in radial stress spreads out further than the change in void ratio.

• Radial stresses at the pile toe:
All installation methods produce high horizontal stresses in the region right below the pile toe. The highest stress peaks occur below the jacked pile with the highest increase in dense sands. Smaller stress peaks can be observed for the driven pile and the smallest peaks occur for the vibratory driven pile.

All installation methods show significant differences concerning radial stresses, void ratio and surface settlements during and after pile installation.

From the numerical results it can be stated that the radial stresses depend highly on the degree of densification during the installation process. In all cases, a certain amount of soil is pushed aside by the pile. In the case of the jacked pile there is only a slight densification of the surrounding soil. In dense sand there is even a volume increase due to dilation of the soil close to the pile. Because of the monotonic nature of the installation method the soil is not able to rearrange its grain structure and as a result high stress interlockings occur around the pile. In case of the vibratory pile driving instead, the soil close to the pile undergoes strong cyclic and dynamic shearing which leads to a higher densification of the soil. As a result, the soil is able to relax and the radial stresses decrease. The installation method pile driving can be seen as a mixture of pile jacking and vibratory pile driving. The soil is compacted close to the pile shaft due to small cyclic shearing. In this compacted area the soil grains are allowed to rearrange themselves like for vibratory pile driving. As a result, the radial stresses at the pile shaft decrease. In a larger distance from the pile the shearing is not strong enough to compact the soil. Therefore the soil is just compacted by compression like for the jacked pile and the radial stresses increase.

The radial stress decrease close to the pile shaft during dynamic pile installation has already been observed in various field tests. Ng et al (1988) and Axelsson (1998) for example both investigated the earth pressures acting on a pile after installation by directly installing pressure gauges into the shaft of prefabricated full-scale concrete piles which had been driven into sand. Consistent to this study they measured very low effective earth pressures on the shaft after the end of pile driving of around $K = 0.07$–0.3, although pore pressures had already been equalised.

8 CONCLUSIONS

The numerical analysis of the three presented installation methods show significant differences of the soil state around the pile after installation. For the jacked pile, the soil is only slightly densified or even acts dilatant which results in high radial stresses around the pile. For the dynamic installation methods the densification is much larger, due to the cyclic and dynamic shearing which results in lower radial stresses around the pile shaft.

Generally, it can be stated that if the volume decrease caused by densification of the soil is higher than the volume of soil pushed away by the pile, a relaxation of the radial stresses at the shaft can be expected. If instead the densification is smaller than the displaced soil volume, the radial stresses will increase.

At the pile toe all installation methods cause high lateral stresses due to stress interlocking, which are highest for the jacked pile and lowest for the vibrated pile.

According to these results the generally observed higher bearing capacity of driven piles with respect to vibrated piles can be explained with higher stress interlocking at the pile toe. Further, the higher stiffness and capacity of jacked piles with respect to dynamically installed piles can be explained with the higher radial stresses at the pile shaft and the pile toe.

REFERENCES

Axelsson, G. (1998). Long-term increase in shaft capacity of driven piles in sand. *Proc. IVth International Conference on Case Histories in Geotechnical Engineering.* St. Louis, Missouri.

Cudmani, R. (2001). Statische, alternierende und dynamische Penetration in nichtbindigen Böden. *Karlsruhe:IBF.* Heft 152.

Deeks, A.D., White, D.J. & Bolton, M.D. (2005). The comparative performance of jacked, driven and bored piles in sand. *Proc. XVIth International Conference of Soil Mechanics and Geotechnical Engineering.*, Osaka.

Dierssen, G. (1994). Ein bodenmechanisches Modell zur Beschreibung des Vibrationsrammens in körnigen Böden. *Karlsruhe:IBF.* Heft 133.

Grabe, J., Kelm, M. & Mahutka, K.-P. (2003). Zur numerischen Modellierung der Verdichtung rolliger Böden. *Bodenverdichtung,* Heft 5. Workshop in Hamburg, 23 September 2003: 83–107.

Grabe, J. & König, F. (2004). Zur aushubbedingten Reduktion des Drucksondierwiderstandes. *Bautechnik.* 81(7): 569–577.

Grabe, J. & Mahutka, K.-P. (2005). Finite Elemente Analyse zur Vibrationsrammung von Pfählen. *Bautechnik.* 82(9):632–640.

Gudehus, G. (1996). A comprehensive constitutive equation for granular materials. *Soils and Foundation.* 36(1):1–12.

Hagen, C. (2005). Wechselwirkungen zwischen Bauwerk, Boden und Fluid bei transienter Belastung. *Band 1. Hamburg: Schriftenreihe des Arbeitsbereiches Modellierung und Berechnung der TU Hamburg-Harburg.*

Hartung, M. (1994). Influence of the installation process on the bearing capacity of piles in sand. *Braunschweig: IGB-TUBS.* No 45.

Herle, I. (1997). Hypoplastizität und Granulometrie einfacher Korngerüste. *Karlsruhe:IBF*. Heft 142.

Kelm, M. & Grabe, J. (2004). Numerical simulation of the compaction of granular soils with vibratory rollers. In Triantafyllidis (ed), *Cyclic behaviour of soils and liquefaction phenomena*. Rotterdam: Balkema. 661–664.

Mahutka, K.-P. & Grabe, J. (2005). Numerical investigations of soil compaction and vibration propagations due to strong dynamic excitation. *Electronic Proceedings of Joint ASCE/ASME/SES Conference on Mechanics and Materials – McMat 2005*. Baton Rouge (USA).

Ng, E.S., Tsang, S.K. & Auld, B.C. (1988). Pile foundation: The behaviour of piles in cohesionless soils. *Federal Highway Adm. Report FHWA-RD-88-081*.

Niemunis, A. & Herle, I. (1997). Hypoplastic model for cohesionless soils with elastic strain range. *Mechanics of cohesive-frictional materials*. 2:279–299.

Rodger, A. A. & Litlejohn, G. S. (1980). A study of vibratory driving in granular soils. *Géotechnique*. 30:269–293.

von Wolffersdorff, P.-A. (1996). A hypoplastic relation for granular materials with a predefined limit state surface. *Mechanics of Cohesive-Frictional Materials*. 1:251–271.

Numerical Modelling of Construction Processes in Geotechnical Engineering for
Urban Environment – Triantafyllidis (ed)
© 2006 Taylor & Francis Group, London, ISBN 0 415 39748 0

Bored and screwed piles, continuum and discontinuum approaches

A. Schmitt
Björnsen Consulting Engineers (BCE), Koblenz, Germany

R. Katzenbach
Institute and Laboratory of Geotechnics, Darmstadt University of Technology, Germany

ABSTRACT: The efficiency of piling methods, where soil is retrieved and methods that make use of soil displacement during drilling, results from the interaction between the performance of drilling equipment and the modifications created by the drilling process to soil characteristics. However, little is known yet on what is really happening during the installation of a pile. In Distinct Element models the soil is modelled by an assembly of particles where every single particle is interacting with neighbours by normal and shear forces as well as by transmitting moments. Every single particle can lose contact with neighbouring particles and form new contacts depending on forces acting on the particle and its environment. From the theoretical point of view many aspects of pile installation could therefore be modelled more accurate by using the 3D Distinct Element Method (DEM) than by approaches based on the Finite Element Method (FEM). Both methods have been used exemplarily in the following to assess their capabilities and limitations to simulate pile installation and pile performance.

1 INTRODUCTION

Pile bearing behaviour has intensively been investigated by means of the classical continuum mechanics utilizing the Finite Element Method (FEM). This includes numerical simulations mainly of bored piles by elasto-plastic approaches. These approaches are able to provide an overall satisfying result to estimate bearing capacities of piles in sand (Katzenbach et al. 2003) as well as in clay (e.q. Katzenbach et al. 2002, Schmitt et al. 2002). As many piles are installed with an outer casing, the change of soil characteristics caused by the installation process is generally estimated to be small. Back-analysing pile load tests and re-adjusting parts of the numerical model generally provides a complex but powerful tool for the design and optimization process of pile foundations. Very efficiently this knowledge is used to predict and optimize even more difficult foundation structures such as Combined-Pile-Raft Foundations (Fig. 1). Obviously, in these cases the simulation of the complex and strong interaction between different foundation parts is successful, because the processes of pile installation and the change to soil characteristics are known and relatively well understood.

When it comes to simulate the horizontal soil displacement during the drilling process of typical soil displacement piles (Fig. 2), simulations based on elasto-plastic approaches can only be used when highly sophisticated material laws are used or when the

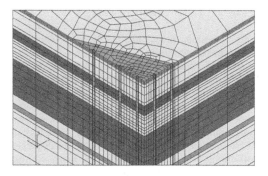

Figure 1. Part of a 3D FE-Model for the high-rise building ERGO-Tower, Mannheim, Germany.

change of soil density during pile installation is monitored and the simulation parameters are later adjusted throughout the simulation in dependency of plastic deformation and density change in the adjacent soil.

Depending on the diameter of the final pile and the amount of soil displacement achieved during installation, a second problem that frequently is encountered in respect of large displacement FE analysis, is the ill-conditioning effect due to severely distorted elements. In areas of intense deformation, straining sometimes can lead to elements with large aspect ratios. Such distortion degrades the accuracy of the analysis and often leads to an early stop of simulations. Only in some

Figure 2. Soil displacement piles (examples).

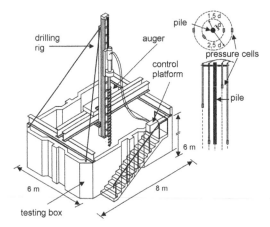

Figure 3. Testing box for different pile installation methods in loose and medium dense sand, pile length 3 m.

cases remeshing procedures can be implemented to overcome these problems.

Although simulations based on FEM have proved to be a reliable tool for predicting the pile resistance of bored piles, they can not entirely clarify the question what really happens when soil particles are partly retrieved and displaced in drilling and screwing processes during the pile installation. The capability of FE simulations to simulate fully or partly displaced soil during drilling or more complex structures that include these types of piles are presently restricted as the discontinuous and inhomogeneous state of granular materials can only insufficiently be considered in large scale.

2 INFLUENCE OF PILE INSTALLATION

The influence of pile installation methods on soil characteristics has intensively been investigated for different types of three meter long soil displacement piles and continuous flight auger piles within the European research project TOPIC (Technically optimized pile concept). Many aspects and results of this work are summarized in Schmitt (2004). The model tests were performed within a large testing box (Fig. 3) filled with sand. All piles were equipped with common measuring devices. During pile installation the

change of horizontal stresses within the adjacent soil was recorded. The soil density before and after pile installation was investigated by CPT tests.

In the following, numerical simulations based on FEM are presented which were used to predict the pile resistance of the model piles. In a further step, numerical particle based simulations are introduced which demonstrate the possibilities and present limitations of the discontinuum approach.

3 CONTINUUM APPROACH

Since the piles introduced before are cylindrical, up to a certain degree isolated and axially-loaded, the models considered within this contribution are assumed to be axisymmetric.

3.1 FE-mesh

The discretisation of the soil and concrete is carried out using four-noded elements. The spatial discretisation of the soil includes about 630 elements. The geometry of the model with the model's dimensions is shown in Figure 4. The pile has a conical shaped base and a shaft diameter corresponding to test pile dimensions. The model consists of up to 6 soil layers which allow to vary densities and initial properties with depth to fit the real test conditions.

3.2 Constitutive models

Within this study the Drucker Prager/Cap model as implemented in the FE-code ABAQUS has been used. In order to capture the phenomenon of densification (plastic volume contraction due to isotropic stressing), a yield cap is used to close the yield surface of the classical Drucker/Prager cone surface (Fig. 5).

A more detailed description of the soil model and how this model is used for other purposes is given in Katzenbach et al. (2003). The hardening rule for sand was derived from triaxial testing under hydrostatic loading and the measurement of corresponding plastic strains (Fig. 6).

3.3 Shearzone between pile and soil (interface elements)

The mobilisation and magnitude of the ultimate shaft friction is controlled by the behaviour of a thin zone, close to the interface between pile shaft and soil. This shear zone is subjected to plastic straining similar to a simple shear mode. Depending on the soil state after pile installation and the soil characteristics within the shear zone, the soil can exhibit dilative or contractive behaviour. The tendency of the shear zone to change volume is influenced by the "elastic" behaviour of

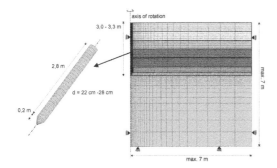

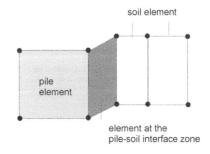

Figure 4. Axisymmetric FE-model.

Figure 7. FE-modeling of pile-soil interaction zone (shear-zone).

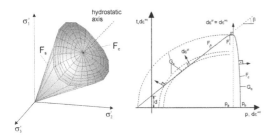

Figure 5. Drucker-Prager/Cap yield surfaces in the principal stress state and in the deviatoric stress plane.

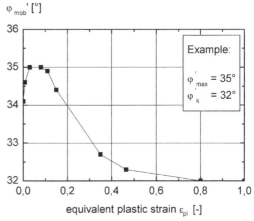

Figure 8. Friction angle φ'_{mob} in dependency of equivalent plastic strain.

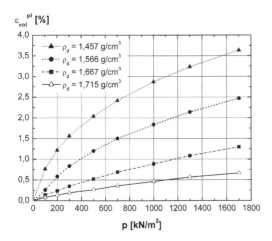

Figure 6. Cap-hardening for the test-box sand.

the surrounding soil which in a certain way, acts as a constraint.

Based on laboratory shear box tests, the shearzone between pile and sand can be extended to a small restricted area adjacent to the pile shaft either than exactly being placed at the materials contact surfaces. Therefore, in the present study a special interface element has been considered as depicted in Figure 7.

The element is located between the pile and the soil with a width chosen 20% of the pile diameter. The classical Mohr-Coulomb constitutive model is customized to account for progressive mobilization (hardening and softening) of the soil shear strength parameter φ' and the dilatation angle ψ within this interface element. This approach is similar to those reported in Potts & Zdravkovic (1999) and Sarri (2001). Each yield criterion (Mohr Coulomb line) can be considered as the current yield surface during a process of hardening and softening. To define the material behaviour, the equivalent plastic strain ε^{pl} was selected as the state variable (Fig. 8). The mobilized shear strength parameter φ' is expressed as function of the equivalent plastic strain magnitude (Hibbitt et al. 1998):

$$\varepsilon^{pl} = \sqrt{\frac{2}{3}\varepsilon_{ij}^{pl}\varepsilon_{ij}^{pl}} \tag{1}$$

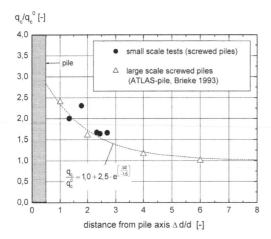

q_c/q_c^0 [-]

distance from pile axis Δ d/d [-]

$$\frac{q_c}{q_c^0} = 1,0 + 2,5 \cdot e^{\left(\frac{\Delta d}{-1,6}\right)}$$

• small scale tests (screwed piles)

△ large scale screwed piles
(ATLAS-pile, Brieke 1993)

pile

Figure 9. Decrease of cone resistance in dependency of the distance from the pile shaft.

The angle of dilatancy is calculated according to Bolton (1986), resulting from the mobilized friction angle ϕ'_{mob} and the friction angle at critical state ϕ'_{crit}.

$$\psi = \frac{\left(\phi'_{mob} - \phi'_{crit}\right)}{0.8} \qquad (2)$$

3.4 Accounting for soil displacement

During the installation of screwed piles, soil is displaced horizontally, densified and prestressed. To quantify this relationship cone penetration testing (CPT) before and after pile installation was used. Figure 9 shows the relation between $q_{c,0}$, representing the measured cone resistance before pile installation and q_c, representing the cone resistance measured after pile installation.

The measured cone resistance q_c after pile installation was used to assess the relative Density D_r by correlations given in DIN 4094. The value of D_r was used to assess the distribution of ϕ', E and ψ in dependency of the distance from the pile shaft and the medium stress level p', using equation (3) according to Bolton (1986) and equation (2) according to Van Impe (1986):

$$\phi' = 3 \cdot \left[D_r \cdot \left(5,4 - \ln\left(\frac{p'}{p_{atm}}\right) \right) - 1 \right] + \phi'_{crit} \qquad (3)$$

with $p_{atm} = 100 \, kN/m^2$

$$E = \alpha_i + \beta_i \cdot q_c \qquad (4)$$

with $\alpha_i = 0$, $\beta_i = 3$ for $q_c \leq 5 \, MN/m^2$ and $\alpha_i = 7.5$, $\beta_i = 1.5$ for $5 \, MN/m^2 \leq q_c \leq 30 \, MN/m^2$. The angle of

dilatancy was calculated according to equation (2) with $\phi'_{mob} = \phi'$.

The modified horizontal stress state due to soil displacement was simulated by adjusting the lateral earth pressure coefficient during the generation of the initial stress state:

$$K = \frac{\sigma_h}{\sigma_v} \qquad (5)$$

The values considered within the simulation were in the range from K = 1.0 to 1.2 and do correspond to values given in Aboutaha et al., 1993 and Rackwitz, 2003 for different types of model tests with soil displacement piles, accounting for the altered stress state around the displacement pile.

Parameters of the sand being used within the numerical modeling are summarized in Schmitt (2004).

3.5 Simulation procedure

Pile installation and testing has been modeled by the following simulation steps:

– initial stress state
– excavate soil according to pile dimensions
– allow soil to expand or relax due to liquid concrete pressure
– reduce concrete pressure to zero and install concrete pile elements with final stiffness
– apply gravity loading on the concrete pile elements
– subsequent loading of the pile (displacement driven)

For the initial horizontal stress distribution a K_0 stress field has been generated for bored piles without soil displacement. The horizontal stresses $\sigma_x = \sigma_y$ are derived from the overburden stresses σ_z by applying the factor of K_0.

$$\sigma_z^{(z)} = \int_0^z \gamma' dz \qquad (6)$$

$$\sigma_x^{(z)} = \sigma_y^{(z)} = K_0 \cdot \sigma_z \qquad (7)$$

K_0 was chosen to be constant for all single soil layers according to

$$K_0 = 1 - \sin \phi' \qquad (8)$$

Simulations where soil displacement is considered start with an already installed pile and an adjusted K value as stated in section 3.4.

3.6 Results of numerical simulation

The load applied at the pile head is split into shaft resistance and base resistance and is compared with

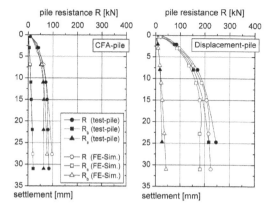

Figure 10. Comparison FE-simulation and test pile behaviour (R: pile resistance, R_b: base resistance, R_s: shaft resistance).

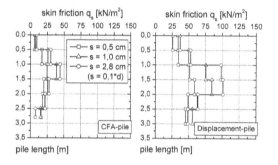

Figure 11. Skin friction along the pile shaft (numerical simulation).

the experimental results. In Figure 10, the results of numerical simulations are exemplarily given for a conventional CFA pile and a soil displacement pile. The increase in horizontal stresses due to the installation of a displacement pile were considered within the numerical model by a value of K being equal to 1.2, according to stress measurements during the pile installation within the testing container of model tests. The initial soil density around the two piles was comparable with a cone resistance (CPT) of: $2{,}5\,\mathrm{MN/m^2} \le q_c \le 5\,\mathrm{MN/m^2}$.

In Figure 11 the distribution of skin friction from FE-simulations is shown. The variation of values is influenced by the variation of soil density at the beginning of the simulation.

However, the graphs show that a soil displacement pile is leading to an increase of pile shaft capacity of more than 100% in the lower pile section. The skin friction obtained for CFA piles within the numerical simulations corresponds to:

$$q_s = \alpha_b \cdot q_c \qquad (9)$$

With $\alpha_b = 0.008$ and q_c in $\mathrm{MN/m^2}$ as stated in DIN 1054. The higher values of skin friction for the displacement piles can be derived in the same manner when relying on cone penetration results evaluated from tests performed after the installation of a displacement pile. The procedure of using CPT also after pile installation becomes even more important when installing displacement piles in layered soils and when using a soil displacement that is not kept constant over the entire pile length.

4 DISCONTINUUM APPROACH

Simulations based on FEM have proved to be a reliable tool for predicting the pile resistance of CFA and soil displacement piles. However, they cannot entirely clarify the question what really happens when soil particles are partly retrieved and displaced in drilling and screwing processes during the pile installation. Also the capability of FE simulations to optimize the shape and dimension of drilling tools is restricted as the discontinuous and inhomogeneous state of granular materials can only insufficiently be considered.

For this reason, numerical particle dynamic simulations have been performed with the long term objective to improve our understanding of changing soil characteristics in the soil during drilling of CFA piles and soil displacement piles and to allow for a later optimization of drilling tools. The DEM describes the behaviour of an assembly of particles. It is based on an explicit local equilibrium algorithm in which all contacts are monitored and the translation of each particle is simulated. These simulations do also account for breaking of existing contacts and forming of new contacts throughout the entire simulation. It is therefore possible to consider cutting as well as drilling processes. The micromechanical behaviour of granular materials within the framework of DEM is determined by how discrete grains are arranged in space and by what kind of interactions are operating among them. First simulations were performed using the DEM to evaluate it's capability.

4.1 Numerical concept

The constitutive model utilized within this study to describe the particle interaction is an elastic-plastic contact model which is characterized by normal and shear springs at the particle contacts. The Particle Flow Code (PFC) which is used within this study was developed by Cundall (1971) and is based on the DEM but restricted to modeling rigid spherical particles (Cundall & Strack 1979).

Numerical simulations are therefore based on the following assumptions:

– All particles are treated as undeformable rigid spheres (compare Fig. 12)

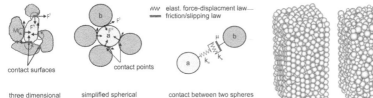

three dimensional
sand grains (cross-
section)

simplified spherical
geometry

contact between two spheres

Figure 12.   Simplified particle shape and contact behaviour.

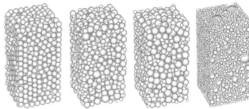

Figure 13.   Example for numerical triaxial testing with spherical particles.

- Contacts only occur at points
- The behaviour at the contacts is ruled by a soft-contact approach in which the rigid particles are allowed to overlap one another at their contact points
- The magnitude of overlap is related to the contact forces by a force displacement law
- The overlap is small compared to the particle size
- All particles are simulated by spheres, overlapping spheres which form a permanent unbreakable clump or by connected (glued) spheres which form a breakable cluster

Slip or plastic deformation occurs when the shear force between two particles exceeds, in comparison to the normal force, a certain level which depends on the dimensionless friction coefficient μ. Every contact is checked for slip conditions by calculating the maximum allowable shear contact force $F_{\max}^s$. In case of

$$\left| F_i^s \right| > F_{\max}^s \tag{10}$$

slip will occur during the next calculation cycle and F_i^s is set to $F_{\max}^s$. A comprehensive overview and detailed calculation scheme is provided by the aforementioned authors.

4.2   Sand investigated

To reduce computational time, all grain diameters of an existing grain size distribution are usually enlarged for the numerical simulation by multiplying (upscaling – USC) the actual diameter with a user defined factor. A steeper slope of the grain size distribution of the simulated sand enables to reduce the upscaling factor from 50 to 100 for different numerical samples of sand to about 40. The effect of particle size distribution and particle friction coefficient on the macroscopic behaviour, especially the macroscopic friction angle, has been investigated by performing triaxial tests on particle assemblies containing spheres as well as clumped particles, representing a more realistic shape of sand grains (Fig. 13). A detailed description of the particle generation process and parameter studies including triaxial and oedometric testing are given in Schmitt & Katzenbach (2003).

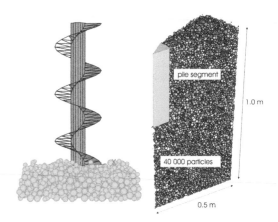

Figure 14.   Simulation of the drilling process and segment-model to simulate the soil-structure interaction of a single pile.

4.3   Pile installation, drilling tools and pile testing

The numerical simulations for studying different piling issues were conducted under gravity loading and real density values. In Figure 14, first models are depicted which were aiming to investigate the simulation capabilities.

Results of theses model tests on a conventional PC were:

- numerical models should be limited to small pile or drilling sections
- simulating the drilling process within an assembly of particles requires too long simulation times and requires certain simplifications also in geometry
- using a particle size which allows for short simulation time leads to unrealistic results, especially under compression loading below the pile base

These results were leading to a model which is presented in Figure 15, a one quarter-shaft section of a pile. In a first stage this model was used to simulate pile installation without horizontal displacement. Due to symmetry, the model size was reduced to one quarter.

The model consists of a cube generated by non-spherical particles and a cylindrical pile shaft that is

252

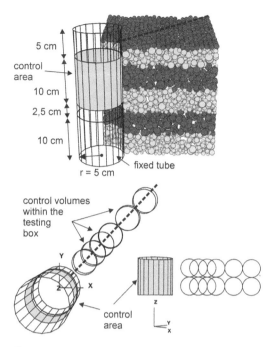

Figure 15. Numerical particle model for a pile section.

placed at one corner of the cube by replacing the particles in this area. The shaft contact area within the soil body is defined by a control area (height 10 cm) where vertical and horizontal forces are monitored. Due to possible boundary effects, upper and lower areas of the pile segment are neglected (height 5 cm and 2,5 cm). Pile loading is modelled by a prescribed vertical translation. To reduce the effects of the model-bottom the lower shaft part is moving into a fixed tube which has a slightly larger diameter and which does not cause any additional friction.

To simulate a deeper soil strata, the top of the model is replaced by a flexible membrane consisting of a layer of spheres which ensure a permanent vertical loading of 100 kN/m² . The process is shown in Figure 16.

The shaft resistance is calculated by the vertical forces on the control area by equation (11).

$$q_s = F_z / \left(\frac{1}{4} \cdot 2 \cdot \pi \cdot r \cdot h \right) \tag{11}$$

An overview on physical parameters and experimental conditions for the generation procedure used in the following is provided in Table 1.

Characteristically, stress values and porosity are recorded throughout the sample generation procedure by spherical control volumes within certain locations inside the model (Fig. 15).

In a second step, the model was modified to simulate the influence of horizontal soil displacement.

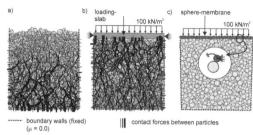

Figure 16. Model after generation procedure (a), during the loading (b) and after the membrane installation (c).

Table 1. Simulation parameters for pile simulation (DEM).

Parameter/Property	Selected value
Number of particles	max. 11 296
Radii of particles	3.5–8.5 mm
Overlapping of particles	0.65 r
Particle density	2 650 kg/m³
Porosity (n)	0.346
Gravity loading	9.81 m/s²
Normal spring constant of particles (k_n)	4.0×10^6 N/m
Stiffness ratio (shear/normal)	1.0
Normal spring constant of walls	4×10^7 N/m
Shear spring constant of walls	0.0 N/m
Shaft friction coefficient $\mu_{S,w}$	tan δ'
Friction coefficient (particles/wall) μ_w	0.0

δ': friction angle between soil and structure.

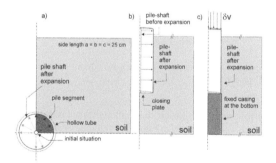

Figure 17. Simulating process for soil displacement.

As shown in Figure 17, the pile shaft within the model was expanded from an initial diameter up to its final size. The initial diameter before expansion was chosen according to the amount of desired displacement.

The measured shaft friction for different soil displacements is shown in Figure 18, in dependency of the vertical shaft displacement for a medium dense packing. Depending on the shaft friction coefficient, the shaft resistance within the model can be increased up to a horizontal displacement of up to 50%. If a higher

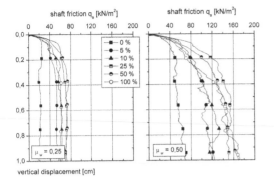

Figure 18.   Measured shaft friction.

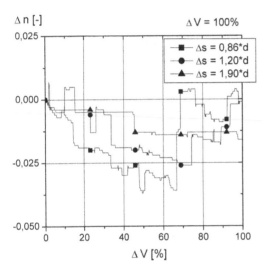

Figure 19. Change of porosity during horizontal displacement.

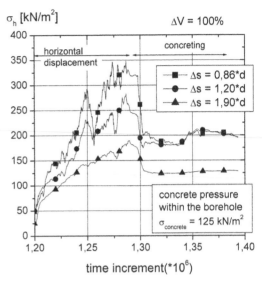

Figure 20.   Stresses during the displacement and concreting stage in different control volumes.

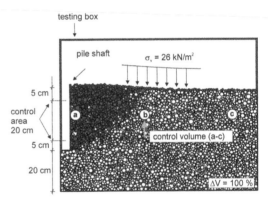

Figure 21.   Modified particle model (length: 1 m, width: 0,75 m, depth: 0,05 m).

displacement is chosen, the shaft resistance is decreasing again. As shown in Figure 19 (change of porosity), the particle assembly is compacted at the beginning of the simulation whereas for a very high displacement the porosity is increasing due to dilatancy effects.

The measured horizontal stresses in different distances Δs from the pile shaft are given in Figure 20. The monitored stresses clearly indicate when the displacement is completed and to which level horizontal stresses decrease again during the concreting stage. Although stresses sharply decrease, they remain higher than the initial stress state and do contribute to a higher shaft friction q_s.

A second model according to Figure 21 was investigated to reduce the effect of the model boundary walls. The testing box with a size of 1 m × 0.75 m × 0.05 m was filled with non-spherical particles. The simulation process was similar to the aforementioned test.

The cylindrical pile was replaced by a rectangular one. The overburden pressure was reduced to about 25 kN/m². In Figure 21, the situation after a 100% pile volume displacement is shown. Dark colored particles represent higher particle velocities during the simulation than white colored ones. The particle size and distribution has not been changed compared to the test with the cubic specimen.

Figure 22 shows the distribution of contact forces after the displacement procedure. It is clearly seen that forces due to expansion are transmitted by preferred force chains straight to the boundary walls of the model. Monitored porosities and stresses are shown in Figure 23 and Figure 24, respectively.

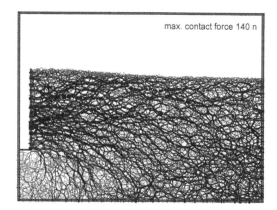

max. contact force 140 n

Figure 22. Distribution of contact forces after a horizontal displacement of about 100% pile volume.

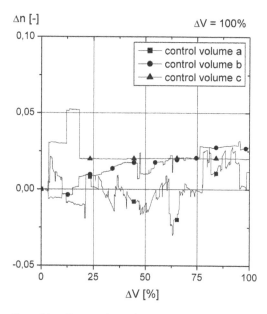

Figure 23. Change of porosity during the pile installation due to horizontal displacement of about 100% in a rectangular testing box.

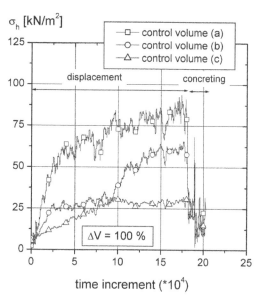

time increment (*10⁴)

Figure 24. Measured stresses within the particle sample.

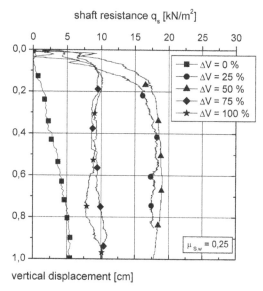

vertical displacement [cm]

Figure 25. Shaft resistance for different pile displacement.

Figure 23 indicates that the scattering of measured porosity values is high and that within the control volume densification is limited to a short phase at the beginning of the simulation. However, the obtained results and the scattering do also lead to the assumption that the chosen size of particles is apparently too large compared to the size of control volumes within the model. Results of Figure 24 are comparable to the results given in Figure 20, although the enlargement of the model has led to a doubtlessly higher decrease of stresses within the concreting stage,

revealing the influence of close boundary conditions within the cubic specimen.

The measured shaft resistance within the modified model is shown in Figure 25. The strong scattering is again a result of the model size and the limited number of contacts at the pile shaft which is leading to a strong influence of every single particle-shaft contact upon the measured result.

5 CONCLUSION

Despite the fact that distinct-element simulations of granular materials such as sand, allow the simulation of a variety of fascinating and interesting phenomena to solve questions related also to piling processes. Unfortunately, these methods remain to be very time consuming.

The application of a distinct element code to the modelling of soil structure interaction, the simulation of pile installation processes and the generation of load settlement curves for different pile types is continued to be pursued. Such meshless modelling with a comparatively small number of material parameters has shown to result in a realistic stress-strain response of soil and is capable of modelling cutting as well as drilling processes. Even when taking into account an exponential growing processor such as speed, the aim to model CFA and soil displacement piles in detail must be considered to be an long-term objective.

Although many aspects of pile installation can be considered within FE-simulations, they will remain a simplification of what really happens during the installation process of piles. A combination of both methods might show a way to overcome limitations of every single method.

REFERENCES

Aboutaha, M., De Roeck, G. and Van Impe, W.F., 1993. Bored versus displacement piles in sand – experimental study. *Deep Foundations on Bored and Auger Piles, (Ed.: Van Impe), Balkema, Rotterdam*, 157–162

Bolton, M.D., 1986. The strength and dilatancy of sand. *Géotechnique 36, No. 1*, 65–78

Brieke, W., 1993. Vergleich der Tragfähigkeit unterschiedlicher Pfahlsysteme. *Franki Grundbau GmbH, Neuss*

Cundall, P.A., 1971. A computer model for simulating progressive, large-scale movements in blocky rock systems. *Proc. Symp. Int. Soc. Rock Mech., Nancy 2, art.8*

Cundall, P.A. and Strack, O.D.L., 1979. A discrete numerical model for granular assemblies. *Géotechnique, 29*, pp. 47–65

Cundall, P.A., Jenkis, J.T. and Ishibashi, I., 1989. Evolution of elastic moduli in a deforming granular assembly. *Powders and Grains, Balkema, ISBN 90 6191 984 3*

DIN 4094, Juni 2002. Baugrund · Felduntersuchungen. *Normenausschuss Bauwesen (NABau) im DIN Deutsches Institut für Normung e.V.*

DIN 1054, Januar 2003. Baugrund · Sicherheitsnachweise im Erd- und Grundbau. *Normenausschuss Bauwesen (NABau) im DIN Deutsches Institut für Normung e.V.*

Hibbit, H.D., Karlsson, B.J. and Sorensen, 1998. *ABAQUS Theory Manual, Version 5.8*

Katzenbach, R., Schmitt, A. and Turek, J., 2002. Three-dimensional modelling and simulation of piled-raft foundations for high rise structures. *9th International Conference on Computing in Civil and Building Engineering, Taipei, Taiwan, 3–5 April 2002*

Katzenbach, R., Schmitt, A. and Turek, J., 2003. Assessing Settlement of High-Rise Structures by 3D Simulations. *Journal of Computer-Aided Civil and Infrastructure Engineering, Special Issue on Computing in Civil and Building Engineering, (submitted for review)*

Katzenbach, R., Schmitt, A., Furmanovicius, L., Nicholson, D., Dinesh, P., Powell, J.M., Skinner, H. and Hamelin, J.-P., 2004. Neue Techniken für Ortbetonverdrängungspfähle. *Vorträge zum 11. Darmstädter Geotechnik-Kolloquium am 18. März 2004. Mitteilungen des Institutes und der Versuchsanstalt für Geotechnik der Technischen Universität Darmstadt, Heft 68*, 99–116

PFC 1999. Theory and Background, ITASCA, www.itascacg.com

Potts, D.M. and Zdrackovic, L., 1999. Finite element analysis in geotechnical engineering theory. *Thomas Telford, ISBN 07-277-27532*

Rackwitz, F., 2003. Numerische Untersuchung zum Tragverhalten von Zugpfählen und Zugpfahlgruppen in Sand auf der Grundlage von Probebelastungen. *Veröffentlichungen des Grundbauinstitutes der Technischen Universität Berlin, Heft 32*

Sarri, H., 2001. Pali trivellati in sabbia soggetti a carico assiale: modellazione fisica e numerica. *Politecnico di Torino, Tesi di dottorato.*

Schmitt, A., 2004. Experimentelle und numerische Untersuchungen zum Tragverhalten von Ortbetonpfählen mit variabler Bodenverdrängung. *Mitteilungen des Institutes und der Versuchsanstalt für Geotechnik der Technischen Universität Darmstadt, Heft 70*

Schmitt, A. and Katzenbach, R., 2003. Particle based modeling of CFA and soil displacement piles. *4th International Geotechnical Seminar, Deep Foundations on Bored and Auger Piles, Ghent-Belgium, June 2–4*, 217–224

Schmitt, A., Turek, J. and Katzenbach, R., 2002. Reducing the costs for deep foundations of high rise buildings by advanced numerical modelling. *5th International Congress on Advances in Civil Engineering (ACE), September 2002, Istanbul*

Van Impe, W.F., 1986. Evaluation of deformation and bearing capacity parameters of foundations, from static CPT-results. *Fourth Int. Geotechnical Seminar – Field Instrumentation and In-Situ Measurements, Nanyang Technological Institute, Singapore, 25–27 November 1986*

Numerical Modelling of Construction Processes in Geotechnical Engineering for
Urban Environment – Triantafyllidis (ed)
© 2006 Taylor & Francis Group, London, ISBN 0 415 39748 0

Numerical simulation of construction-induced stresses around Rammed Aggregate Piers

D.J. White & H.T.V. Pham
Department of Civil, Construction and Environmental Engineering, Iowa State University, Ames, USA

K.J. Wissmann
President and Chief Engineer, Geopier Foundation Company, Inc., Blacksburg

ABSTRACT: This paper describes finite element analysis results that provide new insights into the effects of construction-induced stresses adjacent to isolated Rammed Aggregate Piers (RAPs). An elasto-plastic, strain-hardening constitutive model was selected to characterize the behaviors of the matrix soils and the RAP aggregate. The numerical model is verified by comparing the numerical predictions with the data obtained from full-scale, instrumented load tests performed on three isolated RAPs. Model parameters were estimated from in-situ and laboratory soil tests. The pier installation process was modeled as a cavity expansion problem with boundary conditions determined from field measurements. Interpretations of the numerical results focus on the development of the stress components in the matrix soil adjacent to the piers and the influence of the unique construction-induced stress path on the load-displacement curves and the load transfer mechanism along the piers.

1 INTRODUCTION

Over the past two decades, Rammed Aggregate Piers (RAPs) (also known as *Geopier*[TM] Soil Reinforcement) have been used as a cost-effective solution to provide support for shallow foundations, heavily-loaded slabs, large storage tanks, and a wide range of transportation infrastructures such as box culverts, embankments, and mechanically stabilized earth walls (Lawton & Fox 1994, Wissmann et al. 2000, Wissmann et al. 2002, White et al. 2003). In the United States, annual RAP construction costs will exceed $50 million in 2005. RAPs are constructed by compacting successive thin layers of base-course aggregate in pre-bored cylindrical cavities using a specially designed, beveled tamper. As a result of the high compaction energy, the aggregate in the cavity is compacted and expanded into the soil thus pre-stressing and pre-straining the matrix soil adjacent to the pier.

Finite element (FE) analyses are conducted in this study in an attempt to provide insights into the complex interaction between isolated RAPs and the matrix soil by modeling the unique construction-induced stress regime around the piers. Full-scale, instrumented load tests were performed on three test piers to validate the FE models and modeling parameters used in this study. The pier-soil interaction was investigated by studying the development of interface stresses in the matrix soil due to pier installation and loading, the deformation characteristics of the pier, and the load transfer mechanism within the pier and along the pier-soil contact.

2 INSTALLATION OF TEST PIERS

Figure 1 shows some of the equipment used for installing the RAPs. The installation process involves ramming successive thin layers of base-course aggregate in a drilled cavity using the special beveled tamper. As a result of the high compaction energy and the beveled shape of the tamper head, each layer of aggregate is forced downward and radially into the surrounding soil thus creating an undulatory pier shaft (Fig. 2).

Three isolated RAPs with drilling diameter of 0.76 m were installed at a test site in Neola, Iowa (USA) for the purpose of validating the FE models developed in this study. All test piers were capped with a 0.45-m thick, cast-in-place concrete cap. Test pier P_1 was 3.0 m long and was instrumented with four 23 cm diameter vibrating-wire total stress cells and a tell-tale reference plate. The tell-tale reference plate, which was located at 2.67 m from grade, consisted of a rectangular steel plate attached to two parallel threaded steel bars that were protected by PVC sleeves. Test piers P_2 and P_3 were installed to the depths of 2.74 m and 5.05 m, respectively. Both piers were instrumented with tell-tale reference plates (tell-tale depths of P_2 and P_3 were 2.44 m and 4.75 m, respectively). Load

Figure 1. RAP special beveled tamper and hydraulic rammer.

Rammed aggregate pier

Undulated pier shape

Lift Thickness ~ 0.3 m

Matrix soil

Figure 2. Partially excavated RAP showing undulating pier shape.

tests were performed by applying incremental vertical loads. During testing loads were only increased when the settlement was constant for a period of not less than 15 minutes. Approximately 10 load increments were applied with a total testing time of 5 to 8 hours for each pier.

3 CONSTITUTIVE MODEL

The hardening-soil model (Schanz et al. 1999) was used to characterize the behavior of the matrix soils and the compacted aggregate in this study. The hardening-soil model is essentially an elasto-plastic model developed based on isotropic plastic theory combined with hardening rules (Potts & Zdravkovic 1999). The failure state of the material is defined in accordance with the Mohr-Coulomb failure criterion. The shape and location of the yield locus of the hardening-soil model are shown in Figures 3a and 3b. The nonassociated flow rule (Rowe 1962) is applied for the deviatoric shear yielding whereas the associated flow rule is used to describe yielding on the compression cap.

Constitutive parameters of the hardening-soil model were determined from consolidated-drained (CD) triaxial and oedometer tests. Details on how to estimate the constitutive model parameters from CD and oedometer tests are described in Schanz et al. (1999).

4 MODELING PROCEDURE

An axisymmetric model was developed in this study using the computer program Plaxis (version 8.2). Dimensions and boundary conditions of the model are shown in Figure 3c. The initial diameter of the cavity was 0.76 m with varying depths of 2.74, 3.0, and 5.05 m. The concrete cap was modeled as a linear elastic, non-porous material. Interface elements with perfectly rough condition were introduced along the cap-soil contact. For the pier-soil contact, thin solid continuum elements were used in lieu of interface elements. An un-structured FE mesh was generated which contains 15-node triangular elements.

Based on the *in-situ* test results, the matrix soil profile consisted of a 1 m thick desiccated crust layer overlying 13 m of soft alluvial clay. The initial stress condition (before pier construction) was generated using the K_o approach where element shear stresses are assumed to be zero. For the lower alluvial clay layer, $K_o = 1 - \sin \phi'$ (Jaky 1944) was assumed whereas $K_o = 1.0$ was assigned to the upper desiccated crust layer.

The induced stress regime around the pier element resulted from the pier installation process was simulated by expanding a pre-bored cavity radially into the matrix soil and downward at the base. The cavity expansion modeling was performed using displacement-controlled boundary conditions estimated from field measurements. At the referenced test site, a downward displacement of about 8 cm was measured at the bottom of the cavity after the installation of the pier bottom bulb. The radius of the cavity increased about 4 cm measured near the top of the pier after ramming. To simulate the long-term stress condition after pier installation, the cavity expansion process was modeled as a "drained" process using effective stress parameter values. The modeling procedure is described as follows:

i. Expand the cylindrical cavity by applying incremental outward displacement (4 cm) along the cavity wall and downward displacement at the bottom of the cavity (8 cm). The cavity wall around the pier cap is not expanded;

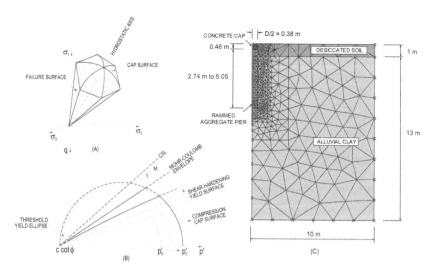

Figure 3. (a and b) Shape and location of yield locus of hardening soil model; and (c) finite element model geometry and mesh.

ii. Back-fill the cavity with compacted aggregate and a concrete cap to create the pier; and

iii. Apply incremental vertical stress on top of the pier until failure is observed.

5 DETERMINATION OF PARAMETERS

Soil properties used in this FE analysis were derived from both *in-situ* and laboratory tests conducted adjacent to the test piers. Cone penetration test (CPT) soundings were carried out prior to pier installation to obtain subsurface information. According to the CPT data, the site stratigraphy consists of a 1 m thick layer of stiff desiccated clay overlying about 13 m of soft alluvial clay (CL), overlying stiff glacial till and then weathered shale bedrock. Correlations from the CPT measurements (see Kulhawy & Mayne 1990) indicate average friction angles of about 22° for the alluvial clay layer and 35° for the desiccated fill. The undrained shear strengths (s_u) of the alluvial clay and the desiccated fill were about 17 kPa and 50 kPa, respectively. The groundwater table was observed at approximately 2 m below the ground surface during pier installation and testing.

Laboratory tests conducted to determine soil parameter values for the analysis included consolidated-drained (CD) triaxial tests and oedometer tests. Undisturbed samples for CD triaxial tests were extracted at 4.2 m from the ground surface and were tested at confining pressures of 25.5, 41, and 60 kPa. The soil exhibits a contractive, strain-hardening behavior under the deviatoric loading condition. Oedometer tests performed on two undisturbed samples collected at 3.74 m and 4.04 m measured from

Table 1. Constitutive model parameters for soils and aggregate.

Parameters*	Unit	Aggregate	Clay	Fill
ϕ'	degrees	47	24	35
c'	kPa	4	2	2
ψ	degrees	12	0	0
γ	ton/m³	2.1	1.924	1.924
E_{50}^{ref}	MPa	61	3	9
E_{oed}^{ref}	MPa	61	1.5	4.5
E_{ur}^{ref}	MPa	183	9	27
m		0.48	1	1
v_{ur}		0.2	0.2	0.2
p^{ref}	kPa	34.5	25.5	25.5
R_f		0.88	0.96	0.96
K_o^{NC}		0.27	0.59	0.43
e_{ini}		0.33	1	1
e_{min}		0.329	–	–
e_{max}		0.393	–	–
$\sigma_{tension}$	kPa	0	0	0

* Parameter descriptions provided in Notation.

the ground surface indicate that the alluvial clay is normally consolidated with the average initial void ratio of 1.0 and a compression index, C_c, of 0.28. Because no laboratory samples from the desiccated layer were collected for analysis, the properties of this layer were estimated from CPT data. Table 1 summarizes the constitutive model parameters for the matrix soils.

Constitutive model parameter values for the compacted aggregate were derived from CD tests conducted under varying confining pressures. Aggregate samples of crushed limestone (classified as GP) obtained from the test site were compacted in a 102 mm diameter by 204 mm high steel split mold to produce specimens for the CD tests. Details of the sample preparation are given by White *et al.* (2002). The aggregate exhibits a strain-softening behavior under deviatoric loading. The FE analysis conducted in this study uses post-peak shear strength parameters to model the aggregate. A summary of the modeling parameters for the compacted aggregate is provided in Table 1.

6 MODEL VERIFICATION

For RAPs installed in homogeneous soft soil, pier bulging and tip movement are two typical deformation mechanisms (Wissmann *et al.* 2001). The development of stresses at the bottom of the pier and tip movement usually occurs when the pier length is less than two to three times the diameter. When the pier length is greater than three times the diameter, bulging deformation often prevails (see White & Suleiman 2004). For RAPs instrumented with a tell-tale reference plate, the deformation mode can be observed. Bulging deformation occurs when the load-displacement curve obtained at the tell-tale location is approximately linear. On the contrary, the development of tip stresses can be recognized when there exists an inflection point on the tell-tale load-displacement curve.

Figure 4 shows the load-displacement curves of three test piers obtained from both FE analyses and full-scale load tests. Based on the displacements recorded at the tell-tale reference plates, at high applied loads, P_1 and P_3 show bulging deformation whereas P_2 develops tip movement (Fig. 4b). As can be seen, the load-displacement curve of P_3 is well captured by the FE model. The maximum design loads (i.e. point of increased curvature) of the three test piers calculated from the FE analyses range from 280 to 330 kN. This range of ultimate load value agrees well with the data obtained from the load tests on P_1 and P_3. Based on the load test performed on P_2, the ultimate load is about 200 kN, which is less than the calculated FE value. Although the FE model overestimated the ultimate load for P_2, the model is capable of capturing the deformation mode of this pier (i.e. movement of the pier tip). The top displacement of P_1, calculated by the FE model, is almost coincident with the load test data until the applied load approaches 350 kN. As the applied load exceeds 350 kN, the FE model predicts greater settlement than shown by the load test data. Moreover, relatively large displacements calculated at the tell-tale location of P_1 (Fig. 3b) suggest the development of tip movement rather than bulging deformation shown by the data. The use of post-peak shear strength parameter values for the aggregate instead of using peak shear strength parameters may attribute to the differences between the measured and the calculated values.

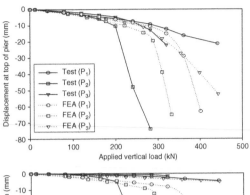

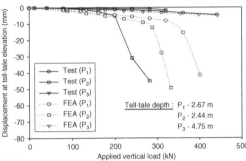

Figure 4. Load-displacement curves of three test piers: (a) at top of the pier and (b) at tell-tale elevation ($P_1 = 3.0$ m; $P_2 = 2.74$ m; $P_3 = 5.05$ m).

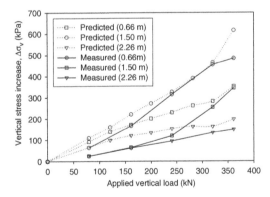

Figure 5. Vertical stress increase along P_1 (L = 3.0 m) as a function of applied load.

Figure 5 shows the variation of the axial stress along P_1 with applied load. The difference between the predicted and the measured values ranges from 0% to 100% but the general trend is acceptable, particularly in light of difficulties that are inherent with total stress measurement (Dunniclif 1993). This agreement lends credibility to the further interpretation of the numerical results.

260

7 STRESSES IN THE MATRIX SOIL

7.1 During pier installation

In axisymmetric analysis, there exist four non-zero stress components – radial stress (σ'_r), vertical stress (σ'_z), tangential stress (σ'_θ), and the shear stress in the z–r plane (τ_{rz}). Figure 6 shows the calculated variation in σ'_r, σ'_z, σ'_θ, and τ_{zr} with radial distance from the edge of pier P_3 at an arbitrary depth of 1.5 m after pier installation and before pier loading, where σ_i is the initial stress prior to cavity expansion. Adjacent to the pier, radial stress σ'_r is greater than σ'_{ri} by a factor of 4.5. The decay of σ'_r with radial distance is approximately logarithmic to a distance of about 20 times the initial pier radius. The calculated tangential stress, σ'_θ, adjacent to the pier is greater than $\sigma'_{\theta i}$ by a factor of 2. The calculated vertical stress, σ'_z, adjacent to the pier slightly increases to about 1.2 times the initial value. The calculated σ'_θ also decreases rapidly with radial distance whereas σ'_z shows an initial drop before returning to the pre-installation value at a distance of about 11 times the pier radius. The trends of σ'_r, σ'_z, and σ'_θ with radial distance are similar to those reported in Randolph et al. (1979) for rigid piles installed in normally consolidated clays.

Results from the numerical modeling also indicate the generation of shear stress, τ_{zr}, along the pier-soil contact after pier installation. The generation of τ_{zr} is a consequence of the non-uniform expansion along the cavity wall and downward displacement applied at the cavity bottom. Moreover, the interface shear stress, τ_{zr}, generated (i.e. the change in shear stress from the initial condition) from pier construction is found to be oriented downward along the pier shaft and decreases rapidly from 13 kPa at the pier edge to zero at a distance of about three times the pier radius.

7.2 During pier loading

The variations of interface stress components adjacent to P_2 and P_3 during the pier loading are presented in Figure 7 for the same soil element shown in Figure 6. For pier settlement less than about 18 mm, σ'_z and τ_{zr} increase and σ'_r decreases. When pier settlement increase over 18 mm, σ'_r adjacent to P_3 increases – indicative of bulging. As bulging occurs, the interface shear stress, τ_{zr}, increases. The calculated σ'_r adjacent to P_2 slightly increases with pier settlement greater than 18 mm, but to a lesser extent as compared to that encountered in P_3. The increase from negative to positive values in τ_{zr}, which occurs for both piers, indicates a reversal of the interface shear stresses during pier loading, which brings about a rotation in the principal stress orientation (Potts and Martin 1982).

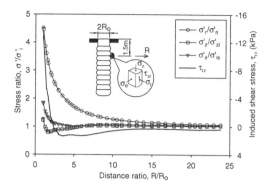

Figure 6. Calculated radial distribution of stresses after the installation of P_3 (L = 5.05 m).

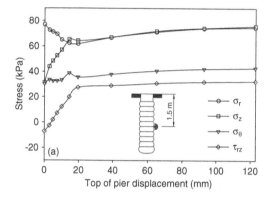

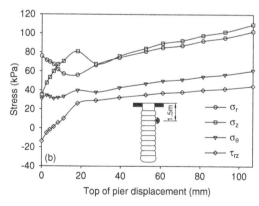

Figure 7. Calculated variation of interfacial stresses during pier loading: (a) P_2 (L = 2.74 m) and (b) P_3 (L = 5.05 m).

8 STRESS DISTRIBUTION IN RAMMED AGGREGATE PIERS

8.1 Vertical stress distribution

Figure 8 shows the distribution of vertical stress with depth through the pier shaft under a range of vertical

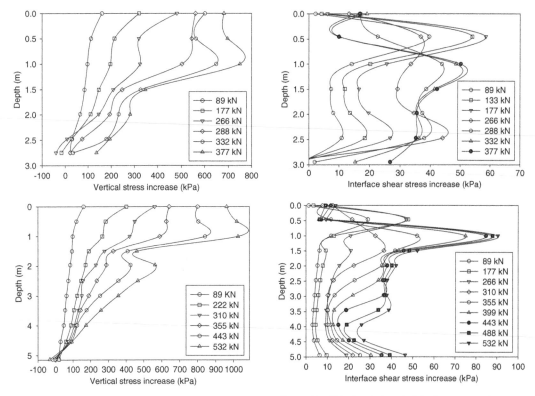

Figure 8. Calculated vertical stress increase: (a) along P_2 (L = 2.74 m); and (b) along P_3 (L = 5.05 m).

Figure 9. Calculated interfacial shear stress increase: (a) along P_2 (L = 2.74 m) and (b) along P_3 (L = 5.05 m).

loads. The results show that RAPs rapidly transfer load to the matrix soil with a large portion of the load transfer occurring in the upper 1.5 m of the pier. In terms of vertical stress transfer, about 75 percent of the stress for P_2 and almost 100 percent for P_3 is dissipated before reaching the bottom of the pier for the range of applied loads.

8.2 Interface shear stress distribution

Figure 9 shows the distribution of pier-soil interface shear stresses for P_2 and P_3 during pier loading. Both piers initially show highly mobilized shear stresses at the upper portion (depth = 0.5 m) of the piers for applied loads that are less than the maximum design values ($P_2 = 200$ kN and $P_3 = 270$ kN). As the applied compressive load increases, the distribution of interface shear stress is affected by bulging deformation, which promotes shaft resistance at the pier-soil contact at a depth of about 1.0 m. As applied compressive loads increase over 280 kN, increased mobilization of shaft resistance is achieved in P_2 which results in a more uniform distribution of the interface shear stress along the pier-soil contact (Fig. 9a).

Conversely, P_3 shows increased bulging with increased compressive loads, which promotes development of interface shear stress at the bulging depth rather than distributing the load down the pier. This behavior is in contrast to rigid piles, where virtually no bulging occurs and the distribution of the interfacial shear can be fairly uniform with depth (Poulos & Mattes 1969).

9 LOAD TRANSFER MECHANISM

As described previously, the unique pier construction process pre-mobilizes tip resistance and shaft friction (i.e. develops residual stresses). These residual stresses develop an upward resultant force acting on the pier prior to pier vertical loading. Figure 10 shows calculated load-deflection responses (top and bottom of pier) for piers P_2 and P_3 under the full range of vertical load conditions. For both piers the construction process results in pre-mobilization of shear stress in the matrix soils and development of tip stresses (left portion of Fig. 10). As the applied vertical load is increased, the pre-mobilized shear stresses must be overcome to induce significant

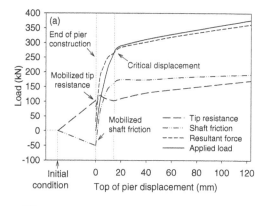

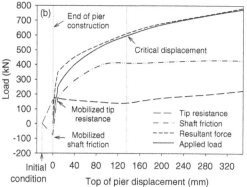

Figure 10. Calculated load transfer as a function of applied load: (a) P_2 (L = 2.74 m) and (b) P_3 (L = 5.05 m).

settlements. Initially, the vertical load is resisted by the shear stresses that develop along the pier-soil interface. As the applied compressive load increases, the shaft resistance becomes fully mobilized first near the top of the pier and then propagates down the shaft as the applied load increases. In this paper the resultant force shown in Figure 10 is the sum of the load carried at the pier bottom plus the load carried by the pier shaft. The displacement at which the interface shear resistance is fully mobilized is herein referred to as the pier critical displacement.

As shown in Figure 10, the critical displacement of P_2 is about 18 mm (top of pier) which corresponds to an applied vertical load of 280 kN. From Figure 3, it can be seen that this load falls in the range of the maximum design load based on the load-displacement curve. As the displacement exceeds 14 mm, interface shear stress in P_2 plateaus giving rise to increased tip stress and tip deformation. The critical displacement for P_3 is much higher at about 110 mm (top of pier). This displacement corresponds to an applied load of about 580 kN, which is much higher than the maximum design load (see Fig. 4). This behavior suggests that the pier is long enough that shear deformation (bulging)

must take place within the pier, rather than in the soil at the top of the pier.

10 CONCLUSIONS

Major conclusions drawn from this study are summarized as follows:

- The results of the FE analyses show that the model successfully captured the load-deflection behavior and stress transfer mechanics of both short and long piers when compared with field instrumentation data. Good correspondence between the field data and the results of the numerical simulations were achieved only when the pier hole was radially and axially expanded to simulate the stress field generated by the expansion of constructed piers. These results suggest that the expansion of the pier cavity is an important mechanism in the behavior of RAPs.
- The expansion of the pier cavity increases the radial stress in the matrix soils surrounding the pier and rotates the principal stresses by introducing interface shear stress that is oriented downward along the pier-soil contact. During pier loading, the interface shear stress reverses which causes a second rotation of the principal stress orientation.
- Much of the applied compressive load is dissipated within the range where bulging deformation occurs. The distribution of the interface shear stress along the piers is highly non-uniform and more complex compared to that reported on isolated rigid piles. The interfacial shear stress becomes relatively uniform in P_2 once the maximum design load is exceeded. Continuing to increase the compressive load on P_3 tends to promote build-up of interface shear stresses at bulging depth rather than distributing to the lower part of the pier.
- The installation of rammed aggregate piers prestresses the matrix soil and creates a pre-mobilized uplift resultant force acting on the pier prior to pier loading. As the compressive load increases, the shaft resistance becomes fully mobilized first at the top of the pier and then propagates down the shaft of the pier. The displacement at which the shaft resistance is fully mobilized is referred to as the critical displacement. The compressive load corresponding to the critical displacement is about the same as the maximum design load estimated for P_2 but is significantly higher than the design load estimated for P_3. This behavior suggests that failure must take place within P_3 rather than in the soil underneath the pier.

ACKNOWLEDGEMENT

This research was sponsored by the Iowa DOT under contract TR-443, Geopier Foundation Company, Inc.

and Iowa State University of Science and Technology. The support of these agencies is greatly acknowledged. Ken Hoevelkamp, Aaron Gaul, Muhannad Suleiman, and Brendan FitzPatrick assisted with conducting the load tests and Peterson Contractors, Inc. installed the test piers.

NOTATION

c' = effective stress cohesion
c_c = compression index
D = pier diameter
e_{min} = minimum void ratio
e_{max} = maximum void ratio
e_{ini} = initial void ratio
E_{ur}^{ref} = reference unloading/reloading modulus corresponding to p^{ref}
E_{50}^{ref} = reference secant modulus corresponding to p^{ref}
E_{oed}^{ref} = reference modulus corresponding to p^{ref}
ϕ' = effective stress friction angle
γ = wet density
K_o = at-rest coefficient of lateral earth pressure
m = power for stress-level dependency
ν_{ur} = unloading-reloading Poisson's ratio
p' = mean effective stress
p^{ref} = reference confining stress
q = deviator stress
R_f = failure ratio
s_u = undrained shear strength
σ_1' = major principal effective stress
σ_3' = minor principal effective stress
$\sigma_{\theta i}'$ = initial tangential stress
σ_{ri}' = initial radial effective stress
σ_{zi}' = initial vertical effective stress
σ_θ' = tangential effective stress
σ_r' = radial effective stress
σ_z' = vertical effective stress
$\sigma_{tension}$ = tension cut-off stress
τ_{rz} = shear stress in $z-r$ plan
ψ = dilatancy angle

REFERENCES

Dunniclif, J. (1993). Geotechnical instrumentation for monitoring field performance. John Wiley and Sons, Inc.

Jaky, J. (1944). *The coefficient of earth pressure at rest.* Journal of the Hungarian Society of Engineers and Architects, 7, 355–358.

Kulhawy, F. H., and Mayne, P. W. (1990). Manual on estimating soil properties for foundation design, EL-6800 Electric Power Research Institute, Paolo Alto, California.

Lawton, E. C., and Fox, N. S. (1994). *Settlement of structures supported on marginal or inadequate soils stiffened with short aggregate piers.* Geotechnical special publication No. 40, ASCE, 2, 962–974.

Potts, D. M., and Martins, J. P. (1982). *The shaft resistance of axially loaded piles in clay.* Geotechnique, 32(4), 369–386.

Potts, D. M., and Zdravkovic, L. (1999). Finite element analysis in geotechnical engineering: Volume I – Theory, Telford Publishing, London.

Poulos, H. G., and Mattes, N. S. (1969). *The behaviour of axially-loaded end-bearing piles.* Geotechnique, 19, 285–300.

Randolph, M. F., Carter, J. P., and Wroth, C. P. (1979). *Driven piles in clay – the effect of installation and subsequent consolidation.* Geotechnique, 29(4), 361–393.

Rowe, P. W. (1962). *The stress-dilatancy relation for static equilibrium of an assembly of particles in contact.* Proc. Roy. Soc. A. 269, 500–527.

Schanz, T., Vermeer, P. A., and Bonnier, P. G. (1999). *Formulation and verification of the hardening-soil model.* Proc., Beyond 2000 in Computational Geotechnics, Balkema, Rotterdam, 281–290.

White, D. J., Gaul, A. J., and Hoevelkamp, K. (2003). Highway applications for rammed aggregate pier in Iowa soils, Final report, Iowa DOT TR-443.

White, D. J., Suleiman, M. T., Pham, H. T., and Bigelow, J. (2002). Constitutive equations for aggregates used in Geopier® foundation construction, Report prepared for the Geopier® Foundation Company, Inc., Iowa State University, Ames, IA, USA.

White, D. J., and Suleiman, M. (2004). *Design of short aggregate piers to support highway embankments.* Journal of the Transportation Research Board, Transportation Research Record, Number 1868, 103–112.

Wissmann, K. J., Fox, N. S., and Martin, J. P. (2000). *Rammed aggregate piers defeat 75-foot long driven piles.* Proc., Performance Confirmation of Constructed Geotechnical Facilities, ASCE, Amherst, MA, 198–210.

Wissmann, K. J., Moser, K., and Pando, M. (2001). *Reducing settlement risks in residual piedmont soil using rammed aggregate pier elements.* Geotechnical Special Publication No. 113, Foundations and Ground Improvement, Blacksburg, VA. 943–957.

Wissmann, K. J., FitzPatrick, B. T., White, D. J., and Lien, B. H. (2002). *Improving global stability and controlling settlement with Geopier soil reinforcing elements.* Proc., 4th International Conference on Ground Improvement Techniques, Kualar Lampur, Malaysia, 753–760.

Numerical Modelling of Construction Processes in Geotechnical Engineering for Urban Environment – Triantafyllidis (ed)
© *2006 Taylor & Francis Group, London, ISBN 0 415 39748 0*

Numerical modelling of the case history of a piled raft with a viscohypoplastic model

F. García & A. Lizcano
Universidad de los Andes, Bogota, Colombia

O. Reul
CDMAG, Germany

ABSTRACT: Hypoplasticity and viscohypoplasticity belong to more powerful constitutive equations for geotechnic. In order to verify the prediction capabilities of the constitutive model one case history of a piled raft foundation – Messeturm tower in Frankfurt, Germany – has been studied using a viscohypoplastic constitutive law in a three dimensional finite element analysis with the program ABAQUS and the user subroutine UMAT developed by A. Niemunis and the geotechnical group of the Karlsruhe University in Germany. In the finite element analysis the construction process of the Messeturm tower is modelled. The calculated results will be compared with the *in situ* measurements with the purpose to verify the viscohypoplastic law in a boundary value problem.

1 INTRODUCTION

Soils with soft particles reveal rate dependence, creep and relaxation. The non-linear viscosity is described by viscohypoplasticity. An extension from hypoplasticity, the viscohypoplastic model, can describe a non-linear stress dependent stiffness, with higher values for unload as for load, using only one tensorial equation. The parameters of the viscohypoplastic equation for monotonic load and unload are physically meaningful and could be determined from routine laboratory test.

The piled raft foundation is a composite construction consisting of three bearing elements, piles, raft and subsoil. Unlike the traditional design of foundation where the load is carried either by the raft or by the piles, in the design of a piled raft foundation the load share between the piles and the raft is taken into account. In this foundation the piles usually are not required to ensure the overall stability of the foundation but to reduce the magnitude of settlements, differential settlements and the resulting tilting of the building and guarantee the satisfactory performance of the foundation system.

The bearing behaviour of a piled raft foundation is characterized by complex soil-structure interactions (Katzenbach et al. 1998). The modelling of these interactions requires a reliable and powerful analysis tool, such as the Finite Element Method in combination with a realistic constitutive law.

Different researches have realized investigations about the numerical modelling of piled rafts with FE analysis and different constitutive models. For example, Prakoso & Kulhawy (2001) used a elastic law with a 2D modelling, or Reul & Randolph (2003) used a elastoplastic law with 3D modelling. Moreover for overconsolidated clays, small-strain non-linearity can have an important influence on the simulated ground movements (e.g. Atkinson 2000). The present research combines the 3D modelling with a modern and realistic constitutive model like the viscohypoplastic law that can model the small-strain non-linearity in clay-like soils.

Key questions that arise in the design of piled raft concerns the relative proportion of load carried by the raft and piles, and the effect of the additional pile support on absolute and differential settlements (Randolph 1994). Therefore three coefficients are introduced to quantify the performance of piled rafts:

(a) The piled raft coefficient, α_{pr}, describes the ratio of the sum of all pile loads, ΣP_{pile}, to the total load on the foundation, P_{tot}:

$$\alpha_{pr} = \frac{\Sigma P_{pile}}{P_{tot}} \tag{1}$$

A piled raft coefficient of unity indicates a free-standing pile group, whereas a piled raft coefficient of zero describes an unpiled raft.

(b) The coefficient of maximum settlement, ξ_s, is defined as the ratio of the maximum settlement of piled raft, s_{pr}, to the maximum settlement of the corresponding unpiled raft, s_r:

$$\xi_s = \frac{s_{pr}}{s_r} \qquad (2)$$

(c) The coefficient of differential settlement, $\xi_{\Delta s}$, is defined in the same way. Unless otherwise stated, this is the differential settlement between the centre and a corner of the raft.

In the scope of this paper one case history has been studied by means of a three dimensional visco-hypoplastic finite element analysis using the program ABAQUS, and the calculated results are compared with the *in situ* measurements with the purpose to verify the viscohypoplastic model in a boundary value problem.

2 CONSTITUTIVE MODEL

Niemunis (1996, 2003) developed a viscohypoplastic model for clay-like soils which can describe the viscous behaviour of soft soils like creep, relaxation and rate dependence. He modified the hypoplastic equation proposed by Wolffesdorff (1996) in order to describe the stiffness upon oedometric/isotropic unloading or reloading, and introduced an expression that depends on true time increment.

The basic assumptions of the viscohypoplastic constitutive law are:

- The state of the soil is defined by the stress state and the void ratio.
- Consider fully saturated clay soils at slow rates of deformation.
- The overconsolidated ratio must be smaller than 2. For larger values of OCR the formula of creep intensity is less precise.
- Change of temperature and ion concentration in pore water are ignored.
- The primary consolidation is left out because the consolidation in not the subject of constitutive modelling.

The viscohypoplastic model has some advantages compared to other constitutive models, that turn it into a very useful tool for the geotechnical engineer. The main advantages are that the material constants of the model are closely related to standard soil parameters and have a physical meaning, and that the model could be easily implemented to a FE program (Niemunis 1996, 2003).

2.1 *Reference model*

The basic viscohypoplastic equation is:

$$\overset{\circ}{\mathbf{T}} = f_b \mathsf{L} : \left(\mathbf{D} - \mathbf{D}^{vis}\right) \qquad (3)$$

therein $\mathbf{T}$ is the actual stress state; L is the hypoplastic fourth order stiffness tensor and $\mathbf{D}$ is the rate deformation.

The barotropy factor f_b was modified by Niemunis in order to describe the stiffness upon unloading and reloading for isotropic and oedometric conditions, and volume changes for constant modulus of strain rate. The barotropy factor is defined according to the condition of the experiment.

$$f_b = -\frac{\mathrm{tr}\,\mathbf{T}}{[1 + a^2/3]\kappa} = -\beta_b\,\mathrm{tr}\,\mathbf{T} \qquad (4)$$

for isotropic conditions.

$$f_b = -\frac{\mathrm{tr}\,\mathbf{T}}{[1 + a^2/(1 + 2K_0)]\kappa^\circ} = -\beta_b\,\mathrm{tr}\,\mathbf{T} \qquad (5)$$

for oedometric conditions.

The parameters κ, κ°, a, K_0 are necessary to calculate the barotropy factor. Using the Butterfield (1979) compression law, the parameters κ and κ° are the unloading or reloading slope of the isotropic and oedometric test respectively. The parameters a and K_0, defined as the earth pressure coefficient, are calculated from

$$a = \frac{\sqrt{3}\left(3 - \sin\varphi_c\right)}{\sqrt{2} \cdot 3\sin\varphi_c} \qquad (6)$$

and

$$K_0 = \frac{-2 - a^2 + \sqrt{36 + 36a^2 + a^4}}{16} \qquad (7)$$

wherein φ_c is the critical friction angle. A critical state is defined by shearing of soil going on indefinitely without change of effective stress and volume.

The viscous rate $\mathbf{D}^{vis}$ is represented by the following equation which is described analogously to Norton's law:

$$\mathbf{D}^{vis} = D_r \vec{\mathbf{B}} \left(\frac{1}{\mathrm{OCR}}\right)^{1/I_v} \qquad (8)$$

therein D_r is the reference creep rate; $\vec{\mathbf{B}}$ is the direction of the creep rate and can be seen as a hypoplastic flow rule; I_v is the viscosity index of Leinelkugel (1976); the OCR is the overconsolidated ratio that can be calculated from $\mathrm{OCR} = p_e/p'$ where p' represents the current effective mean stress and p_e is the equivalent isotropic pressure by Hvorslev (1960).

2.2 *Extended model (intergranular strain)*

In order to improve the small strain behaviour of soils under cyclic loads, or the behaviour after abrupt changes of direction of stress or strain paths,

Table 1. Viscohypoplastic reference model parameters.

Reference model	Laboratory test
e_{100}	oedometric test
λ	oedometric test
κ	oedometric test
β_R	undrained test
Iv	oedometric creep test
Dr	oedometric test
φ_c	triaxial test
OCR	oedometric test

Table 2. Viscohypoplastic extended model parameters.

Extended model	Laboratory test
m_T	Static test with strain path reversals
m_R	Static test with strain path reversals
R	Static test with strain path reversals
β_r	Static test with strain path reversals
χ	Static test with strain path reversals

Niemunis & Herle (1997) have modified the reference equation and proposed:

$$\overset{\circ}{\mathbf{T}} = \mathbf{M} : \mathbf{D} - \mathbf{L} : \mathbf{D}^{vis} \qquad (9)$$

wherein the fourth tensor **M** represents the increased stiffness which is calculated from hypoplastic tensor.

2.3 Model parameters

The viscohypoplastic model has 13 parameters: 8 from the reference model and 5 from the extended model with intergranular strain. In Table 1 the parameters of the reference viscohypoplastic model are presented, and in Table 2 the parameters of the intergranular strain.

For more details of the viscohypoplastic model refer to Niemunis (2003), and for the calibration of the viscohypoplastic model refer to Punlor (2004) or Garcia (2005).

3 CASE HISTORY OF A PILED RAFT FOUNDATION

On its completions in 1991 the Messeturm in Frankfurt, Germany, with a height of 256 m, was the tallest office building in Europe. The piled raft of the Messeturm consists of 64 bored piles, located in three concentric rings and a square raft of 58.8 m with variable thickness. The length of the piles varies from 26.9 m (outer ring) trough 30.9 m (middle ring) to 34.9 m (inner ring). The diameter of the piles is 1.3 m and is the same for all the rings. The thickness of the

rafts decrease from the core area to the edge. In the core area the thickness is 6 m and in the edge 3.8 m. The foundation level lies 11–14 m below ground level.

The construction of the building started in 1988 and was finished in 1991. The behaviour of the foundation was monitored by means of geodetic and geotechnical measurements with 12 instrumented piles, 13 contact pressure cells, one pore pressure cell and 3 multi point borehole extensometers. The positions of the measurement devices are plotted in the ground plan of the raft (Fig. 1b).

The groundwater level is situated 4.5–5 m below ground level. For the construction of a subway tunnel with a station 47 m east of the Messeturm, groundwater had to be drawn down more than 12 m at the tunnel construction site. As a result, the groundwater level in the vicinity of the Messeturm sank by about 10 m, which lead to changes of the uplift on the raft of 287 MN. The maximum load amounts to $P_{eff} = 1885$ MN for the simulation of the groundwater drawdown.

3.1 Finite element model

Thanks to the three symmetry planes only one height of the complete three dimensional model had to be modelled. The thickness of the rafts decrease from the core area ($t_r = 6$ m) to the edge ($t_r = 3.8$ m) (Fig. 1c), and it was modelled in three steps. The depth of the Frankfurt clay has been assumed to be 74.8 m below the bottom of the raft, and the depth of the Frankfurt limestone has been assumed to be 55.2 m below Frankfurt clay (Fig. 1d).

The soil and piles are represented by first-order solid finite elements of hexahedron (brick) and triangular prism (wedge) shape. For the modelling of the raft, first-order shell elements of square and triangular shape with reduced integration have been used. Only the soil below the foundation level is modelled with finite elements. The soil above the foundation level is considered through its weight. The circular piles have been replaced by square piles with the same shaft circumference. For the modelling of the contact zone between soil and raft, between soil and the large-diameter bored piles, thin solid continuum elements have been applied instead of special interface elements. The contact between structure and soil was described as perfectly rough. This means that no relative motion takes place between the nodes of the finite elements that represent the structure, and those of the finite elements that represent the uppermost layer of soil. The material behaviour in the contact area was simulated by material behaviour of the soil (Reul 2000). The properties of the raft and the piles used in the analysis are presented in the Table 3.

The FE analysis began when the construction of the building started in July of 1988 and finished in December of 1998 when the last *in situ* measurement was reported. Approximately it was modelled for nine

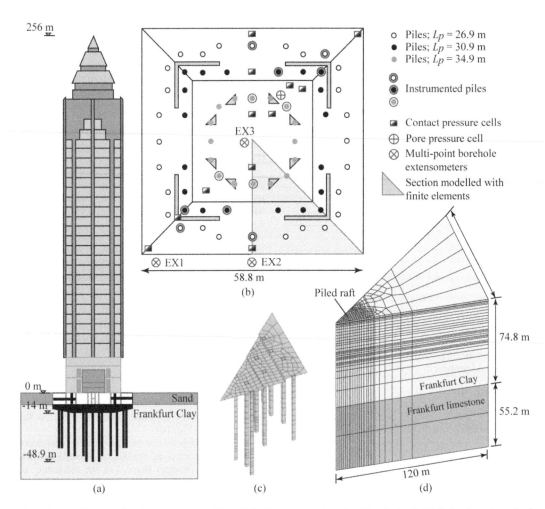

Figure 1.   (a) Cross section; (b) ground plan of the raft; (c) finite element mesh of the piled raft; (d) finite element mesh of the system.

Table 3.   Material parameters used in finite-element analysis.

Parameter	Raft	Piles
Young's modulus, E: MPa	34000	25000
Poisson's ratio, ν	0.2	0.2
γ: kN/m³	25	25
γ': kN/m³	15	15

and a half years. Table 4 summarizes the step by step analysis the construction process in the FE analysis.

3.2   Frankfurt clay properties

The subsoil condition in Frankfurt am Main, Germany, is characterized mainly by tertiary soils and rocks. They consist of Frankfurt clay at on of the underlying rocky Frankfurt limestone. The Frankfurt clay is stiff and overconsolidated. Assuming a maximum previous vertical stress of 450 kPa at its top surface, the analysis is dominated by the soil stiffness rather than the soil strength. Sand and limestone bands of varying thickness are embedded in the Frankfurt clay, which results in a non-homogeneous arrange of the layer as a whole. The compressibility of the Frankfurt limestone is small compared with that of the Frankfurt clay.

The viscohypoplastic parameters of the Frankfurt Clay were calibrated with oedometric and triaxial tests for the reference model. The parameters of the extended model were chosen as standard values. Table 5 presents the viscohypoplastic parameters used in the FE analysis.

The constitutive model used for the Frankfurt limestone is an elastoplastic cap model (Reul 2000).

Table 4. Step by step analysis of construction process of the Messeturm in finite-element analysis.

Step	Applied load, P_{eff} [MN]	Time [days]
1. *In situ* stress state	–	–
2. Excavation to a depth of 7.5 m below ground level	–	21
3. Installation of piles + GW lowering	–	51
4. Excavation to a depth of 14 m below ground level	–	54
5. Application of weight of raft minus uplift due to pore pressure as uniform load on subsoil (zero stiffness of raft) + GW rise	123.0	9
6. Installation of raft	123.0	9
7. Loading of raft 1	1365.9	390
8. Loading of raft 2 + GW lowering	1797.2	140
9. Loading raft 3 (building finished)	1885.0	440
10. Creep	1885.0	210
11. GW rise	1598.0	220
12. Creep	1598.0	580
13. GW lowering	1885.0	160
14. Creep	1885.0	720
15. GW rise	1598.0	440

Table 5. Viscohypoplastic parameter of Frankfurt Clay.

Depth (m)	12–32	32–43	43–87
e_{100}	1.28	0.97	0.92
λ	0.062	0.014	0.010
κ	0.0099	0.0022	0.0016
I_v	0.042	0.043	0.041
β_R	0.95	0.95	0.95
(D_r)	(10^{-6})	(10^{-6})	(10^{-6})
φ_c	23.2	27.5	30.6
OCR	2.1	1.7	1.6
(m_T)	(4.5)	(4.5)	(4.5)
(m_R)	(4.5)	(4.5)	(4.5)
(R_{max})	(10^{-4})	(10^{-4})	(10^{-4})
(β_r)	(0.2)	(0.2)	(0.2)
(χ)	(6.0)	(6.0)	(6.0)

* Standard values may be assumed for parameters in parenthesis.

4 RESULTS

The last documented settlement measurement in the centre of the raft gives a value of 144 mm in December 1998, the calculated settlement with the FE analysis is 120 mm. The calculated settlement is lower than the measured one, but it is a good approximation, considering that the consolidation process in the Frankfurt clay was not modelled in the FE analysis. The settlement of the building was monitored during the construction, and thanks to the viscohypoplastic model it is possible to compare the calculated and measured settlement at different points of time. The settlement profile along the depth was compared in two points of time, when the building shell is finished in March 1990 and when the construction of the building

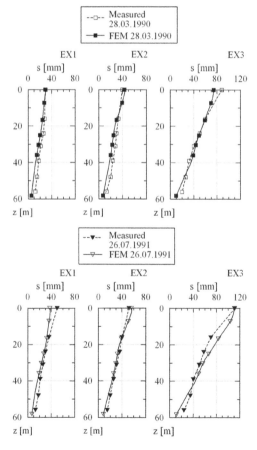

Figure 2. Settlement profile along depth – Measurement after Sommer & Hoffmann (1991) and FE calculation at 28.03.1990 and 26.07.1991.

269

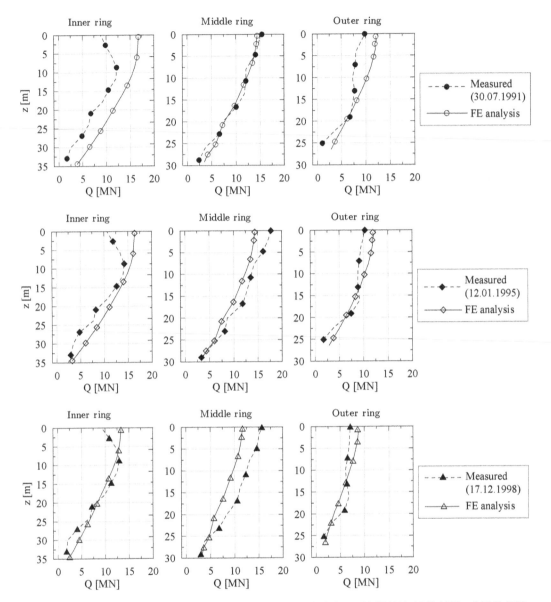

Figure 3. Pile load distribution along shaft – Measurement and FE calculation at 30.07.1991, 12.01.1995 and 17.12.1998.

is finished in July of 1991. In March 1990 the measured settlement in the centre of the raft was 85 mm and the calculated settlement 75 mm, 12% of error, and in July 1991 the measured settlement 111 mm and the calculated settlement 110 mm, only 0.4% of error. Figure 2 shows the settlement profile along the depth for the three extensometers at two different points of time. For both times good agreement between *in situ* measurements and FE analysis are achieved.

Based on the assumption that the average pile load can be derived from the twelve instrumented piles,

the piled raft coefficient in July 1991 was $\alpha_{pr} = 0.41$ and the calculated $\alpha_{pr} = 0.47$. The last measurement reported in the literature of the piled raft coefficient is $\alpha_{pr} = 0.43$, the calculated piled raft coefficient is equal to the one measured.

Figure 3 shows the average pile load profile for the three types of piles in the piled raft at different points of time. Good level of agreement was achieved for the middle and outer ring in 1991. For the inner ring the tendency was good but the calculated values were approximately 25% bigger than the measured ones. In 1995 good agreement was achieved for the three types of piles. For the middle ring in 1998 the tendency was good, but the calculated values were approximately 25% lower than the measured ones. In general, the calculated values were very similar to the ones measured.

To study the influence of the piles on the behaviour of the foundation system, an analysis of the model without the installation of the piles was realized. The calculated settlements for the piled raft and the unpiled raft are related to the stage when the building was finished, in July 1991. The calculated unpiled raft's settlement in the center of the raft is $s_r = 240$ mm and the calculated differential settlement is $\Delta s = 27.2$ mm. The shallow foundation has bigger average settlement and differential settlement, verifying the advantages of the addition of piles and the load share between the raft and the piles. The corresponding settlement coefficients, defined in the first section from the finite element analysis are $\xi_s = 0.46$ and $\xi_{\Delta s} = 0.66$, implying that the ratio of settlement between the piled raft and raft is lower than one, and that the ratio of differential settlement to maximum settlement has actually increased as a result of the addition of piles.

5 CONCLUSIONS

The present paper shows that the viscohypoplastic model can predict to a good level of agreement the behaviour of a piled raft foundation. Also the viscohypoplastic model was verified to predict the behaviour of the boundary value problem with time, the settlement profile with depth (Fig. 2) and the load pile profile (Fig. 3), verifying the great prediction potential of the constitutive law.

The advantages of the piled raft were verified, like the reduction of the total settlement and the differential settlement when piles are added to a shallow or unpiled raft foundation.

It is important to explain, that to be able to use the viscohypoplastic model it is very important to know the parameters of the model and the construction process of the boundary value problem, specially the load distribution with time. Due to the fact that the model is time dependent, the application of the load in time is going to have effects on the results of the numerical modelling.

The present research leaves open questions to future researches, like the effects of the process of consolidation and water flow (fluctuation of the groundwater level) on the behaviour of a piled raft foundation or any boundary value problems in geotechnical engineering. In this research the fluctuations of the groundwater level were modelled by means of the change in the stress level (change from effective to submerged

specific weight or vice versa) and changes in the load applied to the model due to the uplift force of the groundwater. In future works the water flow combined with the process of consolidation could be used to study these effects in the piled rafts foundations.

REFERENCES

Atkinson, J. H. 2000. Non-linear soil stiffness in routine design. *Géotechnique*. 50(5): 487–508.

Butterfield, R. 1979. A natural compression law for soils. *Géotechnique*. 29(4): 469–480.

Garcia, F. 2005. *Viscohypoplasticity applied to piled raft foundation*. M.Sc. thesis, Universidad de los Andes, Colombia (in Spanish).

Hvorslev, M. 1960. Physical components of the shear strength of satured clays. *Proc. ASCE Res, Conf. on Shear Strength of Cohesive Soils. Boulder, Colorado*.

Katzenbach, R., Arslan, U. & Moorman, C. 1998. Design and safety concept for piled raft foundation. *Proc. of the conference on deep foundations on bored auger piles, Ghent*: 439–448. Rotterdam: Balkema.

Leinelkugel, H. (1976). *Deformations- und Festigkeitsverhalten bindiger Erdstoffe. Experimentelle Ergebnisse und ihre physikalische Deutung*. PhD thesis, Institut für Bodenmechanik und Felsmechanik, Universität Karlsruhe, Heft 66.

Niemunis, A. 1996. A visco-plastic model for clay and its FE-implementation. In E. Dembicki, W. Cichy, L. Balachowski (eds), *Resultants Recents en Mechanique des Sols et des roches, Politechnika Gdanska*: 151–162.

Niemunis A. 2003. *Extended hypoplastic models for soils. Monografia Nr. 34*, Politechnika Gdanska.

Niemunis A. & Herle, I. 1997. Hypoplastic model for cohesionless soils with elastic strain range. *Mechanics of Cohesive-Frictional Materials 2*: 279–299.

Prakoso, W. A. & Kulhawy, F. H. 2001. Contribution to Piled Raft foundation Design. *Journal of the Geotechnical and Geoenviorement Engineering Division, ASCE*. 127(1):17–24.

Punlor, A. 2004. *Numerical Modelling of the Visco-Plastic Behaviour of Soft Soils*. PhD thesis, Institut für Bodenmechanik und Felsmechanik, Universität Karlsruhe, Heft 163.

Randolph, M. F. 1994. Design methods for pile groups and piled rafts. *Proc. XIVth ICSMFE, New Dehli*, 5: 61–82. Rotterdam: Balkema.

Reul, O. 2000. *In situ measurements and numerical studies on the bearing behaviour of piled rafts*. PhD thesis, Darmstadt University of Technology, Germany (in German).

Reul, O. & Randolph, M. F. 2003. Piled raft in overconsolidated clay: compression of *in situ* measurements and numerical analyses. *Geotechnique*. 53(3): 301–315.

Sommer, H. & Hoffmann, H. (1991). Last-Verformungsverhalten der Gründung des Messeturmes Frankfurt/Main. *Festkolloquium 20 Jahre Grundbauinstitut Prof. Dr.-Ing. H. Sommer und Partner*, 63–71.

Wolffesdorff, P. A. 1996. A hypoplastic relation for granular materials with predefined limit state surface. *Mechanics of Cohesive-Frictional Materials 1*: 251–279.

Numerical Modelling of Construction Processes in Geotechnical Engineering for
Urban Environment – Triantafyllidis (ed)
© 2006 Taylor & Francis Group, London, ISBN 0 415 39748 0

Failure of a micro-pile wall during remodelling of a hotel: a backanalysis

Pere C. Prat

Technical University of Catalonia (UPC), Barcelona, Spain

ABSTRACT: The paper presents a backanalysis performed to clarify the mechanisms and causes leading to the failure of a micro-pile wall that collapsed while being constructed. Two series of numerical analysis have been performed, a) to determine the most likely strength parameters of the rock mass; b) to evaluate the influence of the actual topography on the earth pressure acting on the micro-pile wall; and c) to evaluate the possible influence of a water table on the failure mechanism. Because of the uncertainty of the strength parameters used in the original project, several combinations of cohesion and friction angle have been used in the analysis, leading to the main conclusion that the primary cause of failure was the use of unrealistic values of these parameters and of incorrect topographic data, affecting the actual earth pressure on the wall. Water in the grounds seems to have had little impact on the collapse.

1 INTRODUCTION

On May 7th of 2001 a micro-pile retaining wall that was built as part of the remodelling of an old building collapsed. The work presented in this paper consists of a backanalysis performed on the failure of the retaining wall in order to clarify the mechanisms and causes that led to the failure.

The building had been a luxury hotel, inaugurated in 1925. Very popular in the 1950's and 60's, it started to decline and in 1979 closed its doors. In 2001 a new ownership began a restoration work to bring back the hotel to its former splendour. During the remodelling, a new underground addition was to be built in front of the hotel to house, among other facilities, a parking garage for hotel guests. This required an excavation of 15 to 17 m of material from the existing ground level, involving a total area of about 1400 m².

The original project required a pile wall to stabilize the grounds of a nearby amusement park. During construction it was decided to change this design to a micro-pile wall. It was this micro-pile wall that collapsed while being constructed, after partial excavation had been performed.

2 LOCATION AND GEOLOGICAL DATA

The zone where the collapse occurred is located on a site within the city of Barcelona (Figure 1), limiting an amusement park on a hill NW of the city. The plot has a shape of a somewhat irregular semi-circle, with a total surface area of approximately 5850 m². North, East and South the plot is limited by a local road reaching the top of the hill and the amusement park, with a sharp 180°-bend around the plot. On the West side, the plot is

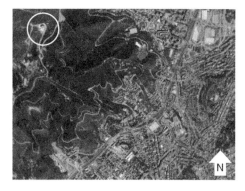

Figure 1. Partial aerial view of the city of Barcelona, showing the area where the collapse occurred.

limited by the amusement park. The plot's topography (Figure 2) is determined by its location, near the top of the Collserola range, at elevations between 478 and 495 m above sea level.

From the geological point of view, Collserola is the SE end of the Catalan Coastal Range, that runs approximately parallel to the coast and is formed by a sequence of low-height ranges. The lithology units outcropping in the Collserola range are the oldest in the area, consisting mainly of meta-sedimentary Palaeozoic rocks, especially dark slates, and hornfels and phyllites as a result of the contact metamorphism produced by the presence of a granitic batholite, currently outcropping at the foothills.

A wide mix of materials, and to a certain extent ages, can be found in the area. The more modern Palaeozoic materials can be found on small Carboniferous outcrops in the region, including the hills around the city, where small-size outcroppings of Silurian-Devonian limestone can be found as well. There is also a considerable amount of Silurian shales and a variety of pre-Silurian materials, probably Ordovician. Among the Silurian and pre-Silurian series there are volcanic rocks and sills of basic rocs. All these materials have been shaped during the Hercinian orogeny, resulting in a superposition of several deformations.

Within the plot limits (see Figure 3), the rock mass consists mainly of fractured slates. From the geomechanics point of view, this rock is characterized by an RMR rating between 21 and 29, thus being classified as "poor" (Bieniawski, 1989; Hoek and Bray, 1981). This rating would suggest preliminary estimates of strength parameters $c' \approx 10$–20 kPa and $\varphi' \approx 15°$–$25°$, much lower than those used in the original design.

However, these values may be only crude estimates of the real values. For instance, using Hoek & Brown's failure criterion (Hoek et al., 2002) the strength parameters obtained are $c' \approx 15$ kPa and $\varphi' \approx 35°$. These are still lower than in the original project, but perhaps closer to the actual values, as will be shown in the following sections.

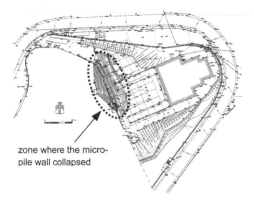

zone where the micro-pile wall collapsed

Figure 2.   Topographic survey conducted after the collapse.

Figure 3.   View of the excavation before the wall was initiated. The collapsed zone will be on the right-hand side of the image.

3   NUMERICAL ANALYSIS

The results obtained in this work are based on two series of numerical backanalysis that will aid in the determination of the causes leading to the collapse of the micro-pile wall.

The first series has been performed with the commercial code *PARATIE* (CEAS, 2003), a FEM-based non-linear computer code specific for flexible earth-support structures. The second series has been conducted with a general-purpose FEM code, *DRAC* (Prat et al., 1993) that has been used for a non-linear plane-strain analysis.

The main objectives of these two series of computations have been:

– Determination, by means of a backanalysis, of the most likely strength parameters of the rock mass, with the knowledge that the wall did fail.
– Evaluation of the influence of the height of the contiguous terrain on the earth pressure on the wall.
– Evaluation of the effect of a possible water table near the surface on the mechanism and causes of failure.

Because of the uncertainty about the strength parameters used in the original design, a set of 36 combinations of cohesion (c') and friction angle (φ') have been used in the analysis (Table 1). Based on the existing geotechnical reports, the maximum values of these parameters have been fixed at 50 kPa and 44°, respectively. These quantities are high enough that they can be taken as an upper bound of the actual values. The minimum values have been fixed at 0 kPa and 20°, respectively.

The remaining parameters (for all materials) have been assumed free from uncertainty and the same values have been used for all combination sets (Table 2). The combination of parameters used in the original project correspond to sets H6 (service state) and H36 (during construction).

Figure 4 shows a typical cross-section used in the plane-strain analysis. Local datum is at elevation +486.83 m a.s.l., two meters below the head of the micro-pile wall. Initial ground level was located at elevation +493.92 m a.s.l., from which 5 m were

Table 1.   Parameter combination set number for different values of cohesion (c') and friction angle (φ').

c' φ'	0 kPa	10 kPa	20 kPa	30 kPa	40 kPa	50 kPa
20°	H1	H7	H13	H19	H25	H31
25°	H2	H8	H14	H20	H26	H32
30°	H3	H9	H15	H21	H27	H33
35°	H4	H10	H16	H22	H28	H34
40°	H5	H11	H17	H23	H29	H35
44°	H6	H12	H18	H24	H30	H36

excavated to reach local reference +2.00, head of the micro-pile wall.

The micro-piles were cast in place, with a steel pipe (Ø114.3 × 7 mm) reinforcement of yield limit 550 MPa. Each micro-pile had a compressive bearing capacity of 1129 kN, and a bending bearing capacity of 28.4 kN × m. There were two micro-piles per meter of wall, for a total bending bearing capacity of 56.8 kN × m.

In the original project, the earth overburden on the wall was assumed to be caused only by the 5 m of earth left after this excavation (because the ground surface was assumed to be horizontal). However, the real topography in the area where the failure occurred presented a certain slope uphill, resulting in an actual overburden equivalent to about 10 m of earth, double the former amount.

Excavation started from level +2.00. Failure occurred approximately when the excavation reached level −3.00, at an excavation depth of about 5 m.

3.1 Using PARATIE

PARATIE (CEAS, 2003) is a non-linear finite element code for the analysis of flexible retaining walls during multiple construction phases. Several components can be activated and/or removed during the analysis, such

Table 2. Fixed parameters.

Parameter	Value
Steel elasticity modulus, E_s	2.1×10^5 MPa
Concrete elasticity modulus, E_c	2.5×10^4 MPa
Virgin modulus of rock, E_v	30 MPa
Unloading/reloading modulus of rock, E_{ur}	50 MPa
Dry specific weight of the rock above the top of the micro-piles, γ_{d1}	20 kN/m³
Dry specific weight of the rock below the top of the micro-piles, γ_{d2}	24 kN/m³
K_0 –coefficient	0.5

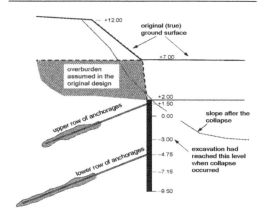

Figure 4. Typical cross-section used in the analysis.

as anchors, struts, fixed or flexible supports, external loadings etc. Water table and seepage forces may also be included.

The program has been used to evaluate forces acting on the wall (bending moment, shear forces) as well as its lateral deformations. The numerical analysis has been carried out with the following premises:

– The problem is assumed to be plane-strain: degrees of freedom are lateral displacements and out-of-plane rotations. Vertical movements are automatically linked, and therefore the axial forces on the wall are not computed.
– The flexible wall is simulated by a series of vertical beam elements.
– The earth pressure can be applied both on the extra-dos or intrados of the wall (active or passive), and it is simulated by a double layer of elasto-plastic springs connected to the nodes on the wall.
– The sustaining elements (anchorages, struts, etc.) are simulated by springs applied to nodes on the wall.

The response of the wall is obtained after numerical simulation of the construction sequence, including construction of the wall, excavation and installation of the rows of anchorages. All phases reproduce accurately the real load history of all structural elements involved in this case. The sequence is as follows:

1. Application of the overburden pressure.
2. Excavation of 1 m, to level +1.00, and construction of the first row of anchorages.
3. Excavation of 1 m, to level 0.00.
4. Excavation of 1 m, to level −1.00.
5. Excavation of 2 m, to level −3.00 (failure occurred at the beginning of this step).
6. Excavation of 2 m, to level −5.00, and construction of the second row of anchorages.
7. Excavation of 2.15 m to level −7.15.

3.1.1 Estimation of strength parameters

From the knowledge that failure did occur, and that it happened after about 5 m of excavation, it is possible to estimate the values of the real strength parameters from the results of the numerical analysis, by recording the combinations for which the program indicates that failure has been reached, or else for which the method does not converge in some of the construction steps (lack of convergence is an indicator of near-failure conditions).

To estimate the most likely strength parameters (cohesion, friction angle), the previous sequence has been applied to four combinations of overburden pressure and water table location, which were magnitudes that were suspect in the original project:

C-1: overburden equivalent to a 10 m high embankment, dry material.
C-2: overburden equivalent to a 10 m high embankment, water table at the head of the micro-piles.

275

C-3: overburden equivalent to a 5 m high embankment, dry material (this combination was the one used in the failed design).

C-4: overburden equivalent to a 5 m high embankment, water table at the head of the micro-piles.

After the collapse, doubts were raised regarding the true position of the water table at the moment of failure, since there was previous evidence that the water table could rise to near-surface levels after periods of intense rain. However, records of rainfall accumulation during the days previous to the collapse show that rain episodes on those days were minimal. Therefore, it seems unlikely that a rain-induced water table existed near the surface at the time of the wall's collapse.

In any case, a micro-pile wall is in essence "discontinuous" and consequently completely permeable. Therefore, water pressures on the wall are automatically cancelled. Similarly, seepage forces are unlikely to play a significant role in this case.

Water, however, can have when present other negative side-effects, mainly the reduction of apparent strength: a partially saturated soil shows an apparent increase in strength (compared to fully saturated or dry conditions) due to suction. Suction vanishes when it becomes fully saturated, e.g. when the water table rises, leading to an effective strength loss, and possibly to failure. Therefore, the possible effects of water on strength should be considered and analysed.

Regarding the actual overburden, a new topographic map made after the collapse shows that the actual height of the ground acting on the wall's extrados was 10 m, instead of the 5 m assumed in the original design.

In conclusion, it seems that the most likely combinations are C-1 (for dry material) and C-2 (if water is present). However, combinations C-3 and C-4, with an incorrect, lower, overburden have also been analysed, for comparison purposes and also to evaluate the impact of each of the variables that played a role in the collapse.

Tables 3–6 show the results obtained from the analysis with the four combinations, and with the 36 parameter sets described in Table 1. The tables indicate the parameter sets that lead to collapse (in **boldface**) or to non-convergence of the numerical method (in *italic*). The numbers point to the construction step at which the event occurred, referring to the sequence described above.

These results show that, for the collapse to occur in the way it actually happened, the strength parameters must have been:

C-1: $c' \approx 0$ kPa $\varphi' \approx 25°$–$30°$
C-2: $c' \approx 0$–10 kPa $\varphi' \approx 30°$–$35°$
C-3: $c' \approx 0$ kPa $\varphi' \leq 20°$
C-4: $c' \approx 0$ kPa $\varphi' \approx 25°$

The last two combinations (C-3 and C-4) are not representative, since the overburden is deterministic,

Table 3. Failure parameters for combination C-1.

c' φ'	0 kPa	10 kPa	20 kPa	30 kPa	40 kPa	50 kPa
20°	**S5**	**S5**	–	–	–	–
25°	**S5**	*S7*	–	–	–	–
30°	*S7*	–	–	–	–	–
35°	–	–	–	–	–	–
40°	–	–	–	–	–	–
44°	–	–	–	–	–	–

Table 4. Failure parameters for combination C-2.

c' φ'	0 kPa	10 kPa	20 kPa	30 kPa	40 kPa	50 kPa
20°	**S2**	**S5**	**S5**	–	–	–
25°	**S5**	**S5**	*S6*	–	–	–
30°	**S5**	*S6*	–	–	–	–
35°	–	–	–	–	–	–
40°	–	–	–	–	–	–
44°	–	–	–	–	–	–

Table 5. Failure parameters for combination C-3.

c' φ'	0 kPa	10 kPa	20 kPa	30 kPa	40 kPa	50 kPa
20°	*S6*	*S7*	–	–	–	–
25°	–	–	–	–	–	–
30°	–	–	–	–	–	–
35°	–	–	–	–	–	–
40°	–	–	–	–	–	–
44°	–	–	–	–	–	–

Table 6. Failure parameters for combination C-4.

c' φ'	0 kPa	10 kPa	20 kPa	30 kPa	40 kPa	50 kPa
20°	**S5**	–	–	–	–	–
25°	**S5**	*S6*	–	–	–	–
30°	–	–	–	–	–	–
35°	–	–	–	–	–	–
40°	–	–	–	–	–	–
44°	–	–	–	–	–	–

and the value used in these two combinations (5 m of earth) is not the real one. However, the results of the analysis with the four combinations illustrate the influence of two main factors that played an important role in the collapse: the total height of earth at the extrados, and the presence or lack of water near the surface. These two variables, that have a considerable effect on the loads acting on the retaining wall, were not taken into account appropriately in the original design of the micro-pile wall.

The results obtained suggest that:

– The presence of water near the surface fully saturates the material and diminishes its apparent strength

- Collapse attributed exclusively to a rise of the water table, would imply strength parameters $c' \approx$ 0–10 kPa, $\varphi' \approx 30°$–35°
- No water influence on the collapse would imply strength parameters $c' \approx 0$, $\varphi' \approx 25°$–30°
- In any case, the estimated strength parameters would be much lower than the ones used in the original design ($c' = 50$ kPa, $\varphi' = 44°$)
- If the actual strength parameters had been the ones used in the original design ($c' = 50$ kPa, $\varphi' = 44°$), collapse would not have happened even with the water table at the surface. In this case, the safety factors would have been 5 (for cohesion) and 1.4 (for friction angle)
- The results indicate that with 5 m of earth overburden collapse would not have happened with reasonable values of the strength parameters in the range indicated above. In this case, and with the water table near the surface, the conditions would be bordering the limit state.

3.1.2 Computed forces on the anchorages

The conditions of the collapse zone were not preserved intact after it occurred, because immediate action was taken to clean the area. Therefore, there is no direct information on the particulars of the failed elements. However, verbal description given by the personnel working in the construction seems to indicate that at most of the anchorages failed due to excessive tension, while some were pulled out with the wall.

The anchorages were made of steel cable with a load capacity of 150 kN per cable. Each anchorage of the upper row consisted of two cables, with a total load capacity of 300 kN, with a separation of 3 m. The total admissible force on the anchorages, per unit length, was therefore equal to 100 kN/m.

Figure 5 represents the change of force acting on the upper row of anchorages at each construction step, for each of the 36 parameter sets defined in Table 1, in the case of combination C-1. This figure shows how the force increases faster for those sets with lower strength parameters, since then the earth pressure on the wall is much larger.

Figure 5 shows that only for six of the parameter sets defined in Table 1 the force remains below the 100 kN/m limit during all construction steps, including those that were never executed because the wall collapsed before: H24 (30 kPa, 44°), H29 (40 kPa, 40°), H30 (40 kPa, 44°), H34 (50 kPa, 35°), H35 (50 kPa, 40°) and H36 (50 kPa, 44°).

Considering only the construction steps previous to collapse (step 5), four more sets would remain below the 100 kN/m limit: H18 (20 kPa, 44°), H23 (30 kPa, 40°), H28 (40 kPa, 35°) and H33 (50 kPa, 30°).

Table 7 shows the maximum values of the force on the upper row of anchorages during all construction steps, including the steps that were never executed (for

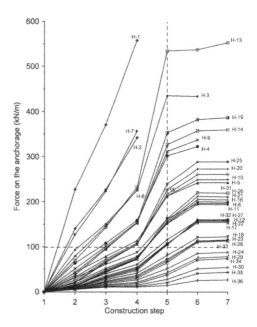

Figure 5. Forces acting on the upper row of anchorages during construction.

Table 7. Maximum load on the upper row of anchorages during all construction steps (kN/m).

c'/φ'	0 kPa	10 kPa	20 kPa	30 kPa	40 kPa	50 kPa
20°	**557.7**	**356.6**	552.2	386.3	288.3	218.9
25°	**341.8**	**226.5**	359.2	272.7	210.7	160.9
30°	**433.8**	**337.0**	260.8	204.2	159.0	113.2
35°	**322.4**	249.6	199.5	157.2	114.9	73.7
40°	242.5	193.8	154.1	115.6	78.0	43.3
44°	196.2	158.2	122.5	87.7	53.9	26.3

Table 8. Load on the upper row of anchorages at the end of step 5, when the collapse occurred (kN/m).

c'/φ'	0 kPa	10 kPa	20 kPa	30 kPa	40 kPa	50 kPa
20°	–	–	534.1	353.1	240.1	162.7
25°	–	–	326.6	227.1	157.3	104.8
30°	435.4	311.6	217.5	154.3	105.8	64.6
35°	301.7	213.3	152.3	107.0	68.7	37.4
40°	212.8	151.7	107.9	72.2	42.3	20.3
44°	162.3	116.4	80.9	51.4	27.8	13.9

some of the parameter sets, the value on the table corresponds to the last converged step, that has not been reached — indicated in **boldface**).

Table 8 shows the value of the force on the same anchorages at the end of step 5, when the collapse occurred (for some of the parameter sets step 5 is not

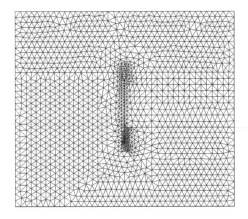

Figure 6. Initial finite element mesh used with DRAC (step 1).

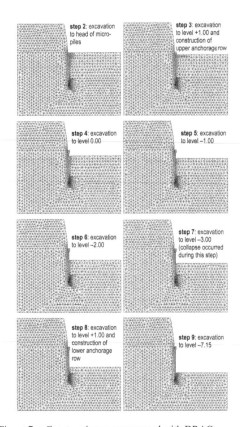

Figure 7. Construction sequence used with DRAC.

reached because collapse occurred in some of the previous steps).

In those tables, the parameter sets for which the limit of 100 kN/m is not reached are indicated in *italic* type. All these sets contain values of the strength parameters much larger than those determined most likely in the previous section. Therefore, since for the remaining parameter sets the load is larger than the maximum admissible, we may conclude that at the moment of collapse the load on the anchorages was larger than its strength, and the cables were fully in the plastic regime.

3.2 Using DRAC

DRAC (Prat et al., 1993) is a general purpose finite element system developed specifically to perform analysis of geotechnical engineering problems. DRAC is a non-linear code allowing 2D and 3D analysis, and includes zero-thickness interface elements needed in solving soil and rock mechanics problems to simulate discontinuities and contact surfaces. Also available are rod elements, used in the simulation of anchorages and struts. For the current analysis, a 2D finite element model in plane strain has been developed (Figure 6, representing step 1).

Computations have been performed for each of the parameter sets described in Table 1. The sequence of excavation and construction of the micro-piles and anchorages has been simulated in 8 additional steps, as depicted in Figure 7, with combination C-1 defined in the previous section.

Although collapse occurred when excavation reached the depth corresponding to step 7, the numerical analysis has been carried out to the end of the described sequence, except when prevented by lack of convergence, an indicator that collapse conditions were reached.

The numerical model is made of triangular and quadrilateral elements. The mesh is denser near the micro-piles wall. The rock mass is modelled with a Mohr-Coulomb material law, with the cohesion and friction angle corresponding to each of the parameter sets defined in Table 1. The rest of the material parameters are taken from Table 2.

The anchorages were simulated by rod elements, fixed at its ends to the wall elements and to the rock mass. No interaction between the rod elements and the surrounding material is modelled. The rod elements, as well as the elements representing the wall were modelled with a linear elastic material law.

The finite element analysis using DRAC has been used as a tool to support the results described before in this paper, and therefore it has not been carried out to the fullest extent possible. A fully three-dimensional analysis of the wall and foundation mass would be necessary to fully understand the collapse circumstances. However, for the purposes of the present study this would not be justifiable because of the computational cost involved.

In general, the results obtained from the finite element analysis corroborate the results obtained with the method described in section 3.1. Especially illustrative are the results showing the plastic deformation contours.

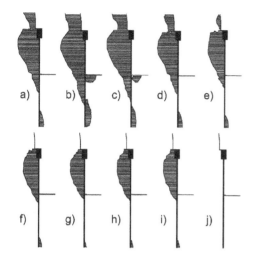

a) b) c) d) e)

f) g) h) i) j)

Figure 8. Plastic points obtained with DRAC near the micro-pile wall at the end of step 7 for parameter sets *a*) H4, *b*) H7, *c*) H8, *d*) H9, *e*) H10, *f*) H14, *g*) H16, *h*) H18, *i*) H27 and *j*) H36.

High values of plastic deformation indicate the zones where the material has reached its yield surface. When this zone, where plastic deformation is high extends to a large volume this indicates that the material has reached a global collapse condition.

Figure 8 shows the extent of plastic points at the end of step 7 for some selected parameter sets. The figure shows that for low values of cohesion and friction angle, the volume of the plastic zone is large, whereas this volume decreases when the values of these parameters increase. In particular, for the values of c' and φ' used in the original design of the retaining wall (50 kPa and 44° respectively, parameter set H36, Figure 8*j*) no plastic zone appears to develop during loading.

The main conclusion of this analysis is that with the strength parameters used in the original design, collapse cannot happen. For collapse to occur, the values of these parameters must be considerably lower, in the order of magnitude indicated in section 3.1.1 above. The results of this 2D finite element analysis support those obtained in section 3.1, strengthening the conclusions presented there.

4 ANALYSIS OF THE RESULTS

The three main issues regarding the collapse of the wall are: the strength of the rock mass, the influence of the water table (if present), and the earth pressure on the wall due to varying slope height overburden. Comparing the results obtained in this work with the original project, the following points can be made:

4.1 About the strength parameters of the rock mass

The geotechnical report on which the original design was based specified zero cohesion and a friction angle of 44°. This was based on a laboratory analysis of a single sample taken from a depth of 2.3 to 2.6 m, about 35 m from where the wall was being constructed.

It seems rather risky to have adopted, as representative of all material, the values obtained from a single sample taken at a considerable distance, since the geological and geotechnical characteristics are very different: the sample used for testing is a clayey gravel (GP-GC) with a natural water content of 6% and a dry density of 20 kN/m³, while the rock mass in the collapsed zone is a slate with a natural water content of 0.05% and dry density of 26 kN/m³. It seems unlikely that the strength of the latter material be the same as the one used in the laboratory tests.

It seems also risky to have taken values as high as the ones in the original project, even for a temporary wall, without taking into account relevant warnings issued in the geotechnical report.

The present analysis shows that the strength parameters would be $c' \approx 0–10\,kPa$, $\varphi' \approx 30°–35°$ in the most unfavourable case, with water table at the surface. In the case of a deep water table, these parameters would be $c' \approx 0\,kPa$, $\varphi' \approx 25°–30°$. In any case, the friction angle is considerable smaller than the value used in the original design. The values of the cohesion deduced from the backanalysis are in all cases smaller than 10 kPa, even for the temporary wall that collapsed. This value is only a small fraction of the 50 kPa assumed in the original project. With this value of cohesion, the backanalysis never reaches failure conditions.

4.2 About the water table

The geotechnical report indicates that the water table is located at depth, in stationary conditions. This assumption was made in the original project, without considering that after periods of intense rain the water table was know to rise to shallower positions. Research on rainfall records show, nevertheless, that during the days preceding the collapse there were no significant precipitations, and therefore rise of the water table due to rain is unlikely.

There is also the question whether seepage from a nearby pond might have caused water to flow to the collapsed zone. Although there is evidence that such seepage did occur, it seems unlikely that the minor volume of water involved could have caused a rise of the water table. However, the presence of water might have caused a reduction of the apparent strength if the material was partially saturated, because the additional strength due to suction vanishes once it becomes fully saturated.

In our case, it seems unlikely that water (from rain or from seepage) was determinant of the collapse. Only in the case of a "dry" situation bordering a limit state, with safety factors near unity, a variation of water content might by the single collapse trigger.

4.3 About the earth overburden on the wall

The original topographic survey did not provide information of the real topography beyond the property limits. Because of that, in the original project the ground surface was taken as horizontal and the overburden on the wall equivalent to the weight of 5 m high embankment.

After the collapse, a new topographic survey extending beyond the property limits showed an uphill sloping surface, with an overburden equivalent to the weight of 10 m high embankment, double of what was used in the original project.

The numerical analysis shows that, assuming an overburden of 5 m of earth, and using $c' = 0$ kPa and $\varphi' = 30°$, the safety factor of the friction angle ranges from 1.2 to 1.5 depending on the presence or non-presence of water in the ground.

5 SUMMARY AND CONCLUSIONS

A simple backanalysis using two numerical methods (one specific for flexible earth-retaining structures, the second a general purpose finite element analysis system) has been used to determine the causes and mechanisms that lead to the failure of a micro-pile wall during construction. Both methods of analysis have been used to simulate the actual construction sequence, and to determine the most likely strength parameters needed for the collapse to occur. The analysis serves also to see the impact on the collapse of other agents, such as the magnitude of earth overburden on the wall or the presence of water in the ground. The main conclusions can be summarized as follows:

1. The most probable cause of the collapse of the micro-pile wall was the use of incorrect values of the strength of the rock mass. The friction angle used in the original project was about 50% larger than the most likely value derived from the present analysis. The cohesion adopted on that project (50 kPa) was much too excessive, since laboratory tests provided a near-zero value for that parameter. The present backanalysis also shows that the most likely value for the cohesion was less than 10 kPa.
2. Collapse appears to be more sensitive to the friction angle than to cohesion.
3. The limitations of the original topographic survey, not extending beyond the property limits, lead to an erroneous estimation of the overburden pressure on the wall which was taken to be only 50% of the actual one. If the strength of the rock mass had been

the one assumed in the original project, this error probably would have had no consequences, according to the results of the present work. However, it did have an impact, since the actual strength was much lower than the one used in the original project.

4. The presence of water does not seem a determinant agent of collapse. The absence of significant rainfall in the days previous to the failure suggests that the water table was indeed low. Seepage from a nearby pond appears too small to have been the trigger of the collapse. Only in a very limit situation the reduction of apparent strength due to vanishing suction as the material becomes saturated might have had a more substantial impact.
5. Overall, it seems reasonable to assume that the three main factors analysed (strength parameters, water, and overburden) were agents in the collapse, with a different degree of implication. But it is quite difficult to quantify precisely this implication. From the results of the present analysis, it would appear that the primary factor was the over-estimation of strength parameters, followed by the underestimation of the overburden pressure and finally, to a certain extent, the possible strength reduction due to the presence of water.
6. The chain of events that ended in the collapse of the micro-pile wall can be summarized as follows: a) the real strength of the rock mass was much lower than the values used in the project; b) the overburden on the wall was actually twice the value assumed in the project; c) the presence of water might have contributed to an additional loss of strength; d) these three items lead to an earth pressure on the wall much larger than the one used in its design, implying inadmissible stresses on the wall and the anchorages, beyond their bearing capacity.

ACKNOWLEDGMENT

Financial support from the Spanish Ministry of Education and Science, (grant BIA2003-03417) is gratefully acknowledged.

REFERENCES

Bieniawski, Z.T., 1989. Engineering rock mass classification. John Wiley & Sons, New York.
CEAS, 2003. PARATIE for Windows User'a Manual. Centro di Analisi Strutturale, s.r.l. Milano, Italy.
Hoek, E. and Bray, J.W., 1981. Rock slope engineering. Spon Press – Institution of Mining and Metallurgy, London.
Hoek, E., Carranza-Torres, C. and Corkum, B., 2002. Hoek-Brown failure criterion – 2002 edition, Proceedings North American Rock Mechanics Society Meeting, Toronto.
Prat, P.C., Gens, A., Carol, I., Ledesma, A. and Gili, J.A., 1993. DRAC: A computer software for the analysis of rock mechanics problems. In: H. Liu (Editor), Application of computer methods in rock mechanics. Shaanxi Science and Technology Press, Xian, China, pp. 1361–1368.

7. Numerical modelling and miscellaneous

*Numerical Modelling of Construction Processes in Geotechnical Engineering for
Urban Environment – Triantafyllidis (ed)
© 2006 Taylor & Francis Group, London, ISBN 0 415 39748 0*

Incorporation of meta-stable structure into hypoplasticity

D. Mašín

Charles University, Prague

ABSTRACT: An existing approach to constitutive modelling of structured soils is in this paper applied to
hypoplastic models. The proposed approach is based on the modification of the state boundary surface in such
a way that the models predict different stress-dilatancy behaviour of structured and reference materials. The
concept is applied to hypoplastic models for both clays and for granular materials. Good predictive capabilities
of the modified model for clays are demonstrated, model for granular materials requires further modifications
in order to take into account different small- to medium- strain behaviour of structured and reference materials.

1 INTRODUCTION

This paper introduces a conceptual framework for
application of existing approaches to constitutive modelling of structured soils[1], developed usually within the
framework of elasto-plasticity, to hypoplastic models
(Kolymbas 1991). The paper discusses general aspects
of the modelling of structural effects in hypoplasticity, its intention is neither compilation of the available
experimental evidence regarding the behaviour of
structured soils, nor thorough evaluation of predictive
capabilities of the proposed constitutive models.

2 REFERENCE CONSTITUTIVE MODELS

The non-linear tensorial equation of the considered
sub-class of hypoplastic models reads (Gudehus 1996)

$$\mathring{\mathbf{T}} = f_s \mathcal{L} : \mathbf{D} + f_s f_d \mathbf{N} \|\mathbf{D}\| \tag{1}$$

The models are formulated using two tensor-valued
functions $\mathcal{L}$ and $\mathbf{N}$ and two scalar factors f_s and f_d. The
soil state is characterised by the Cauchy stress ($\mathbf{T}$) and
void ratio (e). The scalar functions f_s and f_d are named
barotropy and *pyknotropy* factors (Gudehus 1996),
which incorporate the influence of the mean stress and
relative density (overconsolidation ratio), respectively.

   The models characterised by Eq. (1) are suitable
for predicting behaviour in the medium- to large-
strain range. In order to predict correctly the high
stiffness in the small strain range, Eq. (1) must be modified, for example by the *intergranular strain concept*

---

[1] In the following, the term 'structured soil' will be used
for soil, which due to different fabric or additional bonding behaves differently compared to a 'reference' material.
As the reference material equivalent soil reconstituted under
standard conditions (Burland 1990) is usually considered.

(Niemunis and Herle 1997). The application of the
modified model is, however, outside the scope of the
paper.

2.1 Hypoplastic model for clays

A hypoplastic constitutive model for clays (abbreviated HC) was developed by Mašín (2005a) as a
modification of the hypoplastic model for soils with
low friction angles by Herle and Kolymbas (2004).

   The number of parameters of the model and their
physical meaning corresponds to the parameters of
the Modified Cam clay model, the non-linear character of the model, however, provides a significant
improvement of predictions of the hypoplastic model
with respect to the Modified Cam clay model, as
demonstrated by Mašín et al. (2005).

   The model makes use of the following five consti-
tutive parameters: φ_c, N, λ^*, κ^* and r. φ_c is the critical
state friction angle, the isotropic virgin compression
line has the following formulation (Butterfield 1979):

$$\ln(1 + e) = N - \lambda^* \ln\left(\frac{p}{p_r}\right) \tag{2}$$

with parameters N and λ^* and the reference stress
$p_r = 1$ kPa. The parameters κ^* and r may be considered
as factors that control bulk (κ^*) and shear (r) moduli
of overconsolidated specimens.

2.2 Hypoplastic model for granular materials

A model by von Wolffersdorff (1996) (abbreviated
VW) may be seen as a representative example of
hypoplastic models for granular materials developed at
the University of Karlsruhe. Its calibration is described
in detail in Herle and Gudehus (1999).

   Model is in the stress-void ratio space characterised
by three pressure-dependent limit states, describing the

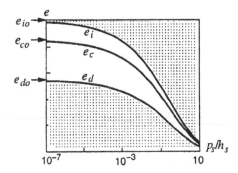

Figure 1. Pressure-dependent limit states of the VW model (Herle and Gudehus 1999).

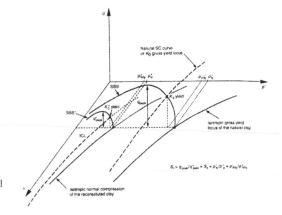

Figure 2. Theoretical framework for structured fine-grained materials (Cotecchia and Chandler 2000).

upper and lower bound of void ratio (e_i and e_d) and a critical state void ratio (e_c) (Fig. 1), defined using a power-law

$$\frac{e_i}{e_{i0}} = \frac{e_c}{e_{c0}} = \frac{e_d}{e_{d0}} = \exp\left[-\left(\frac{-\mathrm{tr}\mathbf{T}}{h_s}\right)^n\right] \qquad (3)$$

with parameters h_s, n, e_{i0}, e_{c0} and e_{d0}. In addition, three parameters are needed, namely critical state friction angle (φ_c) and parameters α and β, which control the influence of *pyknotropy*.

3 CONCEPTUAL APPROACH FOR INCORPORATING META-STABLE STRUCTURE INTO CONSTITUTIVE MODELS

A conceptual framework for the behaviour of structured fine-grained soils was presented, e.g., by Cotecchia and Chandler (2000). Behaviour of structured granular materials, together with the interpretation by means of a constitutive model, was presented by Lagioia and Nova (1995).

Cotecchia and Chandler (2000) demonstrated that the influence of structure in fine-grained soils can be quantified by the different size of the state boundary surfaces[2] of the structured and reference materials (Fig. 2). Assuming a geometric similarity between the state boundary surfaces appears to be a reasonable approximation, although strongly anisotropic natural soils may exhibit SBS which is not symmetric about the isotropic axis. Similar findings were reported for granular materials by Lagioia and Nova (1995), who studied the behaviour of calcarenite. They observed that the state boundary surface of naturally cemented material has similar shape as the reconstituted soil, however SBS of the natural material is bigger and it is

---

[2] State boundary surface (SBS) is defined as a boundary of all possible states of a soil element in the stress-void ratio space.

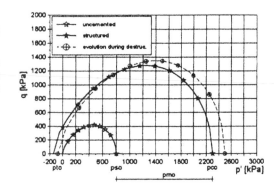

Figure 3. Yield points and theoretical interpretation of natural and uncemented granular material (Lagioia and Nova 1995).

shifted such that the state can access region of tensile stresses (Fig. 3).

These observations are, in principle, applied in most of the currently available constitutive models for structured soils. In general, two additional state variables describing the effects of structure are needed, namely the ratio of sizes of SBSs of natural and reference materials, referred to as 'sensitivity' (s), and the shift of the SBS towards the region of tensile stresses measured along the isotropic axis, denoted by Lagioia and Nova (1995) p_t (with initial value p_{t0}, Fig. 3).

As s and p_t represent natural fabric and degree of bonding between soil particles, they may in the scale of engineering time only decrease or remain stationary. The limit values usually characterise the reference soil ($s = 1$ and $p_t = 0$), although higher values may be reasonable for soils with 'stable' elements of structure caused by natural fabric (Baudet and Stallebrass 2004). s and p_t are usually considered functions of accumulated plastic strain.

4 STATE BOUNDARY SURFACE IN HYPOPLASTICITY

State boundary surface is an important feature of soil behaviour that controls different stress-dilatancy responses of structured and reconstituted materials (Sec. 3). Elasto-plastic constitutive models incorporate state boundary surface, which has a pre-defined shape, explicitly by definition of a yield (or bounding in the case of advanced models) surface and a hardening law. On the other hand, hypoplastic models predict state boundary surface as a consequence of a particular choice of tensorial functions $\mathcal{L}$ and $\mathbf{N}$ and scalar factors f_s and f_d.

Mašín and Herle (2005b) studied the shape of the state boundary surface predicted by the hypoplastic model for clays. Using the concept of normalised incremental response envelopes they demonstrated that although the model predicts the state boundary surface, its explicit mathematical formulation is not available.

The model, however, allows us to derive mathematical formulation for swept-out-memory (limit) states, which may be considered as attractors of soil behaviour (Gudehus 1995). Limit states (defined by constant $\vec{\mathbf{T}} = \dot{\mathbf{T}}/||\dot{\mathbf{T}}||$ and void ratio evolving along normal compression lines) are achieved asymptotically after sufficiently long proportional (constant $\vec{\mathbf{D}}$) deformation paths. Mašín and Herle (2005b) demonstrated that in the stress-void ratio space limit states constitute a surface (named swept-out-memory (SOM) surface) and that this surface is a *close approximation of the state boundary surface*.

Because at swept-out-memory conditions $\dot{\mathbf{T}} \parallel \mathbf{T}$, it is possible to introduce a scalar multiplier γ such that

$$\dot{\mathbf{T}} = \gamma \vec{\mathbf{T}} \tag{4}$$

Eq. (1) therefore, at swept-out-memory conditions, reduces to

$$\gamma \vec{\mathbf{T}} = f_s \mathcal{L} : \mathbf{D} + f_s f_d \mathbf{N} ||\mathbf{D}|| \tag{5}$$

The second condition for SOM states (void ratio evolving along normal compression line) is in considered hypoplastic models described by

$$f_d = 0 \tag{6}$$

SOM conditions are for a particular hypoplastic model fully described if we solve Eqs. (4) and (5) such that for *given* $\mathbf{T}$ and $||\mathbf{D}||$ we find corresponding γ, $\vec{\mathbf{D}}$ and f_d. For HC and VW models solution was derived by Mašín and Herle (2005a), only the main conclusions are reported in the following.

4.1 *SOM surface of the hypoplastic model for clays*

Solution of Eqs. (4) and (5) for the hypoplastic model for clays is relatively straightforward, as the multiplier

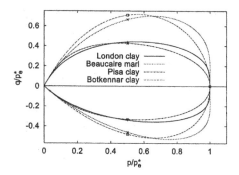

Figure 4. SOM surface of the hypoplastic model for clays for four different sets of material parameters.

γ is independent of void ratio. This follows from the fact that for the HC model the Eq. (1) is for $\dot{f}_d = 0$ positively homogeneous of degree 1 with respect to $\mathbf{T}$ (the model assumes linear normal compression lines in the $\ln p : \ln(1+e)$ space). Solution for the HC model reads

$$f_d = ||f_s \mathcal{A}^{-1} : \mathbf{N}||^{-1} \tag{7}$$

$$\vec{\mathbf{D}} = -\frac{\mathcal{A}^{-1} : \mathbf{N}}{||\mathcal{A}^{-1} : \mathbf{N}||} \tag{8}$$

with

$$\mathcal{A} = f_s \mathcal{L} + \frac{1}{\lambda^*} \mathbf{T} \otimes \mathbf{1} \tag{9}$$

The HC model assumes the following expression for the *pyknotropy* factor f_d:

$$f_d = \left(\frac{2p}{p_e^*}\right)^\alpha \tag{10}$$

where α is a scalar factor calculated from model parameters and p_e^* is Hvorslev's equivalent pressure at the isotropic normal compression line. (2). Therefore it is possible to calculate the value of p_e^* for *given* $\mathbf{T}$ (from (10) and (7))

$$p_e^* = -\frac{2}{3} \operatorname{tr} \mathbf{T} ||f_s \mathcal{A}^{-1} : \mathbf{N}||^{1/\alpha} \tag{11}$$

The SOM surface of the HC model may be conveniently plotted in the normalised space $\mathbf{T}/p_e^*$.

The shape of the SOM surface of the HC model for four different sets of material parameters in the normalised space $p/p_e^* : q/p_e^*$ is plotted in Fig. 4, corresponding parameters are in Tab. 1 (London clay – Mašín 2005a; Beaucaire marl – Mašín et al. 2005; Bothkennar and Pisa clay – Mašín 2005b).

4.2 *SOM surface of the hypoplastic model for granular materials*

Explicit solution of Eqs. (4) and (5) is not available for the hypoplastic model for granular materials. In

Table 1. Parameters of the hypoplastic model for clays.

Soil	$\varphi_c \, [°]$	λ^*	κ^*	N	r
Lond. c.	22.6	0.11	0.014	1.375	0.4
Beau. m.	33	0.057	0.007	0.85	0.4
Pisa c.	21.9	0.14	0.005	1.56	0.2
Both. c.	35	0.119	0.002	1.344	0.05

Table 2. Parameters of the hypoplastic model for granular materials for Zbraslav sand (Herle and Gudehus 1999).

$\varphi_c \, [°]$	h_s [MPa]	n	e_{d0}	e_{c0}	e_{i0}	α	β
31	5700	0.25	0.52	0.82	0.95	0.13	1.00

this case, the *pyknotropy* factor at SOM conditions reads

$$f_d = \left[\|\mathbf{B}\|^2 + \left(\frac{\|\mathbf{C}\| \left(\frac{1+e}{e}\right) \mathrm{tr}\mathbf{B}}{G - \left(\frac{1+e}{e}\right)\mathrm{tr}\mathbf{C}} \right)^2 + \frac{2(\mathbf{B}:\mathbf{C})\mathrm{tr}\mathbf{B} \left(\frac{1+e}{e}\right)}{G - \left(\frac{1+e}{e}\right)\mathrm{tr}\mathbf{C}} \right]^{-\frac{1}{2}} \quad (12)$$

with

$$\mathbf{B} = \mathcal{L}^{-1} : \mathbf{N} \quad (13)$$

$$\mathbf{C} = \frac{\mathcal{L}^{-1} : \vec{\mathbf{T}}}{f_s} \quad (14)$$

$$G = \frac{n}{h_s} \mathrm{tr}\vec{\mathbf{T}} \left(\frac{3p}{h_s} \right)^{(n-1)} \quad (15)$$

Eq. (12) is an implicit equation for f_d, as the hypoplastic model for granular materials assumes

$$f_d = \left(\frac{e - e_d}{e_c - e_d} \right)^{\alpha} \quad (16)$$

and therefore

$$e = f_d^{(1/\alpha)}(e_c - e_d) + e_d \quad (17)$$

In addition, unlike in the case of the HC model, also factor f_s of the VW model is dependent on e. Eq. (12) may be, however, solved numerically.

Due to the particular formulation of factor f_d (16), different constant-volume cross-sections through the SOM surface have different shape (normalisation with respect to p_e^* is not applicable for the VW model). For Zbraslav sand parameters (Tab. 2) these cross-sections are shown in Fig. 5.

5 META-STABLE STRUCTURE IN HYPOPLASTICITY

As mentioned in Sec. 4, Mašín and Herle (2005b) demonstrated that the swept-out-memory surface is a close approximation of the state boundary surface.

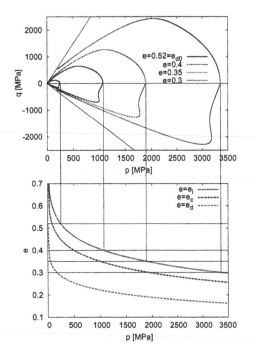

Figure 5. Constant-volume cross-sections through the SOM surface of the VW model for Zbraslav sand parameters.

Therefore, hypoplastic models should, in principle, allow us to modify the size of the state boundary surface in such a way that the models predict different stress-dilatancy behaviour of structured and reference materials, using the approach commonly applied in elasto-plastic models for structured soils (Sec. 3). The incorporation of meta-stable structure into two reference hypoplastic models will be discussed in this section.

5.1 Hypoplastic model for clays

Modification of the reference HC model for predictions of structured soils has been proposed by Mašín (2005b).

First, the reference model will be enhanced for predictions of the behaviour of soil with "stable" structure ($\dot{s} = 0, \dot{p}_t = 0$). A study of the expression for the SOM surface of the HC model (Sec. 4.1) reveals that the size of the SOM surface would be increased by the factor s by a simple replacement of p_e^* in the expression for f_d by sp_e^*:

$$f_d = \left(\frac{2p}{sp_e^*} \right)^{\alpha} \quad (18)$$

Effect of this modification on the SBS in the stress-void ratio space is shown in Fig. 6.

Further, the origin on the SOM surface would be shifted towards negative stresses along the isotropic

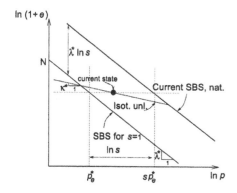

Figure 6. SBS in the $\ln p : \ln (1+e)$ space for $s > 1$.

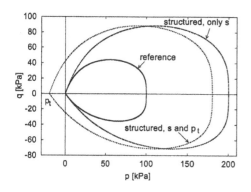

Figure 7. Constant-volume cross-sections through the SOM surface of the HC model with stable structure (reference parameters for London clay).

axis, if the Cauchy stress tensor $\mathbf{T}$ was in the constitutive model replaced by the "transformed stress tensor" $\mathbf{T}^t$ (introduced already by Bauer and Wu (1994) who modified the early hypoplastic model considering $\mathbf{T}$ the only state variable)

$$\mathbf{T}^t = \mathbf{T} - p_t \mathbf{1} \tag{19}$$

Note that this is the most simple modification (constant p_t, $p_t < 0$), which implies that the SOM surface than does not have a unique image in the normalised space $\mathbf{T}/p_e^*$. Effect of the two modifications on the constant volume cross-section through the SOM surface of the HC model is demonstrated in Fig. 7.

Second, the meta-stable structure will be incorporated into the constitutive model by defining the evolution equations for additional state variables and by modifying the *barotropy* factor f_s. For simplicity only one additional state variable s will be considered ($p_t = 0$). Evolution equation for s may read (after Baudet and Stallebrass 2004)

$$\dot{s} = -\frac{k}{\lambda^*}(s - s_f)\dot{\epsilon}^d \tag{20}$$

where s_f is the ultimate value of s (typically for bonded materials $s_f = 1$), k is a parameter controlling the rate of structure degradation and $\dot{\epsilon}^d$ is a "damage" strain rate with formulation (Mašín 2005b)

$$\dot{\epsilon}^d = \sqrt{(\dot{\epsilon}_v)^2 + \frac{A}{1 - A}(\dot{\epsilon}_s)^2} \tag{21}$$

$\dot{\epsilon}_v$ and $\dot{\epsilon}_s$ are rates of the volumetric and shear strains respectively[3] and A is a non-dimensional scaling parameter that controls their relative contribution to the structure degradation.

The *barotropy* factor f_s needs to be modified in order to ensure consistency in model predictions and the pre-defined structure degradation law (20). Formulation of the HC model for the isotropic compression from the isotropic normally compressed state reads

$$\dot{p} = -\left[\frac{1}{3(1+e)}f_s\left(3 + a^2 - 2^\alpha a\sqrt{3}\right)\right]\dot{e} \tag{22}$$

The isotropic normal compression line of the model incorporating structure is given by (see Fig. 6)

$$\ln(1+e) = N + \lambda^* \ln s - \lambda^* \ln\left(\frac{p}{p_r}\right) \tag{23}$$

Time differentiation of (23) results in

$$\frac{\dot{e}}{1+e} = \lambda^*\left(\frac{\dot{s}}{s} - \frac{\dot{p}}{p}\right) \tag{24}$$

The isotropic formulation of the structure degradation law (20–21) is

$$\dot{s} = \frac{k}{\lambda^*}(s - s_f)\frac{\dot{e}}{1+e} \tag{25}$$

Combination of (24) and (25) yields

$$\frac{\dot{p}}{p} = -\left[\frac{s - k(s - s_f)}{\lambda^* s}\right]\frac{\dot{e}}{1+e} \tag{26}$$

which may be compared with (22) to find an expression for the *barotropy* factor f_s of the new hypoplastic model:

$$f_s = \frac{3p}{\lambda^* s}[s - k(s - s_f)]\left(3 + a^2 - 2^\alpha a\sqrt{3}\right)^{-1} \tag{27}$$

The influence of parameter k on predictions of an isotropic compression test is shown in Fig. 8, evaluation of model predictions in Sec. 6.

---

[3] In hypoplasticity, Eq. (21) is defined in terms of total, instead of plastic strain rates. The difference between the two definitions is in the large-strain range insignificant, as high "truly" elastic stiffness causes the elastic part of the total strain increment to be negligible with respect to the plastic part.

287

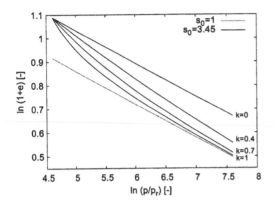

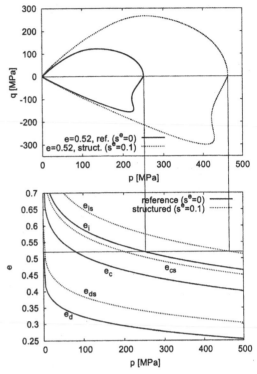

Figure 8. Prediction of an isotropic compression test by the structured HC model, influence of parameter k (parameters for Pisa clay).

5.2 Hypoplastic model for granular materials

Hypoplastic model for granular materials may be modified to incorporate the effects of inter-particle bonding in a similar way as the HC model. It is, however, worth noting that bonding in granular materials influences the small-strain behaviour more significantly than structure effects in clays (by significant enlargement of the elastic range – see, e.g., Coop and Atkinson 1993). Therefore, the VW model for structured soils will in its basic version necessarily perform poorly in the small- to medium-strain range and additional measures (e.g. on the basis of the intergranular strain concept) are needed to ensure its practical applicability.

It follows from Sec. 4.2 that the size of the SBS of the VW model may be increased by modifying the characteristic void ratios e_{i0}, e_{c0} and e_{d0}. Their increase by a factor s^e in the form

$$e_{x0s} = e_{x0} + s^e \quad (28)$$

where x stands for i, c, d and e_{x0s} are characteristic void ratios of a structured soil at $p = 0$ kPa, would have the influence on the size of the SBS as demonstrated in Fig. 9.

Characteristic void ratios e_{is}, e_{ds} and e_{cs} are now calculated by

$$e_{xs} = (e_{x0} + s^e)\exp\left[-\left(\frac{-\mathrm{tr}\mathbf{T}}{h_s}\right)^n\right] \quad (29)$$

where e_{xs} comes in place of e_x in the expression of factors f_d and f_s. Note that e_{xs} are *not* the state limits of the structured soil, as the evolution equation for s^e is a natural component of the modified constitutive equation. Similarly to the HC model, the SBS may be further shifted by introducing the transformed stress tensor $\mathbf{T}'$.

The evolution equation for s^e may read (similarly to the structured HC model)

$$\dot{s}^e = -k^e(s^e - s^e_f)\dot{\epsilon}^d \quad (30)$$

Figure 9. Constant-volume cross-sections through the SOM surface of the VW model for reference ($s^e = 0$) and structured ($s^e = 0.1$) soil (reference parameters for Zbraslav sand).

where typically $s^e_f = 0$ for bonded material, $\dot{\epsilon}^d$ may be calculated according to (21), k^e controls the rate of the structure degradation.

Finally, the factor f_s must be modified to ensure that the consistency condition at the isotropic normally compressed state is not violated. Formulation of the VW model for the isotropic compression from the isotropic normally compressed state is given by

$$\dot{p} = \frac{-f_s}{3(1+e)}\left[3 + a^2 - a\sqrt{3}\left(\frac{e_{i0} - e_{d0}}{e_{c0} - e_{d0}}\right)^\alpha\right]\dot{e} \quad (31)$$

The isotropic normal compression line of the model incorporating structure reads

$$e = (e_{i0} + s^e)\exp\left[-\left(\frac{-\mathrm{tr}\mathbf{T}}{h_s}\right)^n\right] \quad (32)$$

Time differentiation of (32) results in

$$\frac{\dot{e}}{e} = \frac{\dot{s}^e}{e_{i0} + s^e} - n\frac{3\dot{p}}{h_s}\left(\frac{3p}{h_s}\right)^{n-1} \quad (33)$$

The isotropic formulation of the structure degradation law (30) and (21) is

$$\dot{s}^e = k^e(s^e - s^e_f)\frac{\dot{e}}{1 + e} \quad (34)$$

288

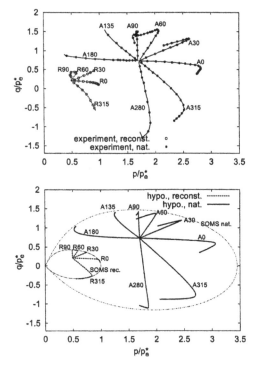

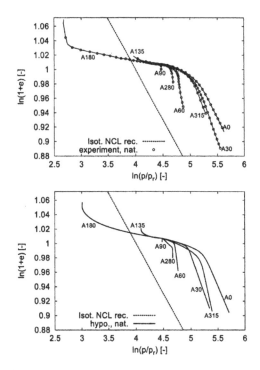

Figure 10. Normalised stress paths of the natural and reconstituted Pisa clay (data from Callisto and Calabresi 1998) and predictions by the structured HC model (from Mašín 2005b).

Figure 11. Experiments on natural Pisa clay plotted in the $\ln(p/p_r) : \ln(1 + e)$ space (data from Callisto and Calabresi 1998) and predictions by the structured HC model (from Mašín 2005b).

Combination of (33) and (34) yields

$$n\frac{3\dot{p}}{h_s}\left(\frac{3p}{h_s}\right)^{n-1} = \left[\frac{k^e(s^e - s_f^e)}{(1+e)(e_{i0} + s^e)} - \frac{1}{e}\right]\dot{e} \qquad (35)$$

The factor f_s of the structured VW hypoplastic model follows from the comparison of (31) and (35) and unlike the factor f_s of the HC model, it contains also the *pyknotropy* component $(e_{is}/e)^\beta$ (for details see Gudehus 1996).

$$f_s = \left(\frac{e_{is}}{e}\right)^\beta \left[\frac{1 + e_{is}}{e_{is}} - \frac{k^e\left(s^e - s_f^e\right)}{e_{i0} + s^e}\right]$$

$$\frac{h_s}{n}\left(\frac{3p}{h_s}\right)^{1-n}\left[3 + a^2 - a\sqrt{3}\left(\frac{e_{i0} - e_{d0}}{e_{c0} - e_{d0}}\right)^\alpha\right]^{-1} \qquad (36)$$

6 EVALUATION

Thorough evaluation of the structured HC model is given in Mašín (2005b). In Figs. 10, 11 and 12 are shown predictions of the experiments on natural Pisa clay reported by Callisto and Calabresi (1998). All the parameters of the structured HC model except the

parameters that quantify the effects of structure (k, A and s_f) were calibrated solely on the basis of experiments on reconstituted clay. Only two experiments (A0 and A90, for normalised stress path see Fig. 10) were used for calibration of parameters k, A and s_f. Although the model was calibrated using a simple procedure with a minimal number of laboratory experiments required, its predictions are satisfactory in the entire range of stress paths directions. Parameters of the structured HC model for natural Pisa clay are in Tab. 3.

7 CONCLUSIONS

The paper presented a conceptual framework for incorporation of meta-stable structure into hypoplastic constitutive models. The approach is based on the modification of the state boundary surface predicted by the models. By increasing the size of the state boundary surface it is possible to take into account different stress-dilatancy behaviour of structured and reference soils.

The approach is equivalently applicable to the two reference hypoplastic models. Predictive capabilities of the structured hypoplastic model for clays were

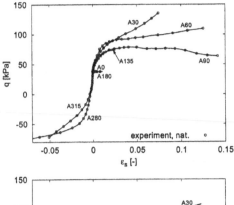

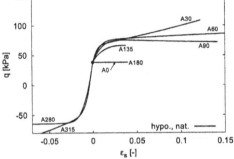

Figure 12. (a) $\epsilon_s : q$ curves from experiments on natural Pisa clay (data from Callisto and Calabresi 1998) and predictions by the structured HC model (from Mašín 2005b).

Table 3. Parameters of the structured HC model for natural Pisa clay.

φ_c	λ^*	κ^*	N	r	k	A	s_f
21.9°	0.14	0.005	1.56	0.2	0.4	0.1	1

demonstrated, structured hypoplastic model for granular materials, however, requires further modification in order to take into account the large quasi-elastic range caused by inter-particle bonding.

ACKNOWLEDGEMENT

The author is grateful for the financial support by the research grants GAAV A200710605 and GAUK 331/2004/B-GEO/PřF and to Dr. Luigi Callisto for providing data on Pisa clay.

REFERENCES

Baudet, B. A. and S. E. Stallebrass (2004). A constitutive model for structured clays. *Géotechnique 54*(4), 269–278.

Bauer, E. and W. Wu (1994). Extension of hypoplastic constitutive model with respect to cohesive powders. In Siriwardane and Zeman (Eds.), *Computer methods and advances in geomechnics*, pp. 531–536. A.A.Balkema, Rotterdam.

Burland, J. B. (1990). On the compressibility and shear strength of natural clays. *Géotechnique 40*(3), 329–378.

Butterfield, R. (1979). A natural compression law for soils. *Géotechnique 29*(4), 469–480.

Callisto, L. and G. Calabresi (1998). Mechanical behaviour of a natural soft clay. *Géotechnique 48*(4), 495–513.

Coop, M. R. and J. H. Atkinson (1993). The mechanics of cemented carbonate sands. *Géotechnique 43*(1), 53–67.

Cotecchia, F. and J. Chandler (2000). A general framework for the mechanical behaviour of clays. *Géotechnique 50*(4), 431–447.

Gudehus, G. (1995). Attractors for granular storage and flow. In *3rd European Symposium – Storage and Flow of Particulate Solids, Paper for the conf. 'Partec 95'*, pp. 333–345.

Gudehus, G. (1996). A comprehensive constitutive equation for granular materials. *Soils and Foundations 36*(1), 1–12.

Herle, I. and G. Gudehus (1999). Determination of parameters of a hypoplastic constitutive model from properties of grain assemblies. *Mechanics of Cohesive-Frictional Materials 4*, 461–486.

Herle, I. and D. Kolymbas (2004). Hypoplasticity for soils with low friction angles. *Computers and Geotechnics 31*(5), 365–373.

Kolymbas, D. (1991). An outline of hypoplasticity. *Archive of Applied Mechanics 61*, 143–151.

Lagioia, R. and R. Nova (1995). An experimental and theoretical study of the behaviour of a calcarenite in triaxial compression. *Géotechnique 45*(4), 633–648.

Mašín, D. (2005a). A hypoplastic constitutive model for clays. *International Journal for Numerical and Analytical Methods in Geomechanics 29*(4), 311–336.

Mašín, D. (2005b). A hypoplastic constitutive model for clays with meta-stable structure. *Canadian Geotechnical Journal (submitted)*.

Mašín, D. and I. Herle (2005a). State boundary surface in hypoplasticity. In *Proc. Int. workshop Modern trends in geomechanics (in print)*. Vienna, Austria.

Mašín, D. and I. Herle (2005b). State boundary surface of a hypoplastic model for clays. *Computers and Geotechnics (in print)*.

Mašín, D., C. Tamagnini, G. Viggiani, and D. Costanzo (2005). Directional response of a reconstituted fine grained soil. Part II: Performance of different constitutive models. *International Journal for Numerical and Analytical Methods in Geomechanics (submitted)*.

Niemunis, A. and I. Herle (1997). Hypoplastic model for cohesionless soils with elastic strain range. *Mechanics of Cohesive-Frictional Materials 2*, 279–299.

von Wolffersdorff, P. A. (1996). A hypoplastic relation for granular materials with a predefined limit state surface. *Mechanics of Cohesive-Frictional Materials 1*, 251–271.

Numerical Modelling of Construction Processes in Geotechnical Engineering for
Urban Environment – Triantafyllidis (ed)
© 2006 Taylor & Francis Group, London, ISBN 0 415 39748 0

Numerical simulation of consolidation and deformation of pulpy fine slimes due to re-contouring and covering of large uranium mill tailing ponds at Wismut

U. Barnekow, M. Haase & M. Paul

WISMUT GmbH, Chemnitz, Germany

ABSTRACT: Approximately $5.7\,km^2$ of uranium mill tailings ponds containing 150 million m^3 of fine-grained tailings (incl. slimes) were left as part of the legacy of the uranium mining and milling of the former Soviet-German company Wismut in East Germany. From 1951 to 1990, uranium mill tailings were disposed in large tailing ponds and settled, forming 25 m to 60 m thick pulpy fine slime layers. The stabilization and covering of such tailing ponds is to be based on a reliable prediction of the time-settlement and time-deformation of the fine tailings, with respect to the step-wise loading during ongoing tailing pond remediation. The paper presents progress achieved for numerical simulations of time-dependent consolidation and deformations of fine tailings and experiences made for calibration and validation of the applied numerical models. A number of numerical modeling methods have been developed. One- and two-dimensional, nonlinear finite large strain models (NLFS-models) were applied to reconstruct the history of the tailings disposal into the pond, to derive the degree of self-weight consolidation, to assess the measured material parameters, to calculate settlement due to placement of embankments and covers, to estimate the development of settlement troughs and to predict the influence of vertical wick drains on consolidation in ponds filled with inhomogeneous tailings. Special compression tests capable of measuring self weight consolidation of the settled slurries have been developed to help define the (a) void ratio vs. effective stress and (b) void ratio vs. hydraulic conductivity relationships. The paper additionally presents experiences made with the numerical simulation of one-dimensional consolidation of fine tailings during and after disposal period including the numerical simulations applied for cost-effective designing of the cover and for controlling and monitoring the remediation progress.

1 INTRODUCTION

Approximately $5.7\,km^2$ of uranium mill tailing ponds containing 150 million m^3 of fine-grained tailings (incl. slimes) were left as part of the legacy of the uranium mining and milling of the former Soviet-German company Wismut in East Germany. From 1951 to 1990, uranium mill tailings were disposed in large tailing ponds and settled, forming 25 m to 60 m thick pulpy fine slime layers. The stabilization and covering of such tailing ponds is to be based on a reliable prediction of the time-settlement and time-deformation of the fine tailings with respect to the step-wise loading during ongoing tailing pond remediation.

The discharged uranium mill tailings can be characterized as graded suspensions that settled inside the pond. Sand fraction often settled above water table near the discharge spots, forming sandy tailing beaches while silt and clay fraction settled below water table forming thick fine slime layers. The transition zone in between is typically composed of an interlayering of silty clay layers and sand layers.

The remediation approach selected in the mid 90's for the tailings ponds of Wismut GmbH is usually realised in the following steps:

1. At first, the pond water is to be expelled and treated before being discharged into the receiving streams. The air-exposed surfaces of fine slimes are then allowed to dry out to enhance trafficability of fine tailing surfaces.
2. Placement of an interim cover on tailing surfaces to reduce dusting and, particularly on fine slime surfaces, to create a stable working platform and initiate consolidation under surcharge loading. On weak or pulpy fine slimes a geotextile and a geogrid and, if needed, a drainmat are placed in advance of the placement of the first earthen layer. Vertical wick drains are stitched in the uppermost 5 to 6 m of the fine slime layer to enhance trafficability for placement of the interim cover.
3. Reshaping of the surrounding dams with respect to their geotechnical and erosional stability to the long term.

4. Re-contouring of the pond surface into a new surface contour with proper surface runoff conditions stable to the long term with respect to time-deformation of the underlying fine slimes.
5. Placement of the final cover to prevent access to tailings and to control infiltration into the tailings.

For prediction of the consolidation behaviour of the fine slimes one-dimensional, nonlinear finite strain models (NLFS-models) have been applied to

1. recalculate the self weight consolidation of the fine slimes for each tailing zone within the pond during and after the discharge period including deriving the degree of consolidation with time,
2. assess the tailings material properties of the different tailing types that settled in different zones and layers within the pond due to the historic discharge pattern,
3. predict the settlement due to surcharge loading during remediation.

A newly developed radialsymmetric two-dimensional model was applied to predict the influence of vertical wick drains on fine tailing consolidation.

In order to model the consolidation of the slimes under self-weight and initial loading, the hydraulic conductivity and stiffness of the material had to be determined as a function of the changing void ratio. For this we used a newly developed slurry oedometer, capable of measuring the consolidation of slimes having a consistency of thick liquids in response to very small loadings from the early stages of loading on.

Beyond laboratory data, the subdivision of the tailing body into zones required in situ measurements of the tailings properties in their present state and the historical record of discharge period.

2 CONSOLIDATION MODELING OF FINE SLIMES

Due to historic discharge pattern, thick fine slime layers settled in WISMUT's tailing ponds distant from the ancient discharge spots. Borehole profiles located in such tailing zones usually show different types of fine slimes with depth. Along an entire borehole profile of up to 30 m to 60 m there were usually pulpy or weak consistency of the fine slimes observed. In addition, we measured void ratios decreasing with depth from values of e = 3 to 5 (e being the void ratio) near surface to e = 1.5 to 2.5 in a depth of 30 m to 60 m, depending on the tailing type and the effective stress at the given depth. Fine slimes from milling and processing of uranium ore are usually of medium or high plasticity and very compressible. Their hydraulic conductivity very much depend on their void ratio. This could unacceptably elongate the consolidation progress of such slimes, which is needed for a sufficient remediation

progress. In order to plan and control the remediation settlement, predictions need the use of numerical models.

The practical need to describe void ratio dependent soil material behaviour, which reacts to even small changes of stress, led to the development of a one-dimensional Nonlinear-Finite-Strain model, NLFS-Model (GIBSON et al. 1967). One of the first internationally available codes based on the NLFS models was ACCUMV developed by Schiffman (SCHIFFMAN et al. 1992). Commercially available NLFS codes, like ACCUMV3 and FSConsol which have been both applied by WISMUT, are capable of describing in one-dimension (a) the self-consolidation of slimes during the filling process of the pond, (b) the behaviour of stratified tailings consisting of heterogeneous material layers and (c) the effect of time-dependent loading. The above described model ACCUMV uses the following material functions:

The void ratio-stress-relationship

$$e(\sigma'): \quad e = e_0 - c_c \log(\sigma') \qquad (1)$$

The value e_0 is the initial void ratio corresponding to a reference stress of $\sigma' = 0$ kpa at the end of the settling and start of "soil's" consolidation.

Derived from (1), the compressive behaviour of the fine slimes can alternatively be expressed by the stiffness modulus depending on the effective stress with respect to the void ratio:

$$E = (1+e)\frac{d\sigma'}{de} \qquad (2)$$

The FS-Consol code applies the hydraulic conductivity-void ratio-relationship k(e):

$$k = C \cdot e^D \qquad (3)$$

Instead of the relationship (3), the ACCUMV code applies the following function:

$$e = c_k \log(k) + c_{kk} \qquad (4)$$

Essential is to determine as exactly as possible the consistency of the void ratio and hydraulic conductivity values. For this purpose, a newly developed compression test has been introduced at WISMUT (see below chpt. 3).

Based on the test results of compression tests and by applying a parameter variation it becomes possible to (a) estimate the representative material functions (e-σ' and e-k_f relationships) and (b) appraise the present state of consolidation in the individual zones. The derived profiles of the depth dependent distribution of the void ratios and of pore pressure can then serve as the initial condition for the coupled differential equations that are used to calculate the settlement.

We use the one dimensional NLFS-models to simulate the time-dependent settlement of heterogeneously layered slimes as affected by the time-dependent loading under consideration of the drainage conditions prevailing at the base of the pond. The results of the computations are compared with the present and past settlement measurements and conclusions are drawn regarding the material functions applied.

Within the used tailings consolidation models it became possible to apply the one-dimensional NLFS-codes for estimation of the extent of settlement troughs, which develop (due to placement of embankment) fills for re-contouring the pond surface.

One of the significant decisions in course of remediation is whether to use and how to layout wick drains to accelerate the dissipation of the pore pressures and thus to speed up the consolidation of thick fine slime layers.

For this purpose, a suitable FEM model code, called Consol2D has been developed at Wismut in cooperation with the Technical University of Chemnitz. The model follows a three dimensional approach, reduced to two dimensions for handling of radial symmetric problems, such as the use of drains. Up to 10 different soil types can be considered in the model. Each soil type can be characterized by an e-σ' and an e-k relationship. Void ratio profiles measured or calculated by the above-mentioned one-dimensional NLFS models are entered into the model as initial conditions. Pore water pressure distribution profiles can be used as initial conditions (REICHEL 1999).

The Consol2D model was validated by analytical solutions and compared with the results of field tests using drains at Wismut. A good agreement was found between the model computations and field test results (WELS et al. 1999).

3 LAB COMPRESSION TESTING

A new compression test apparatus, including a specific testing procedure for deriving all input data needed for non-linear finite strain consolidation calculation, has been introduced and regularly applied at WISMUT (see Fig. 1).

The compression test allows software-controlled testing and continuous measuring of all parameters of interest which are related to the material behaviour of consolidating fine slimes. Samples of 200 mm diameter and up to 250 mm height are tested. The ratio of height vs. diameter of the sample varies during testing from 1 : 0.8 to 1 : 5.

This has to be taken into account for evaluation of the test results. Parameters measured during testing include settlement value, surcharge pressure (load) on top and base pressure at the bottom of the sample as well as pore pressure. Based on these parameters it is possible to eliminate the effect of the geometry of the tested sample. Tailing samples are loaded gently with respect to non-linear consolidation behaviour of the

Figure 1. Automatic compression test apparatus.

fine tailings. Stepwise loading is done with respect to self-weight consolidation and surcharge loading steps during the later covering, commonly in steps of 1, 2, 5, 10, 20, 50, 100, 200 and 300 kPa of surcharge load. Each loading step is carried out until full primary consolidation. Secondary consolidation can be measured as well. During each step pore pressure is measured continuously with time with an accuracy of 0.1 kPa.

4 MATERIAL PROPERTIES OF SLIMES

Table 1 lists values of the compression indices, C_C obtained for the fine tailing types found in some of the tailing ponds of Wismut. The values given for the individual zones may vary for the different types of tailings within a zone, but the presented figures provide the typical ranges. An overview of the material parameters for tailings from 13 various tailing ponds are provided in (WELS et al. 2000).

The hydraulic conductivity as a function of void ratio can be obtained from the same oedometer tests. For illustration, the typical parameters estimated for the slimes of the Helmsdorf and Culmitzsch tailing ponds are in Table 2.

The values in Table 2 indicate that the material function must be estimated for each tailing type individually. Our experience shows that for a range of void

Table 1. Compression indices obtained from specific compression tests on fine slimes from the Culmitzsch A tailings pond. The void ratio has been set identical with e_0 in equation (1).

Tailings type	Compression index C_C	Void ratio e (1 kPa)
Fine slimes	0.60–0.82	3.0–4.2
Transition area	0.40–0.55	2.4–2.9
Sandy tailings	0.06–0.10	0.7…1.2

Table 2. Values for parameters C and D presented with equation (3).

	C in (m/s)	D
Fine slimes TP Helmsdorf	5* 10-9	3.5
Fine slimes TP Culmitzsch A	1.4 *10-10	4.6

ratios from e = 1.5 to e = 4.5 the hydraulic conductivity varies up to more than 2 orders of magnitude for the same material.

5 ZONING OF A TAILINGS POND

The zoning of the tailing pond in principal distinguishes among:

– Fine slime zones consisting only of silty-clayey fine tailings
– Transition zones with mixed layering of silty-clayey fines and sandy tailings and
– Beaches composed of mainly permeable sandy tailings

The relevant tailing zones are identified based on the spatial distribution of different tailing types and the vertical profile of the different tailing layers in each individual zone including the hydraulic boundary conditions varying with time during and after disposal period. To separate the different zones also field measurements are used like cone penetration tests and field shear vane tests. In addition, cone penetration tests with dissipation testing are carried out to measure the vertical static pore pressure distribution with depth and to identify the actual hydraulic boundary conditions in a zone.

6 MODELING RESULTS

Modeling of the filling process is based on the estimation of the discharge rate of the solids in the area of the modeled borehole. Figures 2 and 3 show void ratio profiles for two boreholes calculated on the basis

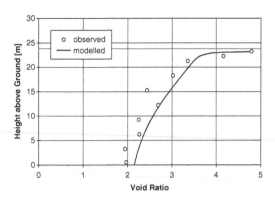

Figure 2. Modeled void ratio profile vs. measured profile for a borehole located in the fine slimes area of the Culmitzsch A tailing pond. The entire profile runs through silty clay material of fine slimes.

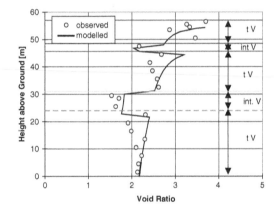

Figure 3. Modeled void ratio profile vs. measured profile for a borehole located in the transition zone of the Culmitzsch A tailing pond.(tV – clayey, int.V – transition zone material).

of the history of filling of the Culmitzsch A tailing pond. The Culmitzsch tailing ponds cover an area of 235 ha (pond A: 160 ha; pond B: 75 ha) and contain in total 85 Million m^3 of tailings (pond A: 62 Mm^3; pond B: 23 Mm^3). The profile of the borehole in Figure 2 is from the central slime zone of the pond, while Figure 3 presents an alternate layering of different tailing types typical for the transition zone next to the slime zone. Based on lab compression test results, both the observed void ratio profile and the achieved thickness can be satisfactorily reproduced. The calibrated hydraulic conductivity -void ratio also matches the laboratory tests very well.

In Figure 3, a calculated void ratio profile from the transition zone in Culmitzsch A tailing pond is compared with the measured values. The stratification consists of an alternate layering of three silty-clayey fine slime layers and two mixed-grained silty-sandy

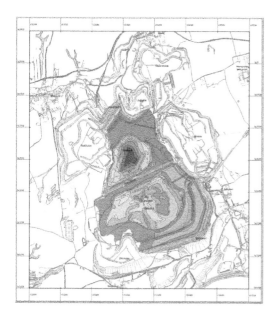

Figure 4. Computed settlement trough under spatially varying loading due to placement of variable thicknesses of interim cover, re-contouring layer and surface cover.

tailing layers, as typically found in the transition zone next to the slime zone.

The numerical implementation involves the calculation of the one-dimensional settlement by the NLFS model with respect to the spatial distribution of the surcharge loading due to interim covering, re-contouring of the pond surface and final covering followed by the calculation of the settlement trough, including the quantitative estimation of the zones and thicknesses. The results are visualized by the EarthVision program system (see Fig. 4).

Figure 4 shows a calculated settlement trough for the two tailing ponds Culmitzsch A and B. Clearly discernable are the high settlements in the area of slimes. The orange and red zones are identical with up to 57 m (southern pond A) to 63 m (northern pond B) thick homogeneous fine slime zones having predicted settlements of up to 6 m (southern pond A) and up to 8.5 m (northern pond B). Consolidation of the thick fine slime layers would need several decades. This fact would elongate the remediation period if no measures would be taken to speed up the fine tailings consolidation.

7 MODELING THE EFFECT OF WICK DRAINS IN STRATIFIED SLIMES

Results of the consolidation modeling presented above show that the consolidation progress of fine slimes must be speeded up during remediation phase to grant

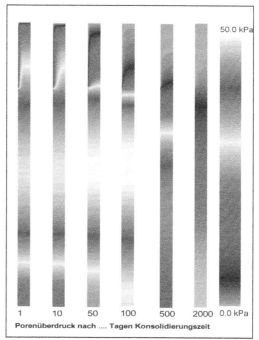

Figure 5. Time dependent development of excess pore pressure in the simulated 22 m high cylinder (Consol2D).

a stable surface runoff regime on and to avoid unacceptable differential deformation of a surface cover. For this vertical wick drains may be applied where needed. The spacing and depth of such wick drains is to be dimensioned with respect to the planned surcharge loading steps with time during the future remediation period. For dimensioning of the spacing and depth of wick drains the FEM model Consol2D has been used.

The input parameters for Consol2D are :

- Void ratio profile vs. depth
- Material functions: void ratio – effective stress and void ratio – hydraulic conductivity
- Loading steps with time
- Drain design: spacing and depth of the vertical drains
- Hydraulic boundary conditions
- if needed also: creep coefficient C_α

The load can be applied stepwise corresponding to the placement of the interim cover, the re-contouring layers and the surface cover. The Consol2D simulation runs are used to investigate the effect of the drains on the displacement, void ratio, pore pressure and stress over the consolidation time and over the depth.

Figure 5 presents the development of the pore water pressure over time in a cylinder 22 m high having a radius of 1.5 m as affected by a 5 m deep drain. Such

295

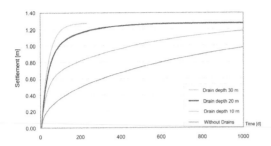

Figure 6. Results of the 2D-radialsymmetric modeling of the settlement over time vs. drain depth (Consol2D).

a drain design is used for improving the trafficability of the fine slime surface in advance of the initial placement of the first thin interim cover layer. The excess pore pressure present at the beginning of the calculation follows from the previously specified void ratio profile. In this case, no load is being applied. The excess pore pressure in the upper strata can be seen dissipating after approximately 100 days. Because of the longer drainage path, the dissipation of the excess pore water pressures from the depth takes considerably longer than 2000 days.

Figure 6 shows how different drain penetration depths affect the time dependent settlement behaviour. One characteristic behaviour becomes apparent: During a first phase (<100 days), consolidation in the sphere of influence of drains goes on with a high settlement rate while the absolute value of settlement depends on the penetration depth of the drain. In a second phase, the settlement rate considerably decreases.

8 CONCLUSIONS

One-dimensional consolidation modeling codes implementing the consolidation behaviour of highly compressible pulpy fine slimes have been successfully applied to backcalculate the filling period of various tailing ponds and to predict the settlement troughs developing under surcharge loading during ongoing remediation. The two-dimensional radialsymmetric

FE-code Consol2D, implementing the nonlinear finite strain consolidation theory, was developed at WISMUT in cooperation with the Technical University of Chemnitz and applied on several tailing ponds for dimensioning of the grid spacing and depths of wick drains for speeding up the consolidation of fine slimes for two different purposes, first for improving trafficability of fine slime surfaces in advance of the placement of the first interim cover layer and secondly to enhance the speed of the consolidation of thick fine slime layers during the ongoing remediation in order to grant a stable surface runoff regime on and the stability of the surface cover to the long term. The relevant material functions of the fine slimes were determined in slurry compression tests developed at WISMUT for measuring the consolidation of pulpy fine slimes even under small loads. Model calculations were validated and calibrated on field measurements of the total tailing thickness, the time settlement rates, the void ratio distribution vs. depth and the pore pressure distribution vs. depth. Currently, re-contouring of the pond surfaces of the two Culmitzsch tailing ponds A and B is being conceptually designed with respect to the predicted consolidation progress with time including the use of vertical wick drains.

REFERENCES

Gibson, R.E., England, G.L., Hussey, M.J.L. Juli 1967: *The theory of one-dimensional consolidation of saturated clays*. Geotechnique 17

Schiffman, R.L., Szavits-Nossan, V., McArthur, J.M. 1992: *ACCUMV – Nonlinear large strain consolidation*. Computer Code Manual

Reichel, U. 20.-25. März 1999: *Two dimensional modelling of consolidation of soft soils using parallel finite element methods, in Proceedings of The Third Euro-Conference on Parallel and Distributed Computing for Computational Mechanics*. Weimar: Civil-Comp Press, ISBN 0-948749-59-8

Wels, C., Barnekow, U., Haase, M., Exner, M., and Jakubick, A.T. *A Case Study on Self-Weight Consolidation of Uranium Tailings.- in E. Oezberk and A. J.Oliver,Eds., Uranium 2000, Proc. of Int. Sym. on the Process Metallurgy of Uranium*. CIM, 9–15 Sept, 2000, Saskatoon, Canada

Numerical Modelling of Construction Processes in Geotechnical Engineering for
Urban Environment – Triantafyllidis (ed)
© 2006 Taylor & Francis Group, London, ISBN 0 415 39748 0

Numerical simulation of underground works and application to cut and cover construction

Th. Zimmermann
LSC-ENAC-EPFL, CH-1015 Lausanne-EPFL and ZACE Services Ltd, Switzerland

A. Truty
Institute of Geotechnics, Cracow University of Technology, Poland

J.L. Sarf
BET J.-L.Sarf, Switzerland

ABSTRACT: Optimal design and construction require analysis tools which reproduce the behavior of media and construction sequences as accurately as possible. The paper discusses some aspects of such a software tool in relation with cut and cover construction procedures, including: large displacement formulation for structures, continua and contact interfaces, and a novel fill algorithm. Validation examples and a case study illustrate the discussion.

1 INTRODUCTION

Underground works often take place in urban environment, a numerical simulation may then be necessary in order to assess safety of existing and planned constructions. Z_Soil.PC (Zimmermann, Truty, Urbański, and Podleś 2005), a finite element software package for geomechanics developed by the authors and colleagues over the past 25 years, offers a unified and generic framework for such computations: analysis of deformation, small and large, flow, steady or transient, in partially or totally saturated media, thermal behavior and continuous stability assessment in soil, rock, and structures, including soil-structure interaction, can be performed within a single environment which allows the modeling of the complete construction time-history, starting with an assessment of the initial state, in two- or three-dimensional media.

The first goal of numerical modeling of mechanical effects associated with underground works is to represent the stress/strain and deformation history in the soil mass as precisely as possible. Constitutive models of soils/rock mass are usually path-dependent; this means that a change of a given state variable (e.g. stress) does not only result from the most recent increment of a conjugate variable (e.g. strain), but also from its previous states. As a consequence, numerical simulation cannot be undertaken as a set of independent computational steps corresponding to the different stages of the construction process; it has to be carried out as a sequential, step-by-step procedure, memorizing

the state of the system at each instant, which is used as a starting point for the subsequent step. In particular evaluation of the initial stress state, existing in the domain prior to the construction process, change of equilibrium conditions caused by excavation, and changes of external loads, change of pore water pressure field, which influences effective stresses, details of construction phases, presence of temporary roofing, proper soil-structure and other interface features, evolution in time of material properties of soil, due to freezing or injection e.g., have to be accounted for.

Numerical simulation in Z_Soil.PC is viewed as a time-dependent evolutionary process, in which time can be real time or just a way of sequencing computations, this depends mainly on the type of phenomena to be simulated: creep or consolidation require real time, an excavation sequence in a dry medium usually not. The overall time-span of the analysis is split into successive sequences, each driven by its own driver algorithm, computational strategies can be specific to one sequence or valid for the whole time-history. A typical sequence could include: 1. non-linear initial state definition, including steady flow conditions (at time $t = to$); 2. evaluation of initial safety of the site (at $t = to$); 3. excavation/ construction sequence, combined or not with consolidation phenomena (from $t = to$ to $t = T$); 4. safety evaluation (at $t = T$). The excavation/construction/safety evaluation sequence can then be repeated as needed.

The initial state algorithm will apply gravity and initial loads progressively, while canceling corresponding deformations. The time-span of this particular analysis is zero and defines a steady state situation. Safety evaluation is also instantaneous in time, it is based on C, ϕ, or (C, ϕ) reduction algorithms for yield criteria based on (C, ϕ), and extensions thereof acting on the deviatoric component of stress, for more advanced criteria. Excavation/construction sequences are standard time stepping sequences during which all acting entities are associated, when meaningful, with a time-evolution function, which governs the evolution in time of amplitudes (of a load e.g.) and an existence function, which governs the birth and death of the entity. Two-phase media may require also an automatic management of boundary conditions, due to their intrinsic evolutionary character at free surface. Constitutive behavior of media and interfaces, is another essential aspect of modeling. While Mohr-Coulomb or Rankine criteria are still favored in many standard situations, more advanced models are often needed. Among these, Modified Cam-Clay, a multilaminate model coupled with the Menetrey-Willam criterion for rock and an extended Aubry-Hujeux for soil are available.

Recent developments in Z_Soil include large rotations and displacements for all structural elements like shells, beams, membranes, anchors and continuum and large deformations at contact interfaces. The corotational approach is used to manage large rotations; its main benefit is that all stresses and strains at the integration points remain the engineering variables although given in a rotated local frame. The new contact formulation, developed to manage really large relative motions of bodies, makes use of a so-called slave-master approach in which the contacting node (slave) cannot penetrate the corresponding master element face (master) by means of a penalty formulation enhanced by an Augmented Lagrangian approach (if needed).

Due to lack of space we concentrate in the following on some aspects of modeling related to cut and cover type construction. In section 2, a corotational formulation for large displacement-small strain analysis is described; in section 3, the large displacement contact formulation is briefly addressed; in section 4, the fill algorithm is presented; in section 5, a real world case study is presented followed in section 6 by concluding remarks.

2 LARGE DISPLACEMENTS-SMALL STRAINS COROTATIONAL APPROACH

The goal is to perform large displacement/rotation analysis re-utilizing standard geometrically linear, possibly materially nonlinear elements for beams, shells, membranes, trusses or anchors and continua. The basic assumption is that displacements and rotations associated with rigid body motion can be arbitrarily large, but "true deformation" remains

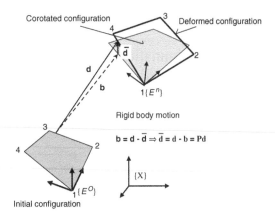

Figure 1. Setting element frames.

within the small strain limit. The implementation requires that the rigid body motion is deducted from total deformation of an element before evaluation of element forces and stiffness emerging from "true deformation". For that purpose, an element reference frame E rigidly attached to the element is introduced and element processing is performed with respect to this frame. Removal of rigid body motion is equivalent to a projection

$$\delta\bar{\mathbf{d}} = \mathbf{P}\,\delta\mathbf{d} \tag{1}$$

$$\mathbf{f} = \mathbf{P}^T\,\bar{\mathbf{f}} \tag{2}$$

$$\mathbf{K} = \mathbf{P}^T\,\bar{\mathbf{K}}\,\mathbf{P} + \frac{\partial \mathbf{P}^T}{\partial \mathbf{d}}\bar{\mathbf{f}} \tag{3}$$

The corresponding mathematics, for truss, beam and shell elements is developed at length in references (Rankin and Nour-Omid 1988), (Rankin and Nour-Omid 1990) and for standard continuum elements in (Crisfield and Moita 1996). It is interesting to mention that large deformation analysis is a fully internal modeling option and does not require any user input.

2.1 Benchmark

Benchmarking needs to be done for each element type separately and in interaction with different element types. A cylindrical shell under point load problem is run as benchmark for shell elements (Fig.(2)). Results are compared with ones given in (Chróścielewski 1998) (Fig.(3)).

3 LARGE DEFORMATION CONTACT

The large deformation frictional contact implementation, in Z_Soil code, follows the approach proposed by Parisch and Lübbing (Parisch and Lübbing 1997). A

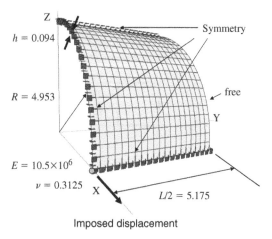

Figure 2. Free edge cylinder. The data.

$h = 0.094$
$R = 4.953$
$E = 10.5 \times 10^6$
$\nu = 0.3125$
$L/2 = 5.175$
Symmetry
free
Imposed displacement

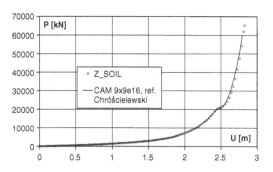

Figure 3. Free edge cylinder. Displacement history.

major benefit of this formulation is that the consistent linearization of the virtual work equations, involving nonlinear contact kinematics, leads to a quadratically convergent Newton iterative scheme when solving finite element equations of the global equilibrium. Moreover, this formulation admits arbitrary large relative movements of the contacting bodies and no difficulty is encountered when a slave node passes through the element edges on the target surface. The 2D representation of the contact kinematics, proposed by Parisch and Lübbing, for a given slave node S and a target surface is shown in the Fig.(4). Two configurations are considered, β_t corresponding to the last configuration of an equilibrium and the current one, possibly not yet in the equilibrium state, $\beta_{t+\Delta t}$. In order to define the normal and tangential gaps, needed later on to compute the interface forces, the motion of the slave node S and the corresponding material point C on the target surface must be traced. The position of points C_t and $C_{t+\Delta t}$ is found by solving the problem of the closest point projection of a point S_t, and $S_{t+\Delta t}$ respectively, on the target surfaces in configurations

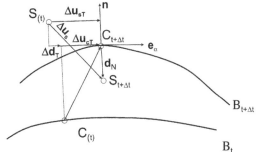

Figure 4. Contact kinematics.

β_t and $\beta_{t+\Delta t}$. The normal and tangential gaps (see Fig.(4)) are defined as follows:

$$g_N = \mathbf{n} \cdot (\mathbf{x}_S - \mathbf{x}_C) \qquad (4)$$

$$g_T = \frac{\Delta \mathbf{d}_T}{\|\Delta \mathbf{d}_T\|} \cdot (\Delta \mathbf{u}_S - \Delta \mathbf{u}_C) \qquad (5)$$

and vector $\Delta \mathbf{d}_T$ is defined as:

$$\Delta \mathbf{d}_T = \Delta \mathbf{u}_{sT} - \Delta \mathbf{u}_{cT} \qquad (6)$$

The penalty approach, enhanced by the Augmented Lagrangian procedure, is used to satisfy the impenetrability condition.

In the large deformations regime, slave nodes may interact with different elements on the target surface during the analysis and for that reason the explicit setting of so-called contact elements is not possible. To avoid rebuilding of the whole profile of the linear system of finite element equations at each iteration, the following strategy is used: at the beginning of the step each slave node is associated with the master element face (on the target surface) and this setup is kept fixed until equilibrium state is achieved. At the state of the equilibrium each slave-master pair is verified (by means of the closest point projection) and if all of them remain unchanged, the step is assumed to be converged, otherwise, after setting new pairs "slave-master", the iteration process is continued until a new equilibrium is reached. The attribute "slave node" can be applied to any nodal point in the mesh while the attribute "master element face" can be associated with any face of the finite element for continuum, shell, beam (only 2D) or a membrane.

3.1 Benchmark

The classical Hertz problem of a deforming disk on a rigid foundation in plane-strain format is analyzed in this section. The reference solution is taken from (Papadopoulos and Taylor 1992). As the foundation

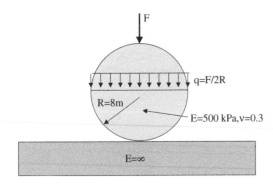

Figure 5. Hertz problem setting.

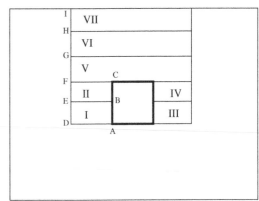

Figure 7. Tunnel construction followed by several fill steps.

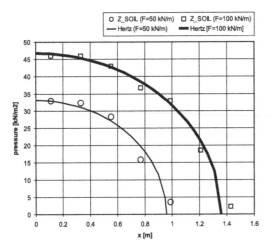

Figure 6. Hertz problem: Distribution of contact stresses.

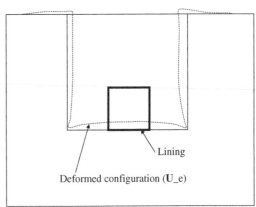

Figure 8. Registering total deformation after excavation.

is rigid, no discretization is necessary; however, in order to facilitate computation of contact stress, the foundation is also discretized here. A comparison of normal stresses on the interface, with the reference solution, for two levels of the force $F = 50$ kN/m and $F = 100$ kN/m (q = (F/(2R))) is shown in the Fig.(6).

4   FILLING

This algorithmic option allows the user to simulate construction or filling of excavations in an automatic manner. New construction stages are detected automatically by the code and an incremental procedure (similar to a standard initial state computation Zimmermann, Truty, Urbański, and Podleś 2005)) is run according to user defined settings. The analysis of constructions evolving in time, especially in the regime of large deformations, generates problems

in finite element modeling, due to violation of kinematic compatibility along the interface between existing deformed domain and newly added undeformed subdomain. Also due to settlement of each layer of the fill, its final elevation will never be achieved and hence the load transferred to the structure can be underestimated.

To explain the general idea, an example of an excavation followed by the construction of a lining and seven stages of filling (see Fig.(7)) should be considered.

After an excavation we get some deformation U_e (see Fig.(8)) which is to be neglected when a tunnel lining is built. This deformation is memorized by the code in an automatic manner. In the next step, the lining is added and its initial configuration is assumed to be undeformed (just before construction) although nonzero deformation at the interface between bottom slab and subsoil exists. This can be handled by memorizing deformation U_e. The major problem

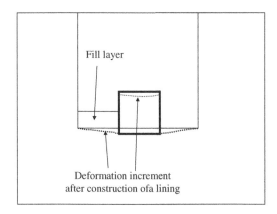

Figure 9. Deformation increment due to construction of the lining.

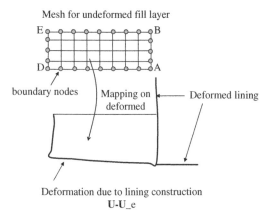

Figure 10. Mapping of the fill subdomain onto deformed configuration.

appears when we begin to add fill material, because newly added fill should satisfy contact kinematics if it touches already deformed lining (deformed due to its own dead weight and/or loading imposed during previous fill steps) and existing deformation on the remaining boundaries (see Fig.(9)).

If we consider the situation shown in Fig.(10) we can notice that the initial undeformed mesh in the zone of fill (stage I) should be mapped onto deformed configuration caused by the construction of the lining. Hence the boundary nodes along section A-B must satisfy contact kinematics (cannot penetrate the lining and cannot be separated from it), nodes along the boundary A-D and D-E must fit current deformation equal to $\mathbf{U} - \mathbf{U}_e$ (the one corresponding to the settlement caused by lining construction), and nodes along section E-B must remain at the initial elevation. This mapping is achieved in the code by

a finite element solution on a sub-domain subject to the imposed boundary displacements. As result we get a shift to the nodal coordinates of all newly added continuum elements (NB. in this finite element sub-problem we assume artificial elastic constants $E = 1.0$ and $\nu = 0.0$). However, to make this mapping, all nodes along the section A-B, being part of the contact interface must be defined as slave nodes. It should also be emphasized here that incremental deformations during single fill step should be small, otherwise strain incompatibilities along the section A-D can cause stress oscillations (although total deformation caused by filling can finally be large). In order to satisfy this condition, the newly added layer of the fill is assumed to behave as a quasi-incompressible material ($\nu = 0.499999$) in the first time step after its appearance. This option is purely algorithmic and activated by the code automatically.

The following general rules are used in this approach:

• structures (beams/truss/membranes/shells) are always added in undeformed configuration; hence, whenever a new structural element is added its initial total deformation (due to presence of its nodes in the previous steps) must be memorized as U_{os} (at the element level); hence the current structure deformation is always equal to $\mathbf{U} - \mathbf{U}_{os}$
• continuum elements are added after mapping onto deformed configuration
• at the end of the excavation time step, the current total deformation corresponding to that state is memorized as $\mathbf{U}_e$ at all nodal points

The proposed fill algorithm is summarized in Box-1.

Box-1: Fill algorithm

1. Given:
   – total deformation $\mathbf{U}$ at t_N
   – total deformation $\mathbf{U}_e$
   – the initial preexisting deformation $\mathbf{U}_{os}$ of structural elements)
2. Perform mapping of continuum elements in the zone of fill according to the algorithm given in Box-2
3. Assume incompressible behavior for all newly added continuum elements in the fill zone (redefine Poisson ratio to $\nu = 0.499999$)
4. Solve nonlinear equation of the global equilibrium $\mathbf{F}_{extN+1} - \mathbf{F}_{intN+1} = \mathbf{0}$ (incremental application of gravity in the fill zone is possible)
5. Retrieve the original Poisson ratio in the fill zone.

NB. The proposed mapping plays a crucial role when contact interfaces are present. Reason being that the initial spurious over-penetrations, related to the violation of contact kinematics, may generate substantial prestress effects in the contacting media and thus may lead to local buckling phenomenona or sudden divergence of the iterative schemes due to huge amplitude of concentrated plastic strains.

5 A THREE-DIMENSIONAL CASE STUDY

A three-dimensional analysis of stability of a tunnel, built in an open trench using cut and cover method, is presented here to validate the algorithm proposed for filling and formulated in the framework of standard finite element technology. The analysis consisted first of an excavation step followed then by construction of a stabilizing concrete wall and a tunnel lining (bottom slab is 1m thick while the thickness of the remaining part is 0.75 m) (see Fig.(12), (13)).

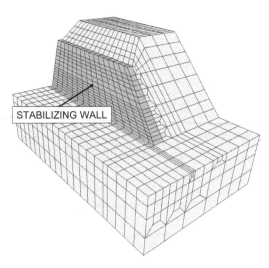

Figure 12.   Excavation and construction of stabilizing wall.

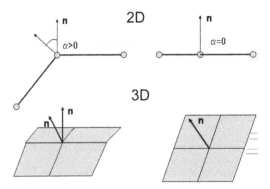

Figure 11.   Different or coincident facet normals.

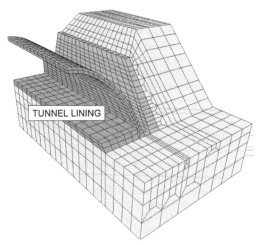

Figure 13.   Construction of tunnel lining.

The fill material was brought-in in four steps according to scheme shown in the Fig.(15). At the end of the construction, a stability analysis was carried out using so-called $c - \phi$ reduction algorithm. To handle the interface between fill and tunnel and fill and stabilizing wall, a large deformations contact interface was generated as shown in Fig.(16). The existing slope was modeled by means of a standard Mohr-Coulomb constitutive model with $E = 1, 00, 000\,\text{kPa}$, $\nu = 0.3$, $\psi = 0°$, $\phi = 35°$, $c = 45\,\text{kPa}$. The same model was used for the fill material ($E = 60, 000\,\text{kPa}$, $\nu = 0.3$, $\psi = 0°$, $\phi = 30°$, $c = 40\,\text{kPa}$) while tunnel lining and stabilizing wall were assumed to be elastic (although geometrically nonlinear) and characterized by $E = 30, 000, 000\,\text{kPa}$ and $\nu = 0.2$.

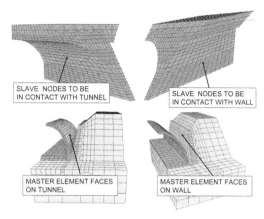

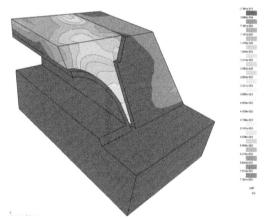

Figure 16.   Setting large deformation contact interface.

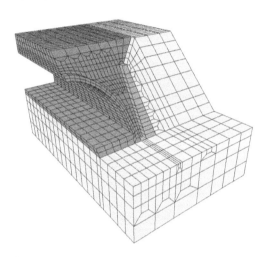

Figure 14.   Final configuration after filling.

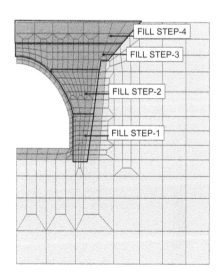

Figure 15.   Filling steps.

Figure 17.   Incremental deformation for SF = 1.85 to SF = 1.95.

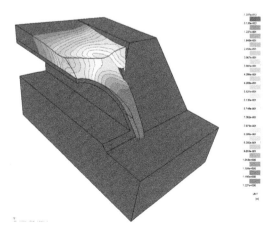

Figure 18.   Failure mechanism at SF = 2.05.

303

The incremental deformation corresponding to the evolution of the safety factor from value SF = 1.85 to SF = 1.95 is shown in Fig. (17), exhibiting formation of a failure mechanism at the external boundary of the bottom part of the fill material. The failure mechanism detected at SF = 2.05 is shown in Fig. (18). It has to be emphasized here that this result does not correspond to a configuration in equilibrium. However, it shows the failure pattern in terms of incremental deformations achieved between steps SF = 1.95 and SF = 2.05.

6 CONCLUDING REMARKS

Optimal design and construction require analysis tools which reproduce the behavior of the medium and construction sequence as accurately as possible. We have presented and validated some aspects of such a tool and illustrated it on a real world cut and cover case study.

REFERENCES

Chróścielewski, J. (1998). Rodzina elementów skończonych klasy C0 w nieliniowej sześcioparametrowej teorii powłok. *Zeszyty Naukowe Politechniki Gdańskiej 53*.

Crisfield, M. A. and G. Moita (1996). A unified co-rotational framework for solids, shells and beams. *International Journal Solids Structures 33, No. 20*, 2969–2992.

Papadopoulos, P. and R. L. Taylor (1992). A mixed formulation for the finite element solution of contact problems. *Computer Methods in Applied Mechanics and Engineering 94*, 373–389.

Parisch, H. and C. Lübbing (1997). A formulation of arbitrarily shaped surface elements for three-dimensional large deformation contact with friction. *International Journal for Numerical Methods in Engineering 40*, 3359–3383.

Rankin, C. and B. Nour-Omid (1988). The use of projectors to improve finite element performance. *Computers & Structures 30*, 257–267.

Rankin, C. and B. Nour-Omid (1990). Finite rotation analysis and consistent linearization using projectors. *Computer Methods in Applied Mechanics and Engineering*.

Zimmermann, T., A. Truty, A. Urbański, and K. Podleś (2005). *"Z_SOIL manual"*. Switzerland: elmepress international & Zace Services Ltd.

Numerical Modelling of Construction Processes in Geotechnical Engineering for Urban Environment – Triantafyllidis (ed)
© 2006 Taylor & Francis Group, London, ISBN 0 415 39748 0

Effects of spatial variability of cement-treated soil on undrained bearing capacity

Kiyonobu Kasama & Kouki Zen
Division of Civil and Structural Engineering, Faculty of Engineering, Kyushu University, Japan

Andrew J. Whittle
Department of Civil and Environmental Engineering, Massachusetts Institute of Technology, U.S.A.

ABSTRACT: Deep mixing methods are gaining popularity as a method for improving the bearing capacity of soft soil foundations. However, spatial variability in the shear strength of the cement-treated ground introduces uncertainties in estimating the bearing capacity for design. This paper presents results of a probabilistic study in which the shear strength of the cement-treated ground is represented as a random field in Monte Carlo simulations of undrained stability for a surface foundation using numerical limit analyses. The results show how the bearing capacity is related to the coefficient of variation and correlation length scale of the soil-cement columns.

1 INTRODUCTION

Wet and dry deep-mixing (DMM; Terashi & Tanaka, 1981; Porbaha et al., 1999) and pre-mixing (Zen et al., 1992) techniques are becoming widely established for stabilizing soft soils in applications ranging from the improvement of foundation properties to mitigation of liquefaction.

This paper considers the undrained stability of foundations on cement-treated ground. Prior studies of these problems have included several series of centrifuge model tests of surface foundations supported by end-bearing, clay-cement columns, with area replacement ratios ranging from 18–80% (e.g., Miyake et al., 1991; Kitazume et al., 2000; Bouassida & Porbaha, 2004). The papers by Miyake et al. (1991) and Kitazume et al. (2000) include details of the interpreted failure mechanisms and finite element simulations of the load-deformation response. Broms (2004) has proposed a semi-empirical method for interpreting the bearing capacity based on the unconfined compressive strength of the deep-mix columns and local shear failure of the adjacent soft clay. A more rigorous theoretical approach is presented by Bouassida and Porbaha (2004). This includes upper and lower bound solutions (based on the earlier yield design theory developed by Bouassida et al., 1995) treating the cement-treated ground as a homogeneous mass. None of these studies have considered the role of spatial variability of the cement-treated ground.

Numerical limit analyses offer a convenient method for analyzing undrained stability problems and can readily be adapted to simulate effects of spatial variability in the shear strength properties of cemented-ground and natural soil layers. The current study presents a probabilistic approach to evaluating the bearing capacity of a surface strip foundation on a spatially random foundation of cemented-treated clay. A statistical interpretation of the bearing capacity is based on Monte-Carlo simulations using the stochastic Upper and Lower bound Numerical Limit Analyses.

2 SPATIAL VARIABILITY OF CEMENT-TREATED GROUND

The main factors influencing the shear strength of the cement-treated ground include the types and amounts of binder/cement (e.g., Clough et al., 1981; Kamon & Katsumi, 1999), physico-chemical properties of the in situ soil (Consoli et al., 2000), curing conditions and effectiveness of the mixing process (Larsson, 2001, Omine et al., 1998). Although there have been significant advances in the equipment and methods used for deep mixing, there remains a high degree of variability in the shear strength properties.

For example, Table 1 summarizes data from a series of construction projects in Japan. In each case, measurements of unconfined compressive strength, q_u, were obtained from core samples cured in the field. The mean values of q_u range from 100–4000 kPa with coefficients of variation, $COV_{q_u} = \sigma_{q_u}/\mu_{q_u} = 0.14$–0.99. These results are consistent with the findings of a recent review of US deep mixing projects by Navin and Filz (2005) who report $COV_{q_u} = 0.17$–0.67. This level of variability is much higher than that expected

Table 1. Project conditions and unconfined strength data for cement-treated soil from construction projects in Japan.

Cement Mixing Method	Depth (m)	Cement Type	Amount Cement (kg/m^3)	Curing Period (day)	w (%)	Sample #	μ_{q_u} (kPa)	COV_{q_u}	Source
Deep Mixing Method (offshore)	$-8.0 \sim -19.0$	NP[1]	140	28	$95 \sim 135$	176	2140	0.358	CDIT[4] (1999)
	$-9.0 \sim -22.0$	NP[1]	150	28, 49	$110 \sim 150$	222	3920	0.353	
	$-12.5 \sim -36.0$	SP[2]	14	28	$55 \sim 110$	182	3760	0.440	
	$-9.0 \sim -39.0$	NP[1]	135	60	$100 \sim 130$	26	3370	0.331	
	$-1.0 \sim -8.0$	SP[2]	150	$28 \sim 52$	$90 \sim 100$	29	3770	0.485	
	–	Cement milk	15	–	–	30	790	0.290	Abe et al. (1997)
Deep Mixing Method (onshore)	$-2.0 \sim -8.0$	NP[1]	74	28	$110 \sim 140$	54	230	0.480	CDIT[4] (1999)
	$0.0 \sim -7.0$	–	150	28	–	47	2360	0.420	Kohinata et al. (1995)
	$0.0 \sim -10.0$	Special cement	250	28	$100 \sim 200$	493	3660	0.140	Noto et al. (1983)
	$0.0 \sim -4.0$	NP[1]	–	–	–	36	2820	0.423	Tamura et al. (1995)
Premixing Method	–	SP[2]	7.5%	91	6.5[3]	32	661	0.470	CDIT[4] (1999)
	–	SP[2]	3%	28	9.4[3]	13	360	0.990	
	–	SP[2]	4%	28	11.8[3]	25	120	0.750	

Notes: 1) Portland cement, 2) Blast furnace slag cement, 3) Water content of original material before mixing cement, 4) Costal Development Institute of Technology.

Table 2. Summary of data on spatial correlation lengths for cement-treated ground.

Type of Ground	Reference Strength	COV	θ_h (m)	θ_v (m)	Reference
DMM Columns	q_u	0.21–0.36 (clay) 0.32–0.40 (sand)	–	0.4–4.0	Honjo (1982)
Cement-Mixed Dredged Fill	CPT	0.114–0.194	2.0	0.5	Tang et al., (2001)
Air-Transported Stabilized Dredged Fill	CPT	–	–	0.22–0.74	Porbaha et al., (1999)
DMM columns	q_u	0.34–0.74	12.0 (wet mix) <3.0 (dry mix)	–	Navin & Filz (2005)
Lime-Cement Columns	Hand-operated penetrometer	<0.6	$\theta_R < 0.15$ $\theta_O < 0.35$	–	Larsson et al., (2005)

Notes: θ_R: correlation length in radial direction, θ_O: correlation length in orthogonal direction.

for the undrained shear strength of natural clays (e.g., Phoon & Kulhawy, 1999; Matsuo & Asaoka, 1977).

Although there is quite extensive data for estimating the coefficient of variation in the unconfined compressive strength, there is much less information available to understand the underlying spatial correlation structure. Table 2 summarizes values of the correlation length, θ, (in both the vertical and horizontal directions) that have been reported from five studies reported in the literature. Two of these are based on q_u data from installed DMM columns, while two others use cone penetration data in dredged fills.

In contrast, the study by Larsson et al. (2005) uses a miniature penetrometer to evaluate the spatial mixing structure within individual, exhumed lime-cement columns. The results show that the vertical correlation length can range from 0.2–4.0 m (this is similar to the range of fluctuation scales quoted for natural clay deposits). Navin and Filz (2005) find that the horizontal correlation length is much larger for wet mix DMM columns (12 m) than for dry-mix (<3 m), while Larsson et al. (2005) find a radial correlation length, $\theta_R < 0.15$ m within 0.6 m diameter columns. Overall, these data suggest that the horizontal correlation length for cement-treated ground is much smaller than for natural sedimentary soil layers.

3 RANDOM FIELD NUMERICAL LIMIT ANALYSIS

The Numerical Limit Analyses (NLA) used in this study were based on 2-D, plane strain linear programming formulations of the Upper Bound (UB) and Lower Bound (LB) theorems for rigid, perfectly plastic materials presented by Sloan and Kleeman (1995) and Sloan (1988a).

The lower bound analyses (Sloan, 1988a) assume a linear variation of the unknown stresses (σ_x, σ_y, τ_{xy}) within each triangular finite element. The formulation differs from conventional displacement-based finite-element formulations by assigning each node uniquely within an element, such that the unknown stresses are discontinuous along adjacent edges between elements. Statically admissible stress fields are generated by satisfying: i) a set of linear equality constraints, enforcing static equilibrium with triangular elements and along stress discontinuities between the elements, ii) inequality constraints that ensure no violation of the linearized material failure criterion. Tresca yield criteria are used to represent both the undrained shear strength of the clay and the cohesive strength of the cement-treated ground. The lower-bound estimate of the collapse load is then obtained through an objective function that maximizes the resultant force, Q, acting on the footing. The linear programming problem is solved efficiently using a steepest edge active set algorithm (Sloan, 1988b).

The upper-bound formulation assumes linear variations in the unknown velocities (u_x, u_y) within each triangular finite element. Nodes are unique to each element and hence, the edges between elements represent planes of velocity discontinuities. Plastic volume change and shear distortion can occur within each element as well as along velocity discontinuities. The kinematic constraints are defined by the compatibility equations and the condition of associated flow (based on an appropriate linearization of the Tresca criterion) within each element and along the velocity discontinuities between elements. The external applied load can be expressed as a function of unknown nodal velocities and plastic multiplier rates. The upper-bound on the collapse load can then be formulated as a linear programming problem, which seeks to minimize the external applied load using an active set algorithm (after Sloan & Kleeman, 1995).

One of the principal advantages of Numerical Limit Analyses is that the true collapse load is always bracketed by results from the upper and lower bound calculations. For example, Ukritchon et al. (1998) were able to achieve estimates of the collapse for footings under combinations of vertical, horizontal and moment

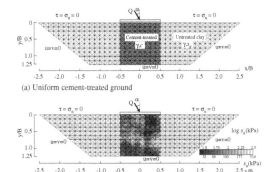

(a) Uniform cement-treated ground

(b) Example of cement-treated ground with spatial variability

Figure 1. Finite element mesh and boundary condition used in UB numerical limit analyses.

loading to an accuracy ±5% for a wide range of undrained strength profiles in the underlying clay.

Figure 1a shows a typical finite element mesh used in the current UB analyses of surface foundations on cement-treated ground. The model considers a clay layer with depth $z/B = 1.25$, where B is the width of the surface strip foundation under conditions of inclined, concentric loading (similar to the geometry used in the centrifuge model tests reported by Kitazume et al., 2000). The clay is underlain by a rigid base, while far-field lateral boundaries of the mesh extend beyond the zone of all potential failure mechanisms. The analyses assume full replacement of the clay beneath the footing such that the zone of cement-treated ground extends to the base of the clay (i.e., acts as an end-bearing zone of ground improvement).

The current simulations also assume that the cement-treated ground and in situ clay have the same total unit weight. The input parameter, $SR = q_u/2s_u$ describes the ratio of the cohesive shear strength of the cement-treated ground to the undrained shear strength of the in-situ clay, where $c = q_u/2$ is the cohesive strength of the cement-treated soil. Hence, the shear resistance along the vertical interface between clay and cement-treated ground is controlled by the undrained shear strength of the clay, while the sliding resistance at the soil-foundation interface is controlled by the shear strength of the cement-treated ground.

The effects of inherent spatial variability are represented in the analyses by modeling the cohesive strength of the cemented-treated soil, c, as a homogeneous random field, while assuming homogeneous undrained shear strength in the clay. The cohesive strength is assumed to have an underlying log-normal distribution with mean, μ_c, and standard deviation, σ_c, and an isotropic scale of fluctuation (correlation length), θ_{lnc}. The use of the log-normal distribution is predicated by the fact that c is always a positive quantity. Following Griffiths et al. (2002) the current

307

analyses present results based on assumed values of the ratio of the correlation length to footing width, $\Theta_{lnc} = \theta_{lnc}/B$.

Spatial variability is incorporated within the Numerical Limit Analyses (both UB and LB meshes) by assigning the undrained shear strength corresponding to the *ith* element:

$$c_i = \exp(\mu_{\ln c} + \sigma_{\ln c} G_i) \tag{1}$$

where G_i is a random variable that is linked to the spatial correlation length, θ_{lnc}.

Values of G_i are obtained using a Cholesky Decomposition technique (e.g., Baecher & Christian, 2003; Kasama & Whittle, 2005a) using an isotropic Markov function which assumes that the correlation decreases exponentially with distance between two points i, j:

$$\rho(x_{ij}) = \exp\left\{-\frac{2x_{ij}}{\theta_{\ln c}}\right\} \tag{2}$$

where ρ is the correlation coefficient between two random values of c at any points separated by a distance $x_{ij} = |x_i - x_j|$ where x_i is the position vector of i (located at the center of element i in the finite element mesh). This correlation function can be used to generate a correlation matrix, K, which represents the correlation coefficient between each of the elements used in the NLA finite element meshes:

$$K = \begin{bmatrix} 1 & \rho_{12} & \cdots & \rho_{1n} \\ \rho_{12} & 1 & \cdots & \rho_{2n} \\ \vdots & \vdots & \ddots & \\ \rho_{1n} & \rho_{2n} & \cdots & 1 \end{bmatrix} \tag{3}$$

where ρ_{ij} is the correlation coefficient between element i and j, and n the total number of elements in the mesh.

The matrix K is positive definite and hence, the standard Cholesky Decomposition algorithm can be used to factor the matrix into upper and lower triangular forms, S and S^T, respectively:

$$S^T S = K \tag{4}$$

The components of S^T are specific to a given finite element mesh (for either UB or LB) and selected value of the correlation length, θ_{lnc}.

The vector of random variables, G (i.e., $\{G_1, G_2, \ldots, G_n\}$, where G_i specifies the random component of the undrained shear strength in element i, eqn. 1) can then be obtained from the product:

$$G = S^T X \tag{5}$$

where X is a vector of statistically independent, random numbers $\{x_1, x_2, \ldots, x_n\}$ with a standard normal

Table 3. Input parameters for current study.

Parameter	Selected Values
α	0°, 15°, 30°, 45°, 60°, 75°, 90°
$SR = \mu_c/s_u$,	1, 2, 4, 8, 20, 40, 80
$COV_c = \sigma_c/\mu_c$	0.1, 0.2, 0.4, 0.8
$\Theta_{lnc} = \theta/B$	0.15, 0.5, 1.0

distribution (i.e., with zero mean and unit standard deviation).

Values of the random variable vector X are then re-generated for each realization in a set of Monte Carlo simulations.

Figure 1b illustrates the spatial distribution of shear strength in the cemented-soil mass for one example simulation with dimensionless input parameters $SR = \mu_c/s_u = 8$, $COV_c = \sigma_c/\mu_c = 0.4$ and $\Theta_{lnc} = 1.0$. The lighter shaded regions indicate areas of higher shear strength.

Based on the literature review of the variability and correlation lengths for cement-treated soil (Tables 1 and 2), a parametric study has been performed using the ranges listed in Table 3.

4 BEARING CAPACITY FOR UNIFORM CEMENT-TREATED GROUND

Figures 2a and 2b compare the UB failure mechanisms for the case of vertical loading ($\alpha = 0°$) for 'untreated' clay and for the case where the foundation soils are uniformly strengthened with strength ratio, $SR = 2.0$, respectively. Each figure shows the deformed mesh, vectors of the UB velocity field, zone of plastic shear distortion (dark shaded region in velocity field), and local rigid body mechanisms (light zones within the failure mechanism). The collapse load is represented by an equivalent bearing capacity factor, $N_c = Q/(Bs_u)$, where Q is the upper or lower bound estimate of the collapse load [F/L]. Results for the untreated soil, show active and passive rigid body wedges within the failure mechanism. The bearing capacity factor is well bounded, $5.00 \leq N_c \leq 5.23$, with errors of $\pm 2.25\%$ compared to the analytical Prandtl solution, $N_c = (2 + \pi)$.

There is a significant increase in the bearing capacity factor when the underlying foundation soils are uniformly cemented with $SR = 2.0$, $7.22 \leq N_c \leq 7.48$. This reflects the mobilization of the shear strength of the cement-treated soil and a subtle change in the mechanism and extent of the failure zone seen in Figure 2b. The failure mechanism in the cement-treated soil can be well-approximated by a three wedge rigid-body mechanism as suggested by Bouassida and Porbaha (2004). These Authors propose simple

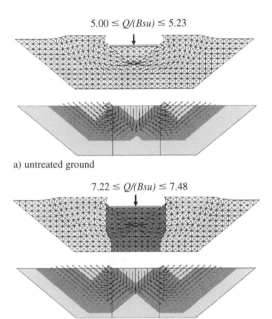

5.00 ≤ Q/(Bsu) ≤ 5.23

a) untreated ground

7.22 ≤ Q/(Bsu) ≤ 7.48

b) uniform cement-treated ground (SR = 2.0)

Figure 2. Deformed mesh and plastic failure zone at failure against vertical loading.

analytical solutions for bounding the bearing capacity factor under vertical loading:

$$2(1 + SR) \leq \frac{Q}{Bs_u} \leq 2\sqrt{2} + 2\sqrt{SR(SR+1)} \qquad (6)$$

Figure 3 confirms that there is very good agreement between the NLA bearing capacity factors and the upper bound solution proposed by Bouassida and Porbaha (2004) for the full range of strength ratios, while the analytical LB solution is conservative for $SR < 10$. Figure 3 also summarizes NLA predictions for inclined concentric loading cases, with inclination angles, $\alpha = 0°–90°$. Figures 4a–d illustrate the different failure mechanisms for inclined loading cases.

Figure 4a shows a rupture mechanism where there is a well defined failure surface passing through the cement-treated zone, and a well defined shear zone forms in the adjacent clay on the 'leading' side of the foundation only. This type of rupture mechanism does not extend to the base of the clay layer/zone of soil treatment.

When the zone of shear failure extends to the base of the layer (at the leading edge of the cement-treated zone), the mechanism of failure is forced to extend back through the trailing clay mass, Figure 4b. This corresponds to a 'leading-edge base rotation mechanism'. It is also possible that failure reaches the rigid base at the trailing edge of the footing. This produces

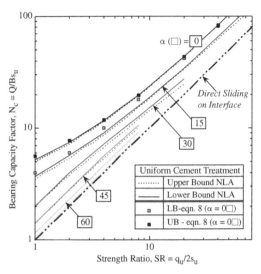

Figure 3. Summary of bearing capacity factors computed for homogeneous soil-cement foundation.

a mechanism that includes a large rigid body mode within the cement-treated soil, Figure 4c.

Finally the lower limit on the bearing capacity for a given inclination angle corresponds to the case where there is direct sliding along the foundation-cemented-soil interface, as illustrated in Figures 4d and 3. In practical terms, it is unlikely that large adhesion forces can develop along the foundation interface, instead the interface shear resistance is more likely constrained by the interface friction angle, δ (i.e., $\alpha \leq \delta$).

5 EFFECT OF SPATIAL VARIABILITY IN STRENGTH OF CEMENT-TREATED GROUND

The effects of spatial variability in the shear strength of the cement-treated soil have been studied by assuming a fixed value of the shear strength ratio, $SR = \mu_c/s_u$, and then varying combinations of the coefficient of variation and correlation length over the ranges shown in Table 3. For each set of parameters, a series of 100 Monte-Carlo simulations have been performed.

Figures 5a–d illustrate the mechanisms of failure under inclined concentric loading from a series of UB calculations with $COV_c = 0.4$ and $\Theta_{Inc} = 0.15$. Each example shows a specific realization of the shear strength field superimposed on the deformed FE mesh, together with the vectors of the computed velocity field. The strength field appears quite ragged due to the small correlation length considered in these examples. The first three examples appear to match closely the results presented for uniform cement-treated soil: Figure 5a shows a one-way rupture plane, Figure 5b

$15.3 \leq Q/(Bsu) \leq 15.7$

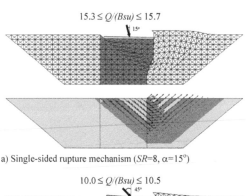

a) Single-sided rupture mechanism ($SR=8$, $\alpha=15°$)

$10.0 \leq Q/(Bsu) \leq 10.5$

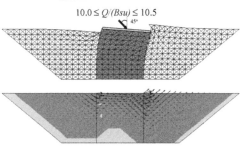

b) Leading edge base rotation ($SR=8$, $\alpha=45°$)

$15.8 \leq Q/(Bsu) \leq 19.2$

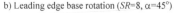
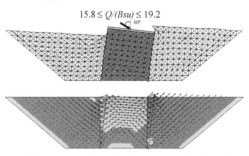

c) Rigid body mode with rotation about lead and trailing edges ($SR=20$, $\alpha=60°$)

$3.77 \leq Q/(Bsu) \leq 4.0$

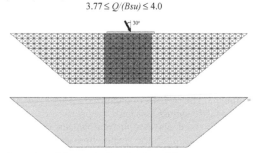

d) Interface sliding mechanism ($SR=2$, $\alpha=30°$)

Figure 4. Failure modes for uniform cement-treated ground under inclined concentric loading.

$1.8 \leq Q/(Bsu) \leq 13.0$

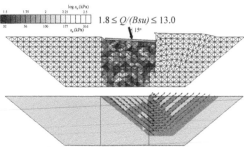

a) One-sided rupture mechanism ($\Theta_{lnc} = 0.15$, $SR = 8$, $\alpha = 15°$)

$9.3 \leq Q/(Bsu) \leq 10.8$

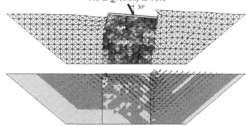

b) Leading edge basal rotation mechanism ($\Theta_{lnc} = 0.15$, $SR = 8$, $\alpha=30°$)

$6.3 \leq Q/(Bsu) \leq 7.0$

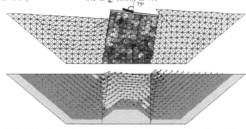

c) Rigid body mode with rotation at leading and trailing edge ($\Theta_{lnc} = 0.15$, $SR = 8$, $\alpha=75°$)

$6.3 \leq Q/(Bsu) \leq 7.0$

d) Shallow horizontal sliding mechanism ($\Theta_{lnc} = 0.15$, $SR = 8$, $\alpha=75°$)

Figure 5. Examples of failure mechanisms for inclined concentric loading accounting for spatial variability of cement-treated ground ($COV_c = 0.4$).

corresponds to a leading edge base rotation and Figure 5c involves a rigid body mode with rotation at the leading and trailing edges. However, close inspection shows that the computed failure mechanisms find paths of least resistance, passing through weaker regions of the cement-treated-soil mass. This result is made clear in Figure 5d where a shallow horizontal sliding plane passes through cemented-soil (i.e., in

310

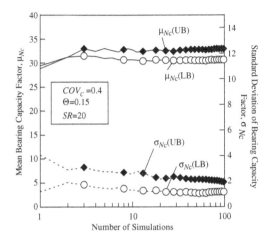

Figure 6. Mean and standard deviation of bearing capacity factor for vertical loading.

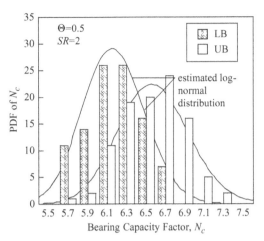

Figure 7. Histogram and estimated log-normal distribution ($SR = 2$, $\Theta_{lnc} = 0.5$, $COV_c = 0.4$, $\chi^2 = 12.7$ for UB, 11.7 for LB).

preference to direct interface sliding). Indeed, the failure mode is no longer strictly a function of SR and α, but is a stochastic property related to the spatial distribution of the shear strength in the cement-treated soil mass.

The computed bearing capacity factor can then be reported for each realization, i, of the shear strength field, N_{ci}. Hence, the mean, μ_{Nc}, and standard deviation, σ_{Nc}, of the bearing capacity factor are recorded through each set of Monte Carlo simulations, as follows:

$$\mu_{N_c} = \frac{1}{n}\sum_{i=1}^{n} N_{ci} \; ; \; \sigma_{N_c} = \sqrt{\frac{1}{n}\sum_{i=1}^{n}(N_{ci} - \mu_{N_c})^2} . \qquad (7)$$

Figure 6 illustrates one set of results for the case with $SR = 20$, $COV_c = 0.4$, $\Theta_{lnc} = 0.15$. The results confirm that the collapse load for any given realization is well bounded by the UB and LB calculations. The mean and standard deviation of N_c both become stable within 100 simulations and hence, reliable statistical interpretation of the data can be obtained from this set of simulations.

Figure 7 shows a 20-bin histogram of the bearing capacity factor from one complete series of Monte Carlo simulations with $SR = 2.0$, $COV_c = 0.2$ and $\Theta_{lnc} = 0.5$ together with the estimated log-normal distribution for N_c. Kasama and Whittle (2005b) show that all of the UB and LB simulations satisfy the χ^2 goodness-of-fit tests at a 5% significance level for both normal and log-normal distributions.

Figure 8 summarizes the mean bearing capacity factor μ_{Nc} for vertical loading as a function of the strength ratio and coefficient of variation, for a fixed correlation length ratio, $\Theta_{lnc} = 0.15$. The results show that μ_{Nc} increases linearly with SR for a given COV_c, while increases in coefficient of variation at a given

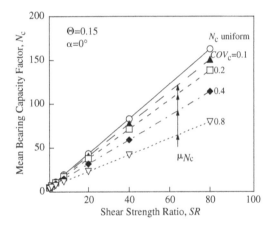

Figure 8. Summary of expected vertical bearing capacity factor as a function of strength ratio, SR ($\Theta_{lnc} = 0.15$, $COV_c = 0.4$).

strength ratio produce a marked reduction in the bearing capacity.

The role of spatial variability in reducing the expected bearing capacity can be more conveniently seen in Figure 9, which reports the normalized mean bearing capacity ratio $\tilde{N}_c = \mu_{Nc}/N_{cuniform}$, where $N_{cuniform}$ is the bearing capacity computed for the case where the shear strength of the cement-treated soil is uniform. The results in Figure 9a show large reductions in $\tilde{N}_c$ as COV_c increases at a given strength ratio, SR, while Figure 9b shows a smaller effect of reducing the correlation length scale, Θ_{lnc}. For typical cement-treated soil masses with $SR = 10$–40 and

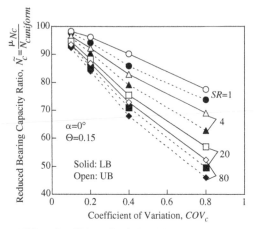

a) Effect of coefficient of variation, COV_c

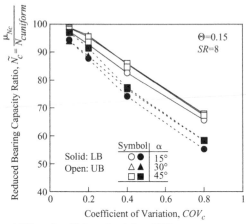

a) Effect of coefficient of variation, COV_c

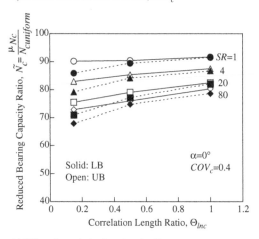

b) Effect of correlation length ratio, Θ_{lnc}

Figure 9. Effect of spatial variability parameters on reduced bearing capacity ratio.

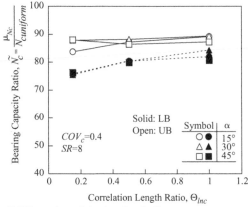

b) Effects of correlation length ratio, Θ_{lnc}

Figure 10. Summary of reduced bearing capacity ratio for inclined concentric loading with $SR = 8$.

$COV_c = 0.4–0.7$, the analyses suggest that the reduced bearing capacity ratio in the range $\tilde{N}_c = 0.5–0.7$.

Finally, Figures 10a and b present exactly analogous results showing the reduction in bearing capacity for inclined concentric loading of footings at a fixed strength ratio, $SR = 8.0$. These data show that the inclination angle has little influence on the reduced bearing capacity ratio needed to account for spatial variability in the cement-treated soil.

6 CONCLUSIONS

This paper has presented initial results from a probabilistic study on the bearing capacity and failure mode for cement-treated ground using numerical limit analyses with random field theory and Monte Carlo simulation. The main conclusions are as follows:

1. From the literature review on the spatial variability of the cohesive shear strength for in situ cement-treated ground, the coefficient of variation indicates wider range than that of naturally deposited soil with $COV_c = 0.42–0.67$. The horizontal spatial correlation length $\theta_h = 0.15\,\text{m}{-}10\,\text{m}$ for cement-treated ground can be at least one order of magnitude smaller than that of naturally sedimented soils.

2. Rigorous upper and lower estimates of undrained stability have been achieved using random field numerical limit analyses. Inherent spatial variability of the shear strength properties of the cement-treated soil mass are simulated using a Chosleky decomposition method.

3. The failure mechanisms for inclined concentric loading of surface foundations on end-bearing, cement-treated ground can be classified as follows; a) single-sided rupture; b) leading edge basal rotation; c) rigid body mode with rotation about the leading and trailing edges, and d) shallow horizontal sliding. The latter case reduces to interface sliding only when strength properties are homogeneous (i.e., uniform cement-treated soil).
4. The bearing capacity factor for inclined concentric loading can be characterized by either normal and log-normal distribution functions.
5. The bearing capacity factor for cement-treated ground decreases with the increased coefficient of variation of the shear strength, which is an influential rather than the spatial correlation length. For typical coefficients of variation of cement-treated soils, the expected bearing capacity factor is 50–70% of that expected capacity for a uniform clay with the same mean cohesive strength.

REFERENCES

Baecher, G.B. & Christian, J.T. (2003): "Reliability and Statistics in Geotechnical Engineering," Wiley & Sons, New York.

Bouassida, M., deBuhan, P. & Dormieux, L. (1995): "Bearing capacity of a foundation resting on a soil reinforced by a group of columns," *Géotechnique*, 45(1), 25–34.

Bouassida, M. & Porbaha, A. (2004): "Ultimate Bearing Capacity of Soft Clays Reinforced by a Group of Columns – Application to a Deep Mixing Technique," *Soils & Foundations*, 44(3), 91–101.

Broms, B.B. (2004): "Lime and lime/cement columns," Chapter 8 of *Ground Improvement* 2nd edition edited by M.P. Moseley and K. Kirsch.

Chew, S.H., Kamruzzaman, A.H.M. & Lee, F.H. (2004): "Physicochemical and Engineering Behavior of Cement Treated Clays," *ASCE J. Geotech. and Geoenvir. Eng.*, 130(7), 696–706.

Clough, G.W., Sitar, N., Bachus, R.C. & Rad, N.S. (1981): "Cemented Sands Under Static Loading," *ASCE J. Geotech. Eng.*, 107(6), 799–817.

Consoli, N.C., Rotta, G.V. & Prietto, P.D.M. (2000): "Influence of curing under stress on the triaxial response of cemented soils," *Geotechnique* 50(1), 99–105.

Costal Development Institute of Technology (1999a): Design and construction manual on Deep Mixing Method for offshore construction, pp.136, (in Japanese).

Costal Development Institute of Technology (1999b): Design and construction manual on Premixing Method, pp.144, (in Japanese).

Griffiths, D.V., Fenton, G.A. & Manoharan, N. (2002): "Bearing capacity of rough rigid strip footing on cohesive soil: probabilistic study." *ASCE J. Geotech. and Geoenvir. Eng.*, 128(9), 743–755.

Honjo, Y. (1982): "A probabilistic approach to evaluate shear strength of heterogeneous stabilized ground by Deep Mixing Method," *Soils & Foundations*, 22(1), 23–38.

Kamon, M. & Katsumi, T. (1999): "Engineering properties of Soil stabilized by ferrum lime and used for the application of road base," *Soils & Foundations*, 39(1), 31–41.

Kasama, K., & Whittle, A.J. (2005a): "Bearing capacity of spatially random cohesive soil using numerical limit analysis," submitted to *ASCE J. Geotech. and Geoenvir. Eng.*

Kasama, K., & Whittle, A.J. (2005b): "Stochastic bearing capacity of footings on cemented-treated soil," submitted to *Soils & Foundations*.

Kitazume, M., Okano, K. & Miyajima, S. (2000): "Centrifuge model tests on failure envelope of column-type Deep Mixing Method improved ground," *Soils & Foundations*, 40(4), 43–55.

Larsson, S. (2001): "Binder distribution in lime-cement columns," *Ground Improvement*, 5(3), 111–122.

Larsson, S., Stille, H. & Olsson, L. (2005): "On horizontal variability in lime-cement columns in deep mixing," *Géotechnique*, 55(1), 33–44.

Matsuo, M. & Asaoka, A. (1997): "Probability models of undrained strength of marine clay layer," *Soils & Foundations*, 17(3), 51–68.

Miyake, M., Akamoto, H. & Wada, M. (1991): "Deformation characteristics of ground improved by a group of treated soil," *Proc. Centrifuge 1991*, 295–302.

Navin, M.P. & Filz, G.M. (2005): "Statistical analysis of strength data from ground improved with DMM columns," *Proc. Deep Mixing '05*, in print.

Omine, K., Ochiai, H. & Yoshida, N. (1998): "Estimation of in-situ strength of cement-treated soils based on a two-phase mixture model," *Soils & Foundations*, 38(4), 17–29.

Phoon, K.K. & Kulhawy, F.H. (1999) "Characterization of geotechnical variability," *Canadian Geotechnical Journal*, 36, 612–624.

Porbaha, A., Tsuchida, T. & Kishida, T. (1999): "Technology of air-transported stabilized dredged fill. Part 2: quality assessment," *Ground Improvement*, 3, 59–66.

Sloan, S.W. (1988a): "Lower bound limit analysis using finite elements and linear programming." *Int. J. Numer. Analyt. Meth. Geomech.*, 12(1), 61–77.

Sloan, S.W. (1988b): "A steepest edge active set algorithm for solving sparse linear programming problems," *Int. J. Numer. Analyt. Meth. Geomech.*, 12(12), 2671–2685.

Sloan, S.W. & Kleeman, P.W. (1995): "Upper bound limit analysis using discontinuous velocity fields." *Comput. Methods Appl. Mech. Eng.*, 127, 293–314.

Tang, Y.X., Miyazaki, Y. & Tsuchida, T. (2001): "Practices of reused dredgings by cement treatment," *Soils & Foundations*, 41(5), 129–143.

Terashi, M. & Tanaka, H. (1981): "Ground Improved by deep mixing method," *Proc. 10th ICSMFE*, Stockholm, 3, 777–780.

Ukritchon, B., Whittle, A.J., & Sloan, S.W. (1998). "Undrained limit analyses for combined loading of strip footing on clay." *ASCE J. Geotech. and Geoenvir. Eng.*, 124(3), 265–276.

Zen, K., Yamazaki, H., Yoshizawa, H. & Mori, K. (1992): "Development of premixing method against liquefaction," *Proc. 9th Asian Regional Conf. SMFE*, 1, 461–464.

313

Author index

Printed and bound by CPI Group (UK) Ltd, Croydon, CR0 4YY

01/11/2024

01782599-0001